Damage and Failure Analysis of Polymer-Based Composites

Damage and Failure Analysis of Polymer-Based Composites

Guest Editors

Xiaoquan Cheng
Wenjun Huang
Qian Zhang

MDPI

Basel • Beijing • Wuhan • Barcelona • Belgrade • Novi Sad • Cluj • Manchester

Guest Editors
Xiaoquan Cheng
School of Aeronautic Science and Engineering
Beihang University
Beijing
China

Wenjun Huang
AVIC China Helicopter Design and Research Institute
Jingdezhen
China

Qian Zhang
School of Mechanical Engineering
Hefei University of Technology
Hefei
China

Editorial Office
MDPI AG
Grosspeteranlage 5
4052 Basel, Switzerland

This is a reprint of the Special Issue, published open access by the journal *Polymers* (ISSN 2073-4360), freely accessible at: www.mdpi.com/journal/polymers/special_issues/DJ209V7257.

For citation purposes, cite each article independently as indicated on the article page online and using the guide below:

Lastname, A.A.; Lastname, B.B. Article Title. *Journal Name* **Year**, *Volume Number*, Page Range.

ISBN 978-3-7258-3258-3 (Hbk)
ISBN 978-3-7258-3257-6 (PDF)
https://doi.org/10.3390/books978-3-7258-3257-6

Contents

polymers

MDPI

Editorial

Damage and Failure Analysis of Polymer-Based Composites

Xiaoquan Cheng [1,*], Wenjun Huang [2] and Qian Zhang [3]

1 School of Aeronautic Science and Engineering, Beihang University, Beijing 100191, China
2 AVIC China Helicopter Design and Research Institute, Jingdezhen 333001, China; chrdihwj@163.com
3 School of Mechanical Engineering, Hefei University of Technology, Hefei 230009, China; zq_hfut@hfut.edu.cn
* Correspondence: xiaoquan_cheng@buaa.edu.cn

1. Introduction

Polymer-based composites are widely used in aerospace, automotive, construction and wind energy fields due to their high specific strength and stiffness, corrosion resistance, and customizable properties. However, the mechanical behaviors of these materials are complex, and various types of damage, such as delamination, matrix cracking, interface debonding and fiber breakage, can occur. These types of damage can significantly affect the mechanical performance and long-term durability of the structure. Therefore, in-depth research on the damage and failure mechanisms of polymer-based composites is of great significance for improving structural efficiency, reducing total life-cycle costs and extending service life.

In recent times, damage and failure analysis of polymer-based composites has remained a key research topic. Researchers focus not only on the influence of the material's microstructure on the damage mechanism [1], but also on developing new numerical methods, damage evolution models and nondestructive testing technologies to achieve more precise failure predictions and performance improvement [2,3]. For instance, mesoscopic and multi-scale models have been widely applied to analyze the damage evolution process of composites under different loading conditions [4]. And experimental methods based on technologies such as Digital Image Correlation (DIC) and X-ray Computed Tomography (CT) imaging have also become effective means for studying the internal damage evolution of composite materials [5]. Furthermore, with the rapid development of artificial intelligence and machine learning technologies, intelligent damage monitoring, analysis and prediction techniques for composite are beginning to show great potential [6].

Based on the research at material level, further investigation into the damage and failure of composite structures is also necessary. Researchers usually focus on typical composite structures, such as sandwich structures, stiffened panels, joints and repaired structures, addressing issues including failure mechanisms, damage tolerance performance, numerical models and design methods [7]. The failure of polymer composite structures is also closely related to loading conditions, external environmental factors, and the manufacturing processes of the composites. Therefore, many researchers are focused on the influence of environmental conditions, such as temperature, humidity, and chemical agents, on the performance of composites, aiming to further improve damage prediction models and design methods [8,9]. Additionally, many studies are dedicated to improving manufacturing processes or developing new material systems, such as novel polymer matrixes, hybrid fibers, natural fiber reinforcements or gradient functional materials to enhance the performance of composite structures, reduce production costs or minimize environmental impact [10–12].

check for updates

Received: 10 January 2025
Accepted: 20 January 2025
Published: 22 January 2025

Citation: Cheng, X.; Huang, W.; Zhang, Q. Damage and Failure Analysis of Polymer-Based Composites. *Polymers* **2025**, *17*, 272. https://doi.org/10.3390/polym17030272

Twenty manuscripts were submitted for this Special Issue and twelve research papers were finally accepted. This Special Issue mainly focuses on the damage and failure mechanisms of different polymer-based composites and their influence factors, while also discussing novel damage assessment methods and numerical models based on various advanced testing technologies. We aim to provide researchers with the latest findings and offer theoretical or practical guidance for the failure analysis and design of composite materials and structures.

2. An Overview of Published Articles

Four articles covered the topics of damage and failure mechanism of polymer-based composites:

Zheng et al. (contribution 1) conducted an investigation into the sand erosion resistance of polyurethane films applied to helicopter rotor blades. Their findings indicated that the impact angle played a crucial role in determining the erosion patterns and damage mechanisms. At lower angles, sand cutting dominated, causing significant surface deformation and eventual shedding of larger polyurethane fragments. However, higher impact angles resulted in concentrated impact damage with smaller debris being shed. Additionally, the study highlighted that volume loss, as a direct measurement approach, provided a more reliable evaluation of sand erosion resistance by eliminating inaccuracies caused by embedded sand particles. The researchers also developed a predictive equation to model volume loss based on experimental observations, offering a robust tool for forecasting erosion behavior.

An et al. (contribution 2) investigated the tensile and compressive properties of woven fabric CFRP laminates containing three-dimensional microvascular channels, which were designed for self-healing applications. The study found that the laminates tend to fail at the location where the thickness-direction micro channels are arranged. While the micro channels minimally affect the overall stiffness, they do lead to a reduction in the strength of the laminate. Several FE models were developed to simulate the mechanical behavior of CFRP laminates containing micro-channels. It was found that resin-rich regions around the microvascular channels, particularly in the warp and weft yarns of the woven fabric composite layers, were identified as a key consideration when establishing FE models. These regions must be accurately represented in the models to predict the mechanical properties of the laminates correctly. Using the validated FE model, the parameter study showed that the reduction in the strength of the laminates exhibited a roughly linear relationship with the diameter of the microvascular channels in the thickness direction, within a typical design range of 0.1 to 1 mm.

Premanand et al. (contribution 3) conducted interrupted constant-amplitude three-point bending fatigue experiments on carbon-fiber satin-fabric-reinforced poly-ether-ketone-ketone (CF-PEKK) composites, using an ultrasonic fatigue testing (UFT) system operating at a cyclic frequency of 20 kHz. The study measured parameters such as the specimen's surface temperature, acoustic activity and ultrasonic generator resonance during cyclic loading. Additionally, stiffness measurements and volumetric damage characterization using 3D X-ray microscopy (XRM) were carried out to trace damage initiation and accumulation. The results showed that UFT significantly shortens the testing duration compared to conventional fatigue tests, proving advantageous for high-cycle fatigue regimes. Damage progression commenced at the surface and progressively spread towards the interior of the specimen, with surface temperature and acoustic activity serving as critical indicators of damage progression. And it was important to consider thermal influences on the material during testing, as variations in temperature can significantly affect the mechanical properties and fatigue life of the composite.

Chen et al. (contribution 4) investigated the mechanical behaviors of Zylon fiber-reinforced polymers (ZFRPs) under the influence pulsed magnets. The study involved mechanical testing of ZFRP laminates and the development of a constitutive model that accounted for plastic deformation and progressive damage under magnetic field stresses. This model was applied to investigate the failure of a 95-T double-coil prototype, with results indicating a significant reduction of approximately 45% in both the radial and axial stiffness of ZFRP composites, particularly in the inner layers. The damage incurred by the end ZFRPs was the primary cause of the failure, which aligned closely with the experimental results. Additional magnet systems of different magnetic fields were also analyzed to explore the performance of ZFRP. The article concluded that ZFRP composites were highly susceptible to damage and stiffness reduction under extreme magnetic field conditions. The end ZFRPs of the magnet were particularly vulnerable, and the failure mechanism was closely tied to axial Lorentz forces and transverse strength of the material.

Three articles were related to performances and failure analysis of composite structures:

Zhang et al. (contribution 5) investigated the energy-absorption characteristics of thin-walled polymer composite C-channels used in civil aircraft structures under low-speed axial compression. An experimental study was conducted on specimens with layups of $[0/90]_{3s}$, $[45/\text{-}45]_{3s}$, and $[45/90/\text{-}45/0]_{3}$, and the thin-walled composite C-channels underwent a stable progressive crushing process, with failure modes including local buckling, fiber breakage, matrix cracks, delamination and corner cracking. Then, stacked shell models were developed after conducting parametric studies on simple single-element models and the numerical results on energy absorption showed less than 6% difference with the test results. It can be seen that thin-walled polymer composite C-channels offered excellent energy-absorbing characteristics, making them suitable for use in energy dissipation applications such as crashworthy structures.

Yu et al. (contribution 6) conducted experiments to study the performance of composite helicopter rotor blades under different ballistic impact conditions. The test results showed that failures were particularly prevalent at the bullet incidence and exit positions of the blade. At the incidence position, the damage was mainly local shear fracture, while at the exit position, fiber tensile fracture was the dominant failure mode. And ballistic damage size decreased as the incident angle of the projectile increased. A numerical model was also established and the verified model was used to conduct the parameter study. It was found that the structural stiffness near the exit position of the blade had a significant impact on the size of the damage. Blades with higher stiffness near the exit experienced larger damage sizes at the exit point. The numerical model provided a reliable tool for predicting ballistic damage and assessing the survivability of composite helicopter blades under attack, which had significant implications for improving the design of composite structures in helicopters.

Shin and Huang (contribution 7) focused on a novel structural health monitoring method for the widely used single-lap composite adhesive joint, doping the adhesive with carbon nanotubes and monitoring the joint's electrical resistance change. Tensile fatigue tests were conducted and electrical resistance of the joints were measured. In addition, the debonding area was recorded using a liquid penetrant technique and the fracture surfaces were examined by scanning electron microscopy (SEM). The results showed that the debonding area showed a linear relationship with the fatigue life expended. However, the relationship between resistance change and debonding was not straightforward, as it followed two different trends, which were associated with different failure mechanisms. Therefore, the percentage voltage change was not a valid quantifying parameter for the fatigue life of a single-lap composite adhesive joint.

Two articles studied the failure of polymer composites under different environments:

Yan et al. (contribution 8) utilized fiber Bragg grating (FBG) sensors to monitor the damage progression of composites, which were embedded in CFRP laminates and can provide real-time data about the deformation and failure progression during the tests. In high-temperature tensile tests, the FBG sensors successfully monitored the damage evolution, including matrix cracking, fiber breakage, and delamination even though the tensile strength of the CFRP composites decreased by 75%. The tests showed the high reliability of FBG sensors in tracking damage under complex conditions and potential of using FBG sensors for long-term structural health monitoring in, for example, aviation insulation layers. In addition, finite element models were established to study the thermal decomposition behavior of the composite materials at high temperatures.

Fan et al. (contribution 9) conducted hydrothermal accelerate aging tests and natural storage tests to study the influence of hydrothermal environments on FRP composites. The results of various testing techniques showed that the plasticization of the resin and interface weakening were major contributors to the decrease in shear strength. Then, a storage life prediction model was established to convert the hydrothermal accelerated aging time to natural storage life. Based on the failure criterion of a specific reduction in shear strength, the storage life of composite can be predicted. This research provided a reliable approach for predicting the storage life of carbon fiber/epoxy composites under real-world conditions. And it was particularly valuable for aerospace applications, whose long-term durability and structural integrity were very critical.

Three articles contribute to the improvement of manufacturing processes or the development of new materials:

Wang et al. (contribution 10) employed three different epoxy resin systems to study the synergistic effects of liquid rubber (CTBN) and thermoplastic particles (PEK-C) for toughening epoxy resin (EP). The structure, curing behavior, thermal properties, impact and bending performances of these EP systems were systematically investigated. The experimental results demonstrated that the combination of PEK-C and CTBN in the epoxy resin led to superior toughness compared to when these agents were used individually, as PEK-C contributed to improved thermal stability while CTBN significantly enhanced the fracture toughness and impact resistance of the resin. Specifically, the synergistic toughening system exhibited improved mechanical properties (including increased tensile strength, elongation at break, and impact strength), enhanced thermal stability (the resin maintained good performance under elevated temperatures, with better resistance to thermal degradation) and better phase morphology (the combination of these toughening agents led to a more uniform distribution of the particles within the resin matrix, which contributed to its improved overall toughness).

Xiao et al. (contribution 11) prepared itaconic acid epoxy resin (EIA) blended with bisphenol A diglycidyl ether (DGEBA) at various ratios, creating bio-sourced composites with different degrees of greenness. Compression, bending and interlaminar shear strength properties of these composites were tested after 48 h of aging at 80 °C under water immersion. The test results showed that the T700/EIA-30 (30% of EIA epoxy) composites showed better dimensional stability and strength retention after hydrothermal aging than traditional petroleum-based composites. The findings highlighted the potential of bio-based epoxy resins to replace petroleum-based counterparts, offering a greener and more sustainable solution with higher performance for future composite material development.

Lee and Yoo (contribution 12) studied the mechanical strength of liquid silicone rubber (LSR) with added carbon black (CB) particles and their resistance to degradation when exposed to oils. The test results showed that the mechanical properties of the LSR/CB composite, including tensile strength and modulus, improved as the CB content increased. When exposed to oil, the swelling rate and oil absorption were significantly reduced in the

LSR/CB composites. And when subjected to high-temperature conditions, the deterioration rate in terms of both mechanical strength and oil absorption was also reduced compared to neat LSR. The improved properties of CB/LSR composites were mainly due to the improvement of the crosslink density and structure of the polymer matrix by CB, making them suitable for various industrial applications, and they are also expected to perform well in oil-exposed environments.

3. Conclusions

A thorough understanding of the damage and failure mechanisms of composites is fundamental to improving the performance of composite structures and achieving optimized designs. In this Special Issue, we present the latest research findings on damage and failure analysis of polymer-based composites, covering a wide range of important topics from the damage mechanisms of composite materials and structures to the enhancement of durability in high-performance materials, utilizing various advanced analytical methods and testing technologies. These studies are derived from practical engineering applications or reflect engineering demands. And the results indicate that through advanced analysis, design or monitoring technologies, the performances of composites are expected to be significantly improved in many aspects, demonstrating their broad potential in applications. These studies provide a technical basis or practical guidance for better design and application of related composite materials or structures.

In the future, we look forward to more studies on nanotechnology, smart materials, intelligent monitoring and analysis technology, bio-derived or eco-friendly materials and other fields. Combined with the evolving damage and failure analysis methods, polymer-based composite materials can play a greater role in more applications and make more contributions to global engineering design and sustainable development.

List of Contributions

1. Zheng, L.; Fan, J.; Gong, Q.; Sun, W.; Jia, X. Sand Erosion Resistance and Failure Mechanism of Polyurethane Film on Helicopter Rotor Blades. *Polymers* **2023**, *15*, 4386. https://doi.org/10.3390/polym15224386.
2. An, Z.; Cheng, X.; Zhao, D.; Ma, Y.; Guo, X.; Cheng, Y. Tensile and Compressive Properties of Woven Fabric Carbon Fiber-Reinforced Polymer Laminates Containing Three-Dimensional Microvascular Channels. *Polymers* **2024**, *16*, 665. https://doi.org/10.3390/polym16050665.
3. Premanand, A.; Prescher, M.; Rienks, M.; Kirste, L.; Balle, F. Online and Ex Situ Damage Characterization Techniques for Fiber-Reinforced Composites under Ultrasonic Cyclic Three-Point Bending. *Polymers* **2024**, *16*, 803. https://doi.org/10.3390/polym16060803.
4. Chen, S.; Peng, T.; Han, X.; Cao, Q.; Xiao, H.; Li, L. Mechanical Behaviors of Polymer-Based Composite Reinforcements within High-Field Pulsed Magnets. *Polymers* **2024**, *16*, 722. https://doi.org/10.3390/polym16050722.
5. Zhang, X.; Mou, H.; Song, S.; Feng, Z. An Investigation of the Energy-Absorption Characteristics of Thin-Walled Polymer Composite C-Channels: Experiment and Stacked Shell Simulation. *Polymers* **2024**, *16*, 2099. https://doi.org/10.3390/polym16152099.
6. Yu, G.; Li, X.; Huang, W. Performance and Damage Study of Composite Rotor Blades under Impact. *Polymers* **2024**, *16*, 623. https://doi.org/10.3390/polym16050623.
7. Shin, C.-S.; Huang, S.-H. Evaluation of Fatigue Damage Monitoring of Single-Lap Composite Adhesive Joint Using Conductivity. *Polymers* **2024**, *16*, 2374. https://doi.org/10.3390/polym16162374.

8. Yan, G.; Wan, B.; Huang, H.; Li, W. Damage and Failure Monitoring of Aerospace Insulation Layers Based on Embedded Fiber Bragg Grating Sensors. *Polymers* **2024**, *16*, 3543. https://doi.org/10.3390/polym16243543.
9. Fan, J.; Zhang, Q.; Chen, X.; He, Y. Interlaminar Shear Strength Change and Storage Life Prediction of Carbon Fiber/Epoxy Composites with Hygrothermal Accelerated Aging. *Polymers* **2024**, *16*, 1109. https://doi.org/10.3390/polym16081109.
10. Wang, Z.; Lai, Y.; Xu, P.; Ma, J.; Xu, Y.; Yang, X. Synergistic Effects of Liquid Rubber and Thermoplastic Particles for Toughening Epoxy Resin. *Polymers* **2024**, *16*, 2775. https://doi.org/10.3390/polym16192775.
11. Xiao, K.; Fang, Y.; Wang, Z.; Ni, N.; Liu, Z.; Kim, S.; An, Z.; Lyu, Z.; Xu, Y.; Yang, X. Bio-Sourced, High-Performance Carbon Fiber Reinforced Itaconic Acid-Based Epoxy Composites with High Hygrothermal Stability and Durability. *Polymers* **2024**, *16*, 1649. https://doi.org/10.3390/polym16121649.
12. Lee, B.-J.; Yoo, H.-M. Effects of Carbon Black on Mechanical Properties and Oil Resistance of Liquid Silicone Rubber. *Polymers* **2024**, *16*, 933. https://doi.org/10.3390/polym16070933.

Author Contributions: Conceptualization, X.C.; writing—original draft preparation, X.C., W.H. and Q.Z.; writing—review and editing, X.C. and Q.Z.; project administration, X.C. All authors have read and agreed to the published version of the manuscript.

Funding: This research received no external funding.

Institutional Review Board Statement: Not applicable.

Data Availability Statement: Not applicable.

Acknowledgments: Ziqian An (anziqian@buaa.edu.cn) provided some references for this paper and participated in the original draft preparation and review.

Conflicts of Interest: The authors declare no conflicts of interest.

References

1. David Müzel, S.; Bonhin, E.P.; Guimarães, N.M.; Guidi, E.S. Application of the Finite Element Method in the Analysis of Composite Materials: A Review. *Polymers* **2020**, *12*, 818. [CrossRef] [PubMed]
2. Wan, L.; Ismail, Y.; Sheng, Y.; Ye, J.; Yang, D. A Review on Micromechanical Modelling of Progressive Failure in Unidirectional Fibre-Reinforced Composites. *Compos. Part C Open Access* **2023**, *10*, 100348. [CrossRef]
3. Bui, T.Q.; Hu, X. A Review of Phase-Field Models, Fundamentals and Their Applications to Composite Laminates. *Eng. Fract. Mech.* **2021**, *248*, 107705. [CrossRef]
4. Fish, J.; Wagner, G.J.; Keten, S. Mesoscopic and Multiscale Modelling in Materials. *Nat. Mater.* **2021**, *20*, 774–786. [CrossRef]
5. Wang, B.; Zhong, S.; Lee, T.-L.; Fancey, K.S.; Mi, J. Non-Destructive Testing and Evaluation of Composite Materials/Structures: A State-of-the-Art Review. *Adv. Mech. Eng.* **2020**, *12*, 168781402091376. [CrossRef]
6. Bishara, D.; Xie, Y.; Liu, W.K.; Li, S. A State-of-the-Art Review on Machine Learning-Based Multiscale Modeling, Simulation, Homogenization and Design of Materials. *Arch. Comput. Methods Eng.* **2023**, *30*, 191–222. [CrossRef]
7. Zimmermann, N.; Wang, P.H. A Review of Failure Modes and Fracture Analysis of Aircraft Composite Materials. *Eng. Fail. Anal.* **2020**, *115*, 104692. [CrossRef]
8. Liu, T.; Liu, X.; Feng, P. A Comprehensive Review on Mechanical Properties of Pultruded FRP Composites Subjected to Long-Term Environmental Effects. *Compos. Part B Eng.* **2020**, *191*, 107958. [CrossRef]
9. Mayandi, K.; Rajini, N.; Ayrilmis, N.; Indira Devi, M.P.; Siengchin, S.; Mohammad, F.; Al-Lohedan, H.A. An Overview of Endurance and Ageing Performance under Various Environmental Conditions of Hybrid Polymer Composites. *J. Mater. Res. Technol.* **2020**, *9*, 15962–15988. [CrossRef]
10. Zuo, P.; Srinivasan, D.V.; Vassilopoulos, A.P. Review of Hybrid Composites Fatigue. *Compos. Struct.* **2021**, *274*, 114358. [CrossRef]

11. Khalid, M.Y.; Al Rashid, A.; Arif, Z.U.; Ahmed, W.; Arshad, H.; Zaidi, A.A. Natural Fiber Reinforced Composites: Sustainable Materials for Emerging Applications. *Results Eng.* **2021**, *11*, 100263. [CrossRef]
12. Ghatage, P.S.; Kar, V.R.; Sudhagar, P.E. On the Numerical Modelling and Analysis of Multi-Directional Functionally Graded Composite Structures: A Review. *Compos. Struct.* **2020**, *236*, 111837. [CrossRef]

polymers

MDPI

Article

Sand Erosion Resistance and Failure Mechanism of Polyurethane Film on Helicopter Rotor Blades

Linfeng Zheng [1], Jinjuan Fan [2,*], Qing Gong [1], Wei Sun [2] and Xinghui Jia [2]

1 China Helicopter Research and Development Institute, Jingdezhen 333001, China; 288288peak@163.com (L.Z.)
2 AECC Beijing Institute of Aeronautical Materials, Beijing 100095, China
* Correspondence: fanjinjuan621@163.com

Abstract: Polyurethane is widely used on the surface of composite materials for rotor blades as sand erosion protection materials. The failure mechanism investigation of polyurethane film under service conditions is useful for developing the optimal polyurethane film for rotor blades. In this article, the sand erosion test parameters were ascertained according to the service environment of the polyurethane film. The sand erosion resistance and failure mechanism of polyurethane film at different impact angles were analyzed by an infrared thermometer, a Fourier transform infrared spectrometer (FTIR), a differential scanning calorimeter (DSC), a field emission scanning electron microscope (FESEM), and a laser confocal microscope (CLSM). The results show that the direct measurement method of volume loss can better characterize the sand erosion resistance of the polyurethane film compared to traditional mass loss methods, which avoids the influence of sand particles embedded in the polyurethane film. The sand erosion resistance of polyurethane film at low-angle impact is much lower than that at high-angle impact. At an impact rate of 220 m/s, the volume loss after sand erosion for 15 min at the impact angle of 30° is 57.8 mm^3, while that at the impact angle of 90° is only 2.6 mm^3. The volume loss prediction equation was established according to the experimental data. During low-angle erosion, the polyurethane film damage is mainly caused by sand cutting, which leads to wrinkling and accumulation of surface materials, a rapid increase in roughness, and the generation of long cracks. The linking of developing cracks would lead to large-scale shedding of polyurethane film. During high-angle erosion, the polyurethane film damage is mainly caused by impact. The connection of small cracks caused by impact leads to the shedding of small pieces of polyurethane, while the change in the roughness of the film is not as significant as that during low-angle erosion. The disordered arrangement of the soft and hard blocks becomes locally ordered under the action of impact and cutting loads. Then, the disordered state is restored after the erosion test finishes. The erosion of sand particles leads to an increase in the temperature of the erosion zone of the polyurethane film, and the maximum temperature rise is 6 °C, which does not result in a significant change in the molecular structure of the polyurethane film. The erosion failure mechanism is cracking caused by sand cutting and impact.

Keywords: polyurethane film; impact angle; sand corrosion resistance; erosion loss mechanism

Citation: Zheng, L.; Fan, J.; Gong, Q.; Sun, W.; Jia, X. Sand Erosion Resistance and Failure Mechanism of Polyurethane Film on Helicopter Rotor Blades. *Polymers* **2023**, *15*, 4386. https://doi.org/10.3390/polym15224386

Academic Editors: Xiaoquan Cheng, Wenjun Huang and Qian Zhang

Received: 25 September 2023
Revised: 4 November 2023
Accepted: 7 November 2023
Published: 11 November 2023

1. Introduction

Helicopter rotor blades are mainly manufactured from fiber-reinforced resin matrix composites. During low-altitude–low-speed flight, hovering, or the takeoff and landing of helicopters, sand impacts the surface of the composite blades, causing fiber fracture, the delamination of composites, and the early failure of composite blades [1–4]. During the Gulf War in the 1990s, the high temperatures and sand dust in Iraq caused severe non-war damage to helicopter rotor blades, and sand erosion became an urgent issue in helicopter design [5,6]. The erosion is mainly caused by sand from two sources: one is the sand in the air, and the other is the sand lifted by the action of rotating rotor blades. From this perspective, the damage of sand erosion for rotor blades is not only related to the natural sand environment but also closely related to the rotation rate of rotor blades [6,7]. Therefore, sand erosion damage on the rotor blades was more severe than that on the general fuselage.

Polyurethane film has a unique micro-phase separation structure containing soft blocks with polymer chain mobility and partially crystalline hard blocks and is, therefore, widely used on the surface of rotor blade composite materials as an anti-sand erosion coating with good impact absorption capacity [6,8–10]. The sand erosion resistance of polyurethane film was closely related to the hybrid ratio of the soft and hard segments and the service conditions. The sand erosion failure mechanism of polyurethane film is very complex and depends on multiple factors, which can be divided into internal and external effect factors. The internal effect factors include the properties and structure of the polyurethane film [11–13], while the external effect factors include sand particle size, rate, erosion angle, impact flow rate, and other environmental conditions [14–16]. It is necessary to understand the sand erosion failure mechanism of polyurethane film under specific service conditions in order to develop the optimal polyurethane film for rotor blades [17–19].

In the research of the inner effect factors, Oka [13] studied the effects of polyurethane toughness on impact performance and found that elastic polyurethane materials showed better anti-erosion properties at lower hardness, which was different from that erosion resistance of polyurethane increased with an increase in the hardness [16]. Vishawa and Dong [20,21] found that the addition of graphite oxide, fibers, and carbon black in the polyurethane improved the energy storage modulus and resistance to solid particle erosion. Cho, Wei, et al. showed that the friction coefficient, tensile strength, and fracture strength of polyurethane could affect its impact resistance [18,22].

In the research of the external effect factors, Liu [23] studied the effects of sand erosion rate (50–100 m/s) and impact angle (15–90°) on the erosion wear of polyurethane films on the surface of high-speed trains. The results showed that the higher the impact rate of sand particles, the higher the erosion rate. With an increase in impact angle, the erosion rate increased in the beginning, followed by a decrease, and finally stabilizing. Acierno [24] studied the effect of sand particle impact angle on the erosion rate of polyurethane films at the impact rate of 160 m/s and found that the impact angle with the maximum erosion loss rate was 15°, and when the erosion angle was higher than 45°, the erosion loss rate remained almost unchanged. Dong [25] studied the erosion resistance of polyurethane nanocomposites to solid particles and pointed out that at the impact rate of 10–30 m/s, when the impact angle of solid particles was 30°, the erosion rate of polyurethane nanocomposites was the highest, and at an impact angle of 90°, the erosion rate was the lowest. Doyle [26] studied the erosion behavior of polyurethane under insulation conditions and found that local deformation and high strain led to the temperature increase of the erosion zone, and the mechanical and thermal stresses led to the degradation and loss of the polyurethane. Barkoula, Qi, and Hao et al. [17,27–29] believe that there are three main ways in which solid particles impact polyurethane, namely elastic impact, plastic impact, and elastic–plastic impact. Because the polyurethane material has a certain elasticity, pure plastic impact rarely occurs in polyurethane materials. However, in high-angle impact, erosion particles will compress the surface of the material to harden the material, and plastic erosion characteristics will appear in the subsequent erosion process. Fan [30] found that there are a large number of microcracks in the front of polyurethane crack propagation, which absorb the dynamic strain energy generated by impacts.

In the present work, the erosion results cannot evaluate the sand corrosion resistance of polyurethane materials for rotor blades because the failure mechanism of polyurethane coatings is closely related to the experimental conditions. In this article, the sand erosion conditions were determined on the basis of analyzing the rotor blades' service conditions, and the sand erosion resistance and failure mechanism of polyurethane film were investigated, which could provide technical support for the application and improvement of polyurethane film on rotor blades.

2. Specimens and Experimental

2.1. Materials and Samples

The polyurethane films with a bi-layer structure were fabricated upon composite substrates to give test samples. The surface layer of the bi-layer structure was a polyurethane layer, the thickness of which was 350 ± 10 μm. The bottom layer was a pressure-sensitive adhesive layer (50 ± 10 μm in thickness) used to bond the polyurethane layer to the surface of the composite substrate. The polyurethane layer was an (AB) n-type linear polymer, mainly formed by the reaction of hard segment dicyclohexylmethane diisocyanate (HDMI) and soft segment polyoxypropylene triol (PTMG), as shown in Figure 1. The pressure-sensitive adhesive layer was mainly acrylic ester materials. The polyurethane and pressure-sensitive adhesive are both colorless and transparent materials.

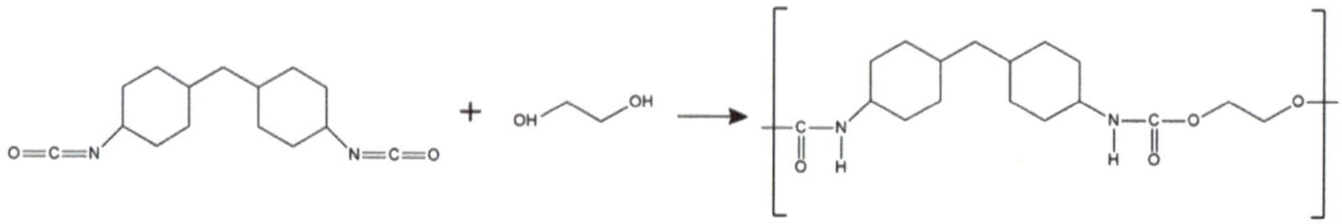

Figure 1. Schematic diagram of polyurethane reaction.

The composite substrate was a 2 mm thick glass fiber reinforced epoxy resin matrix composite board. The type of epoxy resin was medium temperature epoxy resin, 3238A, which is pale yellow. To fabricate the aforementioned samples, the composite substrate was firstly cut into a 75 mm × 50 mm rectangular plate, the surface of which was then ground with 200 μm sandpaper and cleaned with alcohol. The previously described polyurethane film was cut into a 75 mm × 50 mm rectangle and spread and pasted on the surface of the composite specimen. The polyurethane film was pressed with a roller to prevent bubbles from forming between the pressure-sensitive adhesive layer and the composite substrate, which affected the accuracy of the experimental results. An illustrative cross-sectional image of the sand corrosion specimen is shown in Figure 2. The surface of the prepared sample was light yellow, which was identical to the color of the glass fiber-reinforced resin matrix composite.

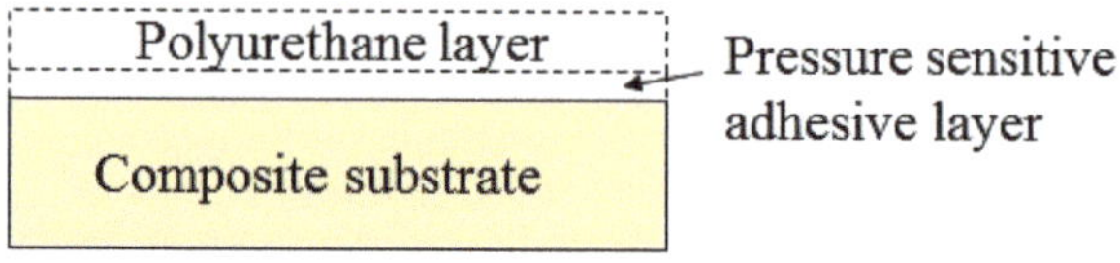

Figure 2. Illustrative cross-sectional image of the sand corrosion specimen.

2.2. Sand Erosion Test

The sands used in the test are SiO_2 particles with rhombohedral angles, purchased from the Minghai quartz sand factory, with a size of 80–120 μm, referring to the standard GJB 1171 "Requirements for Sand and Dust Prevention of Military Helicopters", as shown in Figure 3. The particle size range is measured by AVIC Changcheng Institute of Metrology and Measurement. The results are listed in Table 1.

The sand impact velocity was determined by measuring each condition on actual rotor blades during takeoff and landing, which was equivalent to the helicopter rotor tip speeds and could be calculated by the following Equation [6]:

$$v = 2\pi nR \tag{1}$$

where

v—Sand impact velocity, m/s;

n—Blade rotation angle speed, r/s;
R—Blade length, m.

In this article, the blade length was 3 m, and the blade rotation angle speed was 11.7 r/s. The sand impact velocity was 220 m/s, calculated by Equation (1).

In general, the serious sand erosion damage was on the leading edge of the blades. The sand impact angles were between 30° and 90° due to the fact that the leading edge of the blade is curved [6]. In this article, the typical impact angles of 30°, 45°, 60°, and 90° were selected to investigate the effect of impact angles on the damage mechanism.

Table 1. Sand size distribution.

Sand Size (μm)	**<40**	**40~60**	**60~80**	**80~100**	**100~120**	**120~140**	**>140**
Mass percentage (%)	8 ± 1	12 ± 2	12 ± 3	30 ± 3	23 ± 3	9 ± 2	6 ± 1

Figure 3. Diagram of the sands.

During the experiment, the sand particles were placed in the erodent hopper, then were accelerated in the small diameter nozzle and directed onto the specimen surface illustrated in Figure 4. The impact velocity of sands ejected from the nozzle was controlled by adjusting the pressure of high-pressure air. The particle velocity was measured using a laser Doppler velocimeter, as shown in Figure 5. Firstly, the laser velocity measurement point was adjusted to align with the nozzle of the erosion test device, which made particles pass through the laser. Then, the laser Doppler velocimeter was turned on, and the particle velocity data were collected. The sand velocity was collected three times for each test, and the average velocity value was taken as the final velocity. The test accuracy is ±0.2%. The sand flow was controlled to 6 g/min by adjusting the flowmeter of the hopper. The impact angle was set by adjusting the sample holder. The detailed test conditions are shown in Table 2. The impact time was 2.5 min, 5 min, 10 min, and 15 min respectively. Three samples were tested under each impact angle.

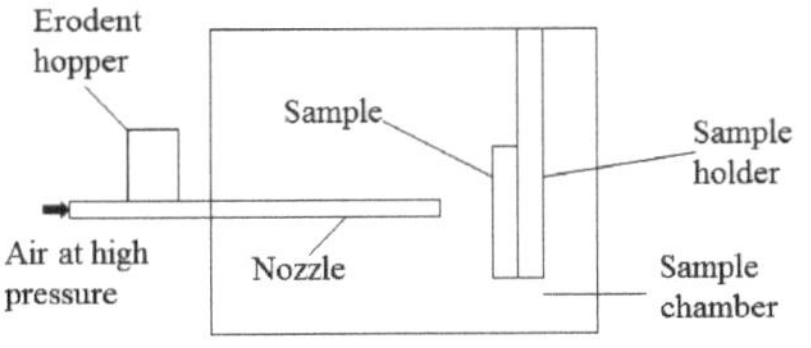

Figure 4. Schematic diagram of sand erosion test device.

Table 2. Sand erosion test conditions.

Sand Diameter	**Impact Rate**	**Sand Flow Rate**	**Impact Angle**	**Distance from Nozzle to Samples**	**Nozzle Diameter**
80–120 μm	220 m/s	6 g/min	30°, 45° 60°, 90°	100 mm	6000 μm

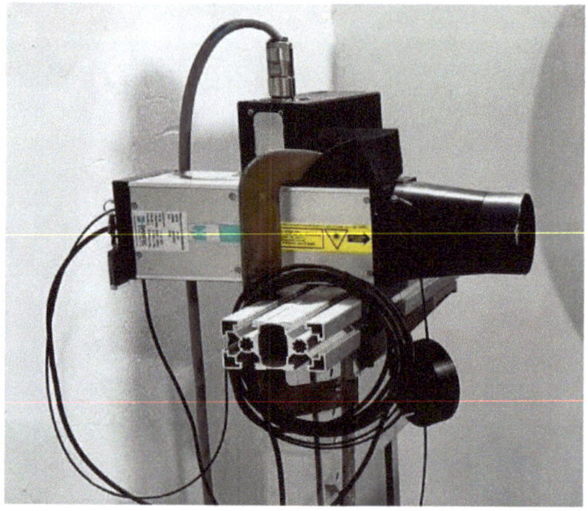

Figure 5. Laser Doppler Velocimetry.

2.3. Analysis of Sand Erosion Loss Mechanism

The temperature of the erosion damage zone was measured by an infrared thermometer after the sand erosion test. The accuracy is 0.5 °C. Each damaged area was tested three times, and the average value was taken. The first measurement was taken at the 10th second after the erosion test finished, and the time interval between the two measurements was 5 s. To determine the characteristic functional groups of polyurethane film before and after erosion, the infrared analysis was conducted on the original surface of polyurethane film and the damage zone after sand erosion test by the Spectrum 100 infrared spectrometer. The glass transition temperatures of polyurethane before and after the sand erosion test were analyzed using the DSC25 differential scanning calorimeter. The test conditions were a nitrogen atmosphere, a heating rate of 10 °C/min, and a testing temperature range of −80~300 °C. The microscopic characteristics and the composition change of the sand erosion damage zone were carried out, respectively, using the field emission scanning electron microscope and the energy spectrometer. The volume loss and roughness changes of the damage zone were measured using the OLS4100 laser confocal microscope. During the testing process, the damaged zone was divided into several small zones of 2 mm × 2 mm. The depth and roughness of every small zone were measured and then combined with computer software Zen lite 2012 to obtain the three-dimensional morphology and size of the damage. Referring to standard GB/T1031 [31], statistical Ra with contour height was selected as the roughness characterization parameter. Three samples were measured under each impact angle, and the maximum volume loss and roughness were taken as the final test results.

3. Results and Discussion

3.1. Erosion Zone Temperature

Doyle's research [26] showed that local deformation caused by solid particles' erosion and high-strain insulation conditions led to the temperature increase at the impact zone on the polyurethane. Before the sand erosion test, the samples were placed in the laboratory and adjusted to 25 °C. After the erosion test, the temperature of the damage zone of the samples was tested. Table 3 shows the temperature of the damage zone on the polyurethane film after sand impact at different angles and times.

It can be concluded that the raised temperature in the damage zone gradually decreased with the increase of the sand impact angle at the same impact time, and it increased with the impact time, which is related to the damage degree of the polyurethane, and the more severe the damage, the greater the temperature increase on the damaged zone. When the polyurethane film is impacted for 15 min at the impact angle of 30°, the raised temperature on the erosion damage zone of the sample is the maximum, and the maximum temperature rise was 6 °C.

Table 3. The damage zone temperature of polyurethane film (°C).

Impact Angle	Damage Zone Temperature (°C)			
	Impact for 2.5 min	Impact for 5 min	Impact for 10 min	Impact for 15 min
30°	26.7 ± 0.21	28.5 ± 0.20	30.2 ± 0.26	31.0 ± 0.25
45°	26.3 ± 0.20	27.0 ± 0.23	28.3 ± 0.22	30.1 ± 0.25
60°	25.8 ± 0.23	26.5 ± 0.24	28.0 ± 0.20	29.3 ± 0.20
90°	25.5 ± 0.21	26.2 ± 0.20	27.0 ± 0.24	28.0 ± 0.23

3.2. Erosion Damage Morphology

The erosion damage morphology and size of polyurethane film at different impact angles are shown in Figures 6–9. It can be seen that the morphologies of erosion damage change with impact angles. At the impact angle of 30°, the plane morphology of the erosion damage is approximately elliptical, and the damage area and depth gradually increase with the increasing impact time. When the impact time is 2.5 min, the long axis of the ellipse is 21,984 μm, and the short axis is 16,246 μm. The damage depth is 72 μm. The damage depth profile appears approximately symmetrical. When the impact time is 15 min, the long axis of the ellipse is 41,562 μm, the short axis is 19,703 μm, and the damage depth is 330 μm. The damage depth profile appears to be asymmetric spoon-shaped. Now, the thickness of the remaining polyurethane film is only about 10 μm.

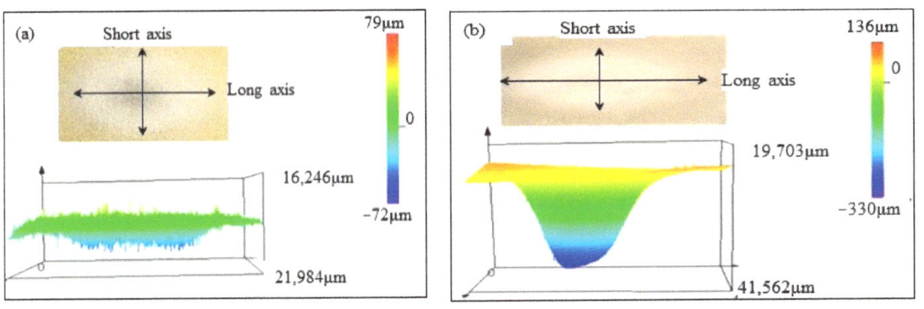

Figure 6. The sand erosion damage morphology impacted for (**a**) 2.5 min and (**b**) 15min at 30°.

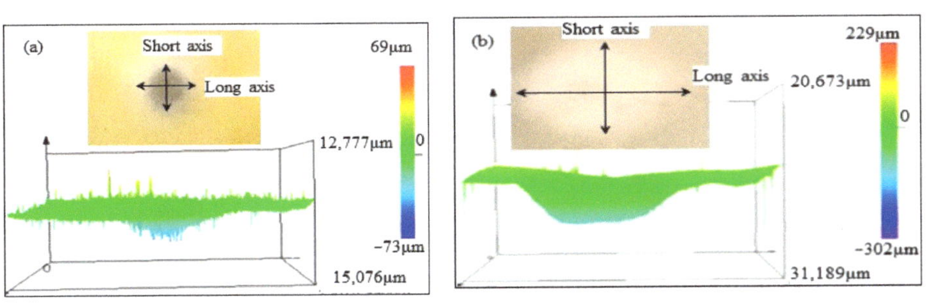

Figure 7. The sand erosion damage morphology impacted for (**a**) 2.5 min and (**b**) 15min at 45°.

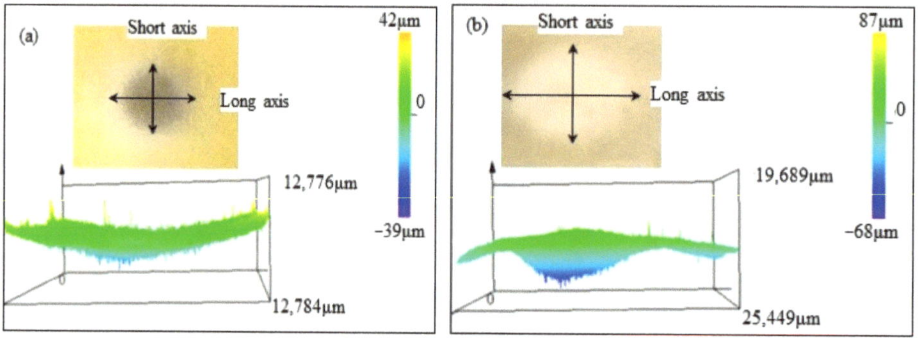

Figure 8. The sand erosion damage morphology impacted for (**a**) 2.5 min and (**b**) 15min at the 60°.

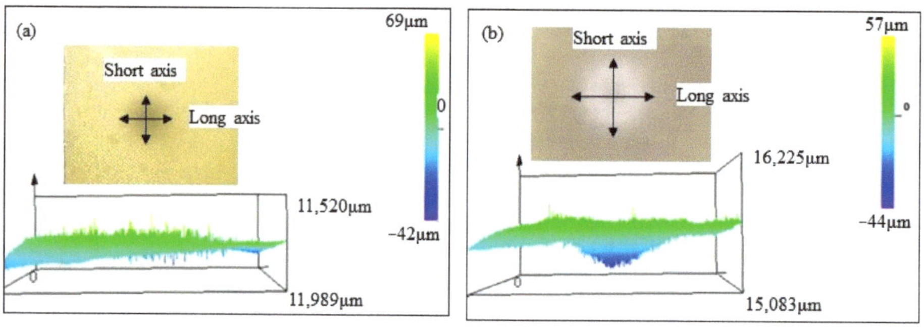

Figure 9. The sand erosion damage morphology impacted for (**a**) 2.5 min and (**b**) 15min at 90°.

With the sand impact angle increasing, the ratio of the long axis to the short axis of the ellipse damage morphology gradually decreases. When the sand impact angle is 90°, the sand erosion damage morphology is approximately circular. The area and depth of the damage zone gradually decrease with the increase of the impact angle. The polyurethane film was impacted for 15 min at an angle of 90°; the depth of the erosion damage is 44 μm, which is much lower than that at the impact angle of 30° (330 μm).

The depth profile shapes of the damage zone change with erosion times. When the impact time is short, the depth profile shape of the damage zone is similar to the symmetrical morphology. With the increase in the sand impact time, the depth of damage gradually changed. Except for the sand impact of 90°, the depth of damage zones gradually develops an asymmetric profile.

3.3. Volume Loss

In erosion and wear tests, mass loss is often used to evaluate the material's resistance to sand erosion [20]. Due to the fact that the polyurethane film used in this article is an elastic material, during the sand erosion test, the sand particles were embedded in the polyurethane film, and the embedded sand mass exceeded the loss mass of polyurethane film, resulting in a significant testing error. Therefore, volume loss is used to evaluate the polyurethane film's resistance to sand erosion. Moreover, volume loss cannot be calculated through mass loss while obtained by direct measurement methods.

Figure 10 shows the variation of polyurethane film volume loss with erosion time at different erosion angles, where the slope of the straight line represents the rate of volume loss. It can be seen that at different impact angles, with the increase in the sand impact time, the volume loss of the polyurethane film increases. The volume loss rate shows a trend of first increasing and then declining with the impact time. At the same impact time, the maximum volume loss occurs at the impact angle of 30°, and after sand erosion for 15 min, the volume loss is approximately 57.8 mm^3. With the sand impact angle increasing, the volume loss gradually decreases. When the impact angle is 90°, the volume loss of polyurethane is the lowest. After the sand erosion for 15 min, the volume loss is only 2.6 mm^3.

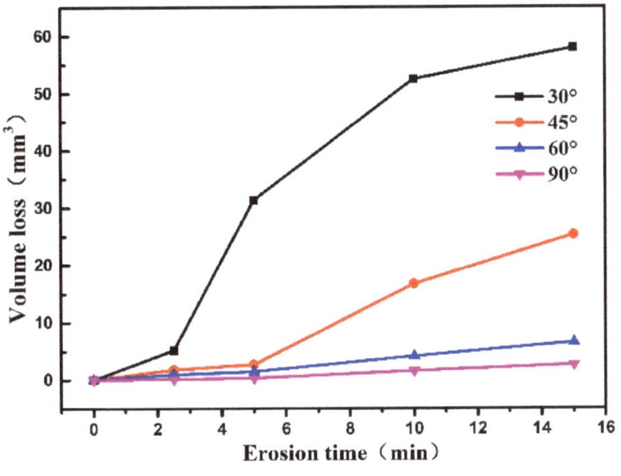

Figure 10. The change of loss volume with sand impact time at different impact angles.

Based on volume loss data from different angles, the relationship between volume loss, erosion time, and erosion angle can be expressed as:

$$
\begin{aligned}
V = {} & 0.003382t^4 + 0.000512t^3\alpha + 0.000636t^2 - 0.000196t^2\alpha^2 + 0.006145t^2\alpha - \\
& 0.024131t^2 - 6.812593 \times 10^{-6}t\alpha^3 + 0.015535t\alpha^2 - 0.716068t\alpha - 0.000715t - \\
& 4.191954 \cdot 10^{-6}\alpha^4 - 0.138532\alpha^3 + 0.913370\alpha^2 + 18.373836\alpha + 6.847228
\end{aligned} \quad (2)
$$

where

V—volume loss, mm^3;

t—erosion time, min;

α—erosion angle, °.

The comparison of experimental and computational results is shown in Figure 11. The goodness of fit R^2 = 0.98 shows that the Equation has a good fitting effect.

When sand particles impact the polyurethane surface at a certain angle, the erosion energy is divided into two components: the component parallel to the erosive surface (tangential energy) and the component perpendicular to the erosive surface (normal energy), as shown in Figure 12. The tangential stress generated by the tangential energy has a cutting effect on the eroded surface, while the normal stress caused by the normal energy impacts the eroded surface. When the sand impact angle changes, the two-component stress also changes accordingly. During the low-angle impact test, the cutting effect of sand is greater than the impact effect, and the damage of erosion particles to the material surface is mainly caused by oblique cutting, while at high-angle impact, the impact effect of sand particles is greater than the cutting effect. The damage of erodent to the polyurethane is mainly caused by direct impact. Based on the damage morphology at different impact angles (from Figures 5–8), it can be seen that the damage area and depth of polyurethane film caused by sand cutting are greater than those caused by sand impact.

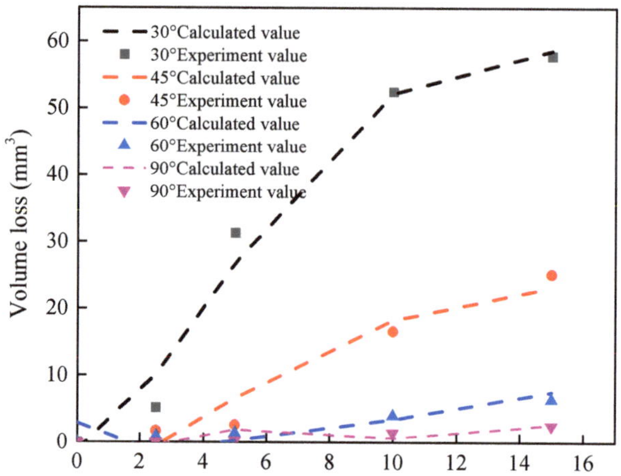

Figure 11. Comparison of experimental and computational results for volume loss.

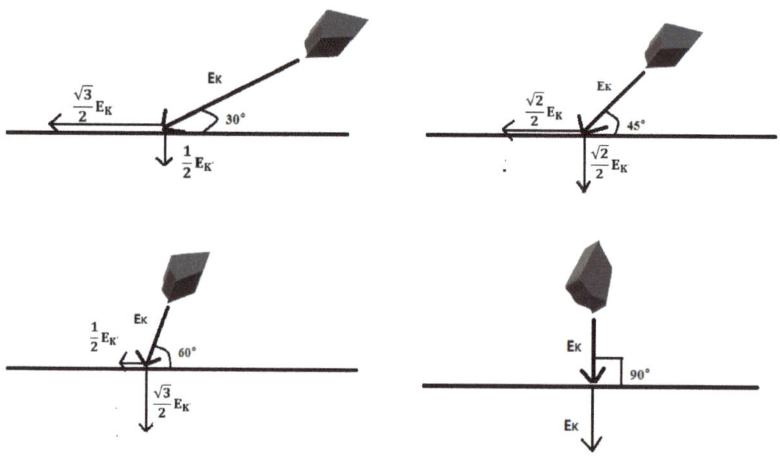

Figure 12. Schematic diagram of erosion load analysis at different impact angles.

3.4. Surface Roughness

Figure 13 shows the surface roughness variation of the polyurethane film with the impact time and impact angles. The surface roughness of the fabricated polyurethane film is 0.052 μm. After sand erosions at the impact angles of 30°, 45°, and 60°, the surface roughness of polyurethane first increases and then decreases with the increase in erosion time, while the roughness continuously increases at the impact angle of 90°. At the same impact time, the roughness of the polyurethane surface gradually decreases with the increase in impact angle. The maximum roughness reaches 8.5 μm at the impact angle of 30°.

During the sand erosion process, when the impact angle is 30°, the shear effect of the tangential load is more likely to cause polyurethane film wrinkling, accumulation, and the formation of peaks and valleys with large height differences on the surface. Therefore, the initial increase rate of roughness is faster. When the surface roughness reaches a certain level, the protruding parts of the surface are easier to be rushed away from the polyurethane surface by the subsequent particle erosion, and a new surface forms. Therefore, there is a slight decrease in the roughness in the later stage of erosion. The variation of surface roughness for polyurethane film with the impact time at the impact angle of 30° is shown in Figure 14. The variation trend of surface roughness at the impact angles of 45° and 60° is similar to that at the impact angle of 30°, and the difference is that the surface roughness changes less.

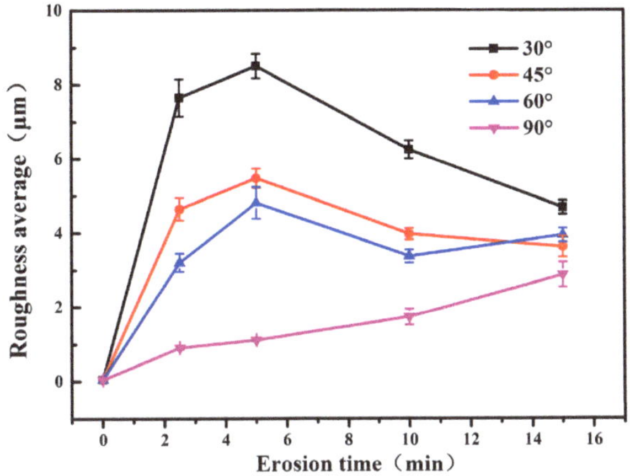

Figure 13. Changes of surface roughness with erosion time at different impact angles.

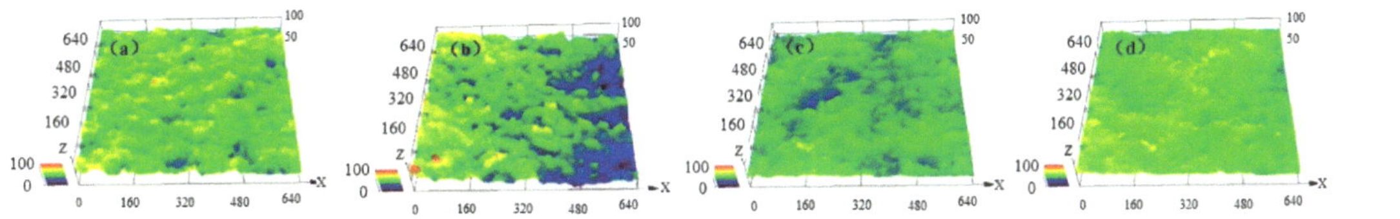

Figure 14. Surface roughness change of polyurethane film at the impact angle of 30° with (**a**) 2.5 min, (**b**) 5 min, (**c**) 10 min, (**d**) 15 min.

At the impact angle of 90°, due to the fact that polyurethane film has good elasticity and some sand particles are directly rebounded, the impact action of erosion particles cannot lead to wrinkling and accumulation of the polyurethane film, and the surface smoothness of the material is maintained well. In the early stage of erosion, the roughness does not change significantly with time. As the erosion time increases, some materials crack and fall off, resulting in an increase in roughness, as shown in Figure 15.

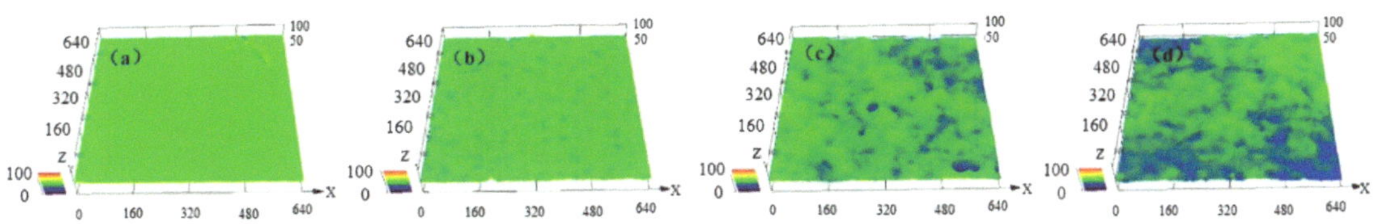

Figure 15. Surface roughness change of polyurethane film at the impact angle of 90° with (**a**) 2.5 min, (**b**) 5 min, (**c**) 10 min, (**d**) 15 min.

3.5. Microscopic Characteristics of Damage

Figure 16 shows the microstructure of the fabricated polyurethane film and 15 min of erosion at the impact angles of 30°, 45°, 60° and 90°. It can be seen that the fabricated coating is flat and smooth, as shown in Figure 15a. After 15 min of erosion at the impact angle of 30°, the polyurethane film surfaces show a plow micro characteristic and become rough. It is mainly micro-cutting marks of sand particles caused by the cut stress parallel to the surface of the polyurethane film. A small amount of microcracks caused by impact stress occurs, as shown in Figure 15b. With the increase in the impact angle, the cutting effect decreases, the plowing marks become shallower, and the cracks on the polyurethane film surface become obvious, as shown in Figure 15c,d. After 15 min of erosion at the impact angle of 90°, the plowing characteristics disappear, and the coating becomes rougher than the original surface. The microcracks caused by impact stress vertical to the film occur, as shown in Figure 15e. It can be concluded that when the polyurethane film is eroded by sand at low angles, the main failure mode is plowing and cutting wear. When it is impacted at high angles by sand, the failure mode is mainly impact cracking. The horizontal and vertical stresses coexist, promoting polyurethane film damage.

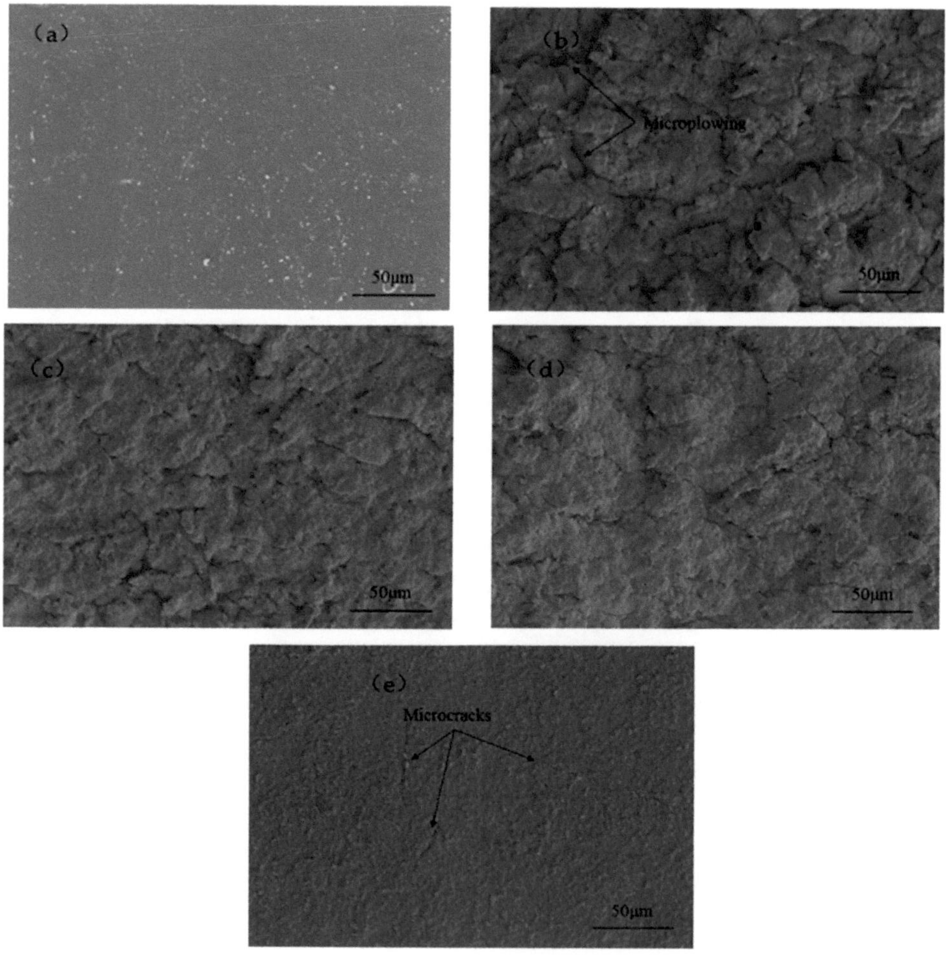

Figure 16. Microscopic morphology of film before and after 15 min of erosion at different impact angles (**a**) as-prepared, (**b**) 30°, (**c**) 45°, (**d**) 60°, (**e**) 90°.

3.6. Fourier Infrared Analysis

To analyze whether the polyurethane film undergoes chemical changes during the erosion process, infrared analysis was performed on the damage zone after 15 min of 30° erosion because under these conditions, the damage of the polyurethane film is the most serious, and its temperature rise is the maximum.

Figure 17 shows the infrared spectrum of the polyurethane film after sand impact at different times. The peak at 3337 cm^{-1} is the N–H peak, the peak at 2923 cm^{-1} is the asymmetric stretching vibration peak of β-CH2, and the symmetric stretching vibration peak of α-CH2; the peak at 2853 cm^{-1} is the asymmetric stretching vibration peak of α-CH2. The C=O stretching vibration peak of carbamate is at 1717 cm$^{-1,}$ and a CHN deformation vibration peak in the amino ester group is at 1525 cm^{-1}. There is a C–O–C stretching vibration peak of carbamate at 1231 cm^{-1}, which is a characteristic peak of polyether polyurethane material. The peak at 1078 cm^{-1} is the stretching vibration peak of C–O. It can be seen that after 2.5 min and 5 min of sand erosion, the position and size of the main characteristic peaks of polyurethane film have not changed significantly. After sand erosions for 10 min and 15 min, the characteristic peak of polyurethane film at 1405 cm^{-1} increased significantly, which is the characteristic peak of the adhesive layer pressure-sensitive glue [32]. The depth of the erosion damage is greater than 300 μm after sand erosion for 10 min and 15 min. The remaining polyurethane film is less, and the pressure-sensitive adhesive at the bottom layer has an effect on the infrared radiation. In addition, the C–H–N deformation vibration peak decreases, indicating that there is a small amount of fracture at the connection between the hard and soft segments of polyurethane.

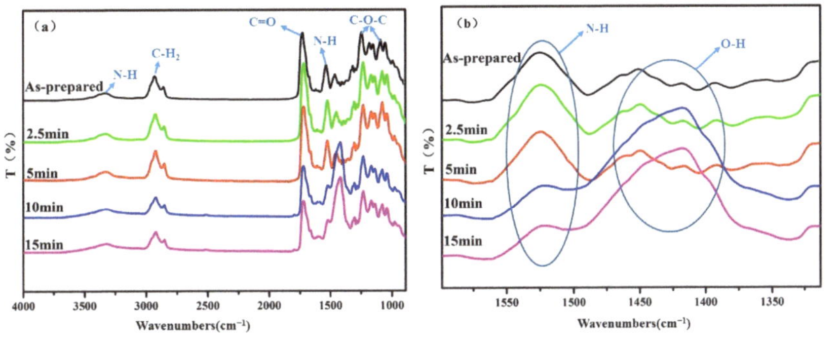

Figure 17. Infrared spectrum of the polyurethane damaged area after 30° erosion at different times with (**a**) full infrared image and (**b**) peak amplification at 1405 cm^{-1}.

During the erosion test, the temperatures of the polyurethane damage zone rise. As the temperature in the most serious damage zone only increases by 6 °C, it would not cause obvious chemical changes to the polyurethane film. Therefore, the sand erosion of polyurethane film is mainly wear failure.

From the above results, it can be seen that when the depth of sand erosion damage is close to the thickness of the polyurethane film, the infrared spectrum of the damaged polyurethane area changes significantly. Therefore, in the later use stage, whether polyurethane film has reached its service life can be inferred from infrared characteristic peak changes.

3.7. DSC Analysis

The glass transition temperatures (Tg) of the soft and hard blocks of the polyurethane film are −21.47 °C and 72.04 °C, respectively. There is no significant change in glass transition temperature after erosion, as shown in Table 4. Furthermore, there was no significant chemical change in the polyurethane film after the erosion test.

Table 4. Glass transition temperature Tg (°C) of the damage zone of polyurethane after erosion under different conditions.

Impact Angle	Impact for 2.5 min		Impact for 5 min		Impact for 10 min		Impact for 15 min	
	Soft Blocks	Hard Blocks	Soft Blocks	Hard Blocks	Soft Blocks	Hard Blocks	Soft Blocks	Hard Blocks
30°	−20.73 ± 0.38	74.06 ± 0.49	−19.72 ± 0.44	74.52 ± 0.51	−21.41 ± 0.51	73.17 ± 0.49	−20.01 ± 0.46	72.88 ± 0.49
45°	−20.31 ± 0.46	73.14 ± 0.56	−20.56 ± 0.45	75.20 ± 0.47	−21.06 ± 0.41	73.26 ± 0.57	−22.97 ± 0.57	71.95 ± 0.49
60°	−21.04 ± 0.36	74.04 ± 0.48	−19.56 ± 0.52	73.86 ± 0.42	−20.35 ± 0.48	73.27 ± 0.49	−19.96 ± 0.44	73.18 ± 0.54
90°	−20.21 ± 0.45	72.43 ± 0.64	−21.13 ± 0.48	72.53 ± 0.60	−20.74 ± 0.50	73.17 ± 0.66	−20.83 ± 0.60	73.40 ± 0.47

3.8. Analysis of Sand Erosion Failure Mechanism

In the initial stage of sand erosion, a small amount of sand particles impact the polyurethane film surface, and elastic deformation occurs in the polyurethane film and absorbs some impact energy. With the impact time increasing, the number of impact particles increases, and the material deformation is insufficient to withstand the continuously increasing impact and shear stresses, resulting in plastic deformation, material wrinkling, accumulation of long cracking, and small cracking. The impact time further increased, and the protrusions on the polyurethane surface were sheared and collided into debris removed from the surface, resulting in material loss.

The sand particles with a certain amount of energy impact the surface of polyurethane film, and the erosion failure mechanism changes with the impact angles. During low-angle (30°) erosion, the cutting action of sand particles is greater than the impact action. In the early stage of erosion, the surface of the polyurethane film becomes wrinkled and accumulates under the cutting action of erosion sands, resulting in a rapid increase of surface roughness. With the erosion time increasing, under the action of sand tangential load, some long cracks appear on the surface of the damage zone. The embedding sands in the polyurethane film can also cause cracking of the damage zone. When the cracks propagate and link, some material sheds off from the surface of the film in large pieces, forming a new relatively smooth surface, and the roughness of the damage zone decreases. Under such cyclic erosion action, the material gradually loses and forms asymmetric spoon-shaped damage pits, as shown in Figure 18a.

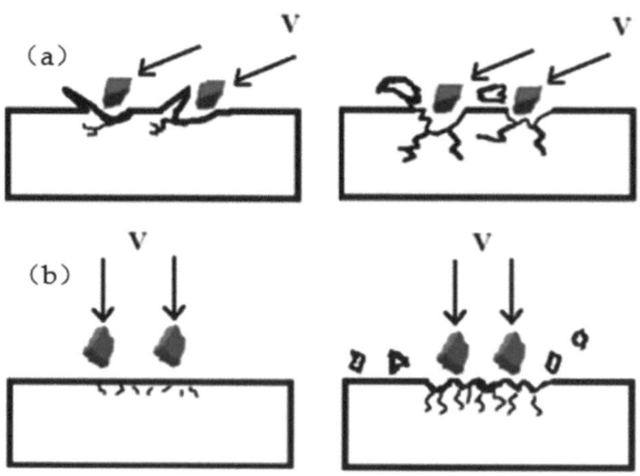

Figure 18. Erosion mechanism of polyurethane film with (**a**) low-angle erosion and (**b**) vertical erosion.

With the increase of the impact angle, the cutting effect of erosion sands on the polyurethane film decreases while the vertical impact effect increases. During the erosion of vertical (90°) impact, in the initial stage of erosion, material loss is difficult on the surface of the elastic polyurethane film, and the roughness changes little. Then, under the continuous impact of erosive sands, dense and short cracks appear on the surface of the polyurethane film. The propagation and connection of cracks cause the surface material to shed off in small pieces, leading to an increase in surface roughness and ultimately forming symmetrical funnel-shaped erosion damage, as shown in Figure 18b.

As the erosion time increased, the polyurethane film became significantly thinner. After erosion of 10 min, the underlying pressure-sensitive adhesive affected the change of characteristic peak at 1405 cm^{-1}. Sand erosion can cause a local temperature rise in the polyurethane film. An increase in temperature may lead to thermal oxygen aging of polyurethane, chemical bond change of molecular chains, and a decrease in physical properties. However, due to the highest temperature only rising to 31 °C, the molecular chain segments had not yet broken or crosslinked, and there was no significant group change in the polyurethane film.

Figure 19 is a schematic diagram of the polyurethane structure before and after load application [33]. For fabricated polyurethane, the rigid blocks randomly orient and disperse in the flexible block matrix (shown in Figure 19a). These ordered regions are kept in the appropriate positions through intermolecular forces (mainly hydrogen bonds). The local molecular orientation changes due to loading action, and the orderliness of these regions increases (shown in Figure 19b). In this article, the loading is the impact and cutting action of sand. A small amount of fracture occurs in the links between the soft and hard segments with the increase in the erosion time.

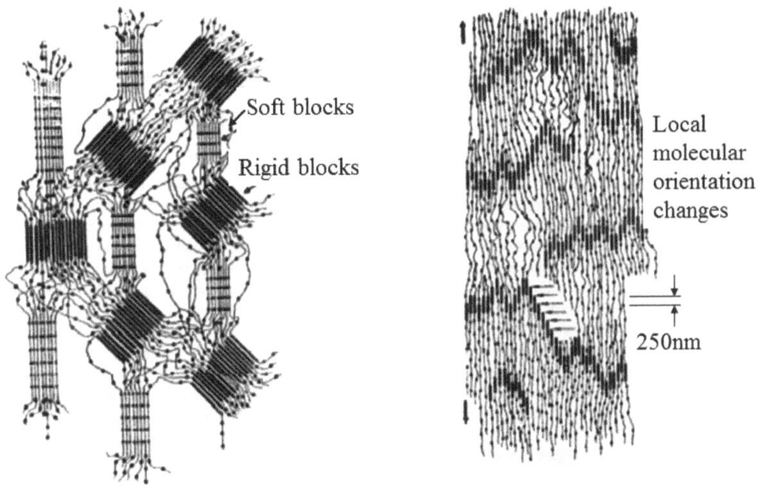

Figure 19. Changes in the arrangement of local soft and hard segments of the polyurethane film during erosion: (**a**) Disordered; (**b**) locally ordered [33].

After the sand erosion test, under the action of intermolecular forces, the molecular orientation is gradually restored, which is also a manifestation of high elastic deformation. Therefore, during the erosion process, there was no significant chemical change in the polyurethane material. The failure mechanism is cracking caused by sand cutting or impact.

4. Conclusions

In this study, the impact velocity of the sand erosion test was determined as 220 m/s according to the service status of polyurethane film on the surface of helicopter rotor blades. The sand erosion resistance and failure mechanism of polyurethane film at different impact angles were analyzed, and the main conclusions are as follows:

The resistance of polyurethane film to low-angle sand erosion is much lower than that to high-angle sand erosion. The erosion damage depth after sand erosion for 15 min at the impact angle of 30° is 330 μm, and the volume loss is 57.8 mm^3. While the impact angle is 90°, the depth of the erosion damage is only 44 μm, and the volume loss is only 2.6 mm^3. The direct measurement method of volume loss avoids the influence of sands embedded in the polyurethane film, which can effectively demonstrate the sand erosion resistance of the elastic polyurethane film. The volume loss prediction equation was established. The calculated results are compared with the experimental data, and good agreements are obtained.

During low-angle sand erosion, the failure of polyurethane film is mainly caused by sand cutting. The cutting effect leads to the wrinkling and accumulation of the polyurethane film, a rapid increase in roughness, and the generation of long cracks. The linking of developing cracks leads to the large-scale shedding of polyurethane film. At high-angle sand erosion, the loss of the polyurethane film is mainly caused by impact. The connection of small cracks by impact leads to the shedding of small pieces of polyurethane, while the change in roughness of the film is not as significant as that during low-angle erosion.

During the erosion process of sands, the disordered arrangement of the flexible and rigid blocks for polyurethane film becomes locally ordered under the action of impact and cutting stress. And after the erosion test, the disordered state is restored. The erosion of sand particles leads to an increase in the temperatures of the damage zone, and the maximum temperature rise is 6 °C. At such a temperature rise during a limited testing period, the molecular structure of the polyurethane film does not change significantly. The erosion failure mechanism is cracking caused by sand cutting or impact. In the future, polyurethane film researchers should pay more attention to the shear performance of the material.

Author Contributions: Formal analysis, W.S.; Investigation, Q.G.; Data curation, X.J.; Writing—original draft, J.F.; Writing—review & editing, L.Z. All authors have read and agreed to the published version of the manuscript.

Funding: This research received no external funding.

Institutional Review Board Statement: Not applicable.

Data Availability Statement: The data presented in this study are available on request from the corresponding author.

Conflicts of Interest: The authors declare no conflict of interest.

References

1. Bousser, E.; Martinu, L.; Klemberg-Sapieha, J.E. Solid particle erosion mechanisms of protective coatings for aerospace applications. *Surf. Coat. Technol.* **2014**, *257*, 165–181. [CrossRef]
2. Gang, D.; Zhang, J. Development of research on solid particle erosion of materials. *J. Mater. Eng.* **2003**, *21*, 307–312.
3. Li, X.; Zhang, Y.; Zuo, Y.; Ou, Y. Wind-blown Sand Resistant Polyurethane Acrylate Coatings for Aerospace Transparency. *Polyurethane Ind.* **2016**, *31*, 40–43.
4. Kaundal, R. Role of process variables on the solid particle erosion of polymer composites: A critical review. *Silicon* **2014**, *6*, 5–20. [CrossRef]
5. Pan, H.; Yang, T. Analysis of sand/particle erosion standards for helicopter blade. *Helicopter Tech.* **2013**, *4*, 51–56.
6. Gong, Q.; Fan, J.; Huang, Y.; Zheng, L. Research progress of sand erosion resistance evaluation technology for protective materials of helicopter rotor blades. *Hi-Tech. Fiber Appl.* **2020**, *45*, 33–38.
7. Pepi, M.; Squillacioti, R.; Pfledderer, L.; Phelps, A. Solid particle erosion testing of helicopter rotor blade materials. *J. Fail. Anal. Prev.* **2012**, *12*, 96–108. [CrossRef]
8. Eslami, R.; Bagheri, R.; Hashemzadeh, Y.; Salehi, M. Optical and mechanical properties of transparent acrylic based polyurethane nano Silica composite coatings. *Prog. Org. Coat.* **2014**, *77*, 1184–1190. [CrossRef]
9. Gong, W.; Zang, Y.; Liu, B.; Chen, H.; Wu, F.; Huang, R.; Wang, S. Effect of using polymeric materials in ecological sand-fixing of Kerqin Sandy Land of China. *J. Appl. Polym. Sci.* **2016**, *133*, 44102. [CrossRef]
10. Zhang, S.W.; He, R.; Wang, D.; Fan, Q. Abrasive erosion of polyurethane. *J. Mater. Sci.* **2001**, *36*, 5037–5043. [CrossRef]
11. Huang, Y.F.; Zhang, Z.C.; Xu, J.Z.; Xu, L.; Zhong, G.J.; He, B.X.; Li, Z.M. Simultaneously improving wear resistance and mechanical performance of ultrahigh molecular weight polyethylene via cross-linking and structural manipulation. *Polymer* **2016**, *90*, 222–231. [CrossRef]

12. Zhang, J.; Tu, W.; Dai, Z. Synthesis and characterization of transparent and high impact resistance polyurethane coatings based on polyester polyols and isocyanate trimers. *Prog. Org. Coat.* **2012**, *75*, 579–583. [CrossRef]
13. Oka, Y.I.; Matsumura, M.; Kawabata, T. Relationship between surface hardness and erosion damage caused by solid particle impact. *Wear* **1993**, *162–164 Part B*, 688–695. [CrossRef]
14. Barkoula, N.M.; Karger-Kocsis, J. Review processes and influencing parameters of the solid particle erosion of polymers and their composites. *J. Mater. Sci.* **2002**, *37*, 3807–3820. [CrossRef]
15. Gao, J.; Zhang, Y.; Wang, J.; Su, L.D.; Wu, G.J.; Qi, G.Q. Study on coatings erosion of wind turbine blades under sand-carrying wind. *ACTA Egergiae Solaris Sin.* **2020**, *41*, 367–371.
16. Zhong, P.; Peng, E.G.; Li, J.; Xiao, X. Study of erosion behavior of polyurethane-urea coating. *Tribology* **2007**, *27*, 447–450.
17. Ma, W.; Tang, P.; Zhou, X.; Li, G.; Zhu, W. Study on the Failure Mechanism of a Modified Hydrophilic Polyurethane Material Pisha Sandstone System under Dry–Wet Cycles. *Polymers* **2022**, *14*, 4837. [CrossRef] [PubMed]
18. Cho, K.; Lee, D. Effect of molecular weight between cross-links on the abrasion behavior of rubber by a blade abrader. *Polymer* **2000**, *41*, 133–140. [CrossRef]
19. Ackland, K.; Anderson, C.; Ngo, T.D. Deformation of polyurea-coated steel plates under localised blast loading. *Int. J. Impact Eng.* **2013**, *51*, 13–22. [CrossRef]
20. Suthar, V.; Asare, M.A.; de Souza, F.M.; Gupta, R.K. Effect of graphene oxide and reduced graphene oxide on the properties of sunflower oil-based polyurethane Films. *Polymers* **2022**, *14*, 4974. [CrossRef]
21. Dong, M.; Li, Q.; Liu, H. Thermoplastic polyurethane-carbon black nanocomposite coating: Fabrication and solid particle erosion resistance. *Polymer* **2018**, *158*, 381–390. [CrossRef]
22. Wei, J.; Kong, F.; Liu, J.; Chen, Z.; Kanungo, D.; Lan, X.; Jiang, C.; Shi, X. Effect of sisal fiber and polyurethane admixture on the strength and mechanical behavior of sand. *Polymers* **2018**, *10*, 1121. [CrossRef] [PubMed]
23. Liu, R.; Zhang, S.; Wang, X. Erosion performance of polyurethane coatings on motor train unit surface. *Mech. Eng.* **2015**, *12*, 60–62.
24. Acierno, D.; Sanguigno, L.; Arena, G.; Friedrich, K.; Padenko, E.; Russo, P. Erosion behavior and mechanical properties of thermoplastic polyurethanes. In Proceedings of the Times of Polymers & Composites, Ischia, Italy, 22–26 June 2014. [CrossRef]
25. Dong, M.; Wang, C.; Liu, H. Enhanced Solid Particle Erosion Properties of Thermoplastic Polyurethane-Carbon Nanotube Nanocomposites. *Macromol. Mater. Eng.* **2019**, *302*, 1900010. [CrossRef]
26. Doyle, R.A.; Ball, A. On thermomechanical effects during solid particle erosion. *Wear* **1991**, *151*, 87–95. [CrossRef]
27. Qi, C.; Bai, Y.; Liu, J.; Bu, F.; Kanungo, D.P.; Song, Z.; He, X. Desiccation cracking behavior of polyurethane and polyacrylamide admixed clayey soils. *Polymers* **2020**, *12*, 2398. [CrossRef]
28. Li, H. Erosion Wear Theory and Influencing Factors. *Light Ind. Sci. Technol.* **2015**, *31*, 31–32.
29. Wang, W.; Shi, G.; Wang, M.; Tao, C.; Bao, J.; Huang, Y.; Xu, G. Effect of isocyanate types on the properties of thermoplastic polyurethane elastomers. *Appl. Chem. Ind.* **2019**, *48*, 811–815. [CrossRef]
30. Fan, J.; Chen, A. Studying a flexible polyurethane elastomer with improved impact-resistant performance. *Polymers* **2019**, *11*, 467. [CrossRef]
31. *GB/T 1031-2009*; Geometrical Product Specifications-Surface texture: Profile method-Surface roughness parameters and their values. Standardization Administration of China: Beijing, China, 2009.
32. Yue, L.; Li, W.; Bai, Y. Synthesis of water-based acrylate pressure sensitive adhesive for electromagnetic shielding and its properties. *Adhesion* **2016**, 22–27. [CrossRef]
33. Sigamani, N.S. Characterization of Polyurethane at Multiple Scales for Erosion Mechanisms under Sand Particle Impact. 2010. Available online: https://www.researchgate.net/publication/45290155 (accessed on 24 September 2023).

polymers MDPI

Article

Tensile and Compressive Properties of Woven Fabric Carbon Fiber-Reinforced Polymer Laminates Containing Three-Dimensional Microvascular Channels

Ziqian An [1], Xiaoquan Cheng [1,*], Dafang Zhao [2], Yihao Ma [3], Xin Guo [1] and Yujia Cheng [1]

1 School of Aeronautic Science and Engineering, Beihang University, Beijing 100191, China; anziqian@buaa.edu.cn (Z.A.); gxdfxxx@163.com (X.G.); yujia.cheng@buaa.edu.cn (Y.C.)
2 Aviation Industry Corporation of China, Ltd. (AVIC) Manufacturing Technology Institute, Beijing 100024, China; zhaodf003@avic.com
3 Research Institute of Navigation and Control Technology, China North Industries Group, Beijing 100089, China; mayihao@buaa.edu.cn
* Correspondence: xiaoquan_cheng@buaa.edu.cn

Abstract: Microvascular self-healing composite materials have significant potential for application and their mechanical properties need in-depth investigation. In this paper, the tensile and compressive properties of woven fabric carbon fiber-reinforced polymer (CFRP) laminates containing three-dimensional microvascular channels were investigated experimentally. Several detailed finite element (FE) models were established to simulate the mechanical behavior of the laminate and the effectiveness of different models was examined. The damage propagation process of the microvascular laminates and the influence of microvascular parameters were studied by the validated models. The results show that microvascular channels arranged along the thickness direction (z-direction) of the laminates are critical locations under the loads. The channels have minimal effect on the stiffness of the laminates but cause a certain reduction in strength, which varies approximately linearly with the z-direction channel diameter within its common design range of 0.1~1 mm. It is necessary to consider the resin-rich region formed around microvascular channels in the warp and weft fiber yarns of the woven fabric composite when establishing the FE model. The layers in the model should be assigned with equivalent unidirectional ply material in order to calculate the mechanical properties of laminates correctly.

Keywords: self-healing composites; woven fabric CFRP; microvascular; experimental test; finite element analysis

Citation: An, Z.; Cheng, X.; Zhao, D.; Ma, Y.; Guo, X.; Cheng, Y. Tensile and Compressive Properties of Woven Fabric Carbon Fiber-Reinforced Polymer Laminates Containing Three-Dimensional Microvascular Channels. *Polymers* **2024**, *16*, 665. https://doi.org/10.3390/polym16050665

Academic Editor: Roberto De Santis

Received: 5 January 2024
Revised: 14 February 2024
Accepted: 26 February 2024
Published: 29 February 2024

1. Introduction

Due to the poor interlaminar performance of fiber-reinforced polymer composite laminates and the shortcomings of existing non-destructive testing and repair methods for composite structures [1], self-healing structural polymers and fiber-reinforced composites have been proposed, which can be divided into intrinsic and external approaches [2]. The intrinsic self-healing system is mainly based on the reversible chemical reaction of the matrix material itself, which does not affect the structural integrity but is only suitable for repairing small damages or scratches [3]. The external approaches, inspired by the self-healing characteristics of organisms after injury, use microcapsules, hollow fibers, or micro channels to transport healing agents to promptly repair damages [4]. Among these approaches, microvascular self-healing is more promising for it allows multiple efficient repairs of delamination damage or matrix cracks [5–9]. It can effectively reduce maintenance costs, improve safety, and extend structure service life if implemented.

Luterbacher et al. [10] incorporated microvascular self-healing into a composite skin-stringer structure to deliver the healing agent and found that the structural performances could be fully restored by using microvascular self-healing to repair the debonding interface between the stringer and skin panel. Sakurayama et al. [11] conducted impact and compression tests on composite stiffened panels containing microvascular networks and repaired the impact damage using them. The results showed that the repaired stiffened panel could recover 50% of its compression strength compared to the unrepaired specimens. These all confirm the potential of microvascular systems under practical conditions. Additionally, there are relatively mature technical routes in terms of manufacture processes and healing agents, such as the vaporization of sacrificial component (VaSC) method [7] and the epoxy resin system [12]. However, the microvascular channels can also be regarded as initial damage that affects the mechanical properties of the structure. Therefore, it is necessary to thoroughly study the mechanical performance of laminates containing microvascular channels so as to determine appropriate design parameters in actual structures.

A number of experimental studies have now been conducted to address the issue of the mechanical performance of microvascular composites. Kousourakis et al. [13] tested the tensile and compressive properties of laminates containing micro channels located in the mid-plane of the laminate. As the diameter of the microvascular channels increased from 0.3 mm to 3 mm, the strength and stiffness of the specimens with longitudinally oriented channels decreased by less than 10%, while the performance of the specimens with transversely oriented channels decreased significantly, with a maximum reduction in tensile strength of 50%. The main reason for this significant performance loss was the bending of fibers around the channels which resulted in a change in the stress state. Devi et al. [14] also reached similar conclusions. Saeed et al. [15] conducted three-point bending and short beam strength tests on laminates containing in-plane microvascular channels and found that both the bending strength and short beam strength of the specimens linearly decreased as the diameter of the channels increased. With a channel diameter of 1.5 mm, the short beam strength decreased by about 33%, and the bending strength decreased by about 15%. Coppola et al. [16] investigated the tensile properties and damage propagation of 3D orthogonally woven glass fiber composites containing straight and undulating wave-shaped micro channels and found that reductions in strength and modulus only occurred when channels distorted the fiber architecture. Norris et al. [17] found that cutting the fibers around the channel can prevent the formation of a resin-rich region, but this will lead to a more significant decrease in the mechanical performance of the laminate.

Some researchers have also attempted to conduct studies using finite element methods. Nguyen and Orifici [18] first conducted experiments on laminates containing micro channels with a diameter of 0.68 mm. They found that the tensile stiffness of the laminate perpendicular to the microvascular channels could decrease by up to 7.5%, and the compressive strength could decrease by 4.9%, while the performance decrease along the direction of the channel was not significant. They further established a representative volume element (RVE) model of the microvascular channel. The composite plies were modeled using continuous shell elements, and the two-dimensional Hashin criterion was used to determine the damage of the composite material. The resin-rich region was considered and the numerical results of mechanical performance and failure modes were in good agreement with the experimental results. Huang et al. [19] established a plane strain model, while Shawk et al. [20], Demiral et al. [21], and Zhao et al. [22] established three-dimensional models to study the influence of in-plane microvascular channels on different mechanical properties of laminates. Ran et al. [23] also considered the variation in fiber volume fraction in the area around the microvascular channels where fibers are bent in an FE model. Compared to models that do not consider this factor, the calculated results of laminate strength and stiffness were more accurate.

It is evident that current researchers primarily focus on laminates with a one-dimensional microvascular channel arranged between composite layers. A few studies have demonstrated research on three-dimensional microvascular channels but lack simulation analysis. Delamination damage can occur at any position within the laminate during the structures' manufacturing and service period [24]. In-plane micro channels can only repair damage between specific layers, and arranging channels in multiple layers would lead to a significant decline in the laminate's mechanical performance. Therefore, in a three-dimensional microvascular configuration, the in-plane microvascular channels are used to transport healing agents and the z-direction microvascular channels are used to repair delamination damage at different positions. This may be the way to make self-healing structures available for engineering applications. In order to provide a reference and basis for the design of microvascular composite structures, studies on the mechanical performance of laminates with such microvascular configurations should be conducted.

In this paper, the tensile and compressive properties of woven fabric CFRP laminates containing three-dimensional microvascular channels were investigated experimentally. New detailed finite element models with a resin-rich region and variations in fiber volume fraction around the z-direction microvascular based on the actual structure were established, which were employed to study the damage propagation of laminates under tensile and compressive loads. The effects of microvascular parameters, including diameter, spacing and volume fraction, on the tensile and compressive properties of the laminates were discussed using the FE model. Finally, the design criteria for microvascular self-healing composite structures were summarized based on the parameter study results.

2. Experiment

2.1. Material

The laminates in this study were all made of CF3031 carbon fiber fabric and 5284 epoxy resin, with the mechanical properties listed in Table 1. The materials and the nominal properties were all provided by AVIC Manufacturing Technology Institute.

Table 1. Mechanical properties of CF3031/5284 ply.

Property	Value	Property	Value
E_1, E_2/GPa	55.0	X_t, Y_t/MPa	550
E_3/GPa	8.4	X_c, Y_c/MPa	593
G_{12}/GPa	3.64	Z_t/MPa	80
G_{13}, G_{23}/GPa	3.0	Z_t/MPa	180
ν_{12}	0.051	S_{12}/MPa	84
ν_{13}, ν_{23}	0.15	S_{13}, S_{23}/MPa	80

2.2. Specimen Design and Manufacture

The tensile and compression specimens containing microvascular channels were designed according to the ASTM D3039 [25] and D6641 [26] standards. The specimens contained two parallel, three-dimensional microvascular channels, as illustrated in Figure 1. The layup of the specimens was $[(0,90)/\pm45/(0,90)/\pm45/(0,90)]$ s, with a nominal thickness of 0.25 mm per layer. The width of the specimens was 24 mm, and the spacing of the z-direction channels was 12 mm. The diameter of the channel was 0.5 mm, with the in-plane channels located two layers beneath the surface of the specimen, as illustrated in Figure 2. For compression specimens, care was taken during preparation to ensure the presence of z-direction channel within the gage section. Blank specimens were also prepared as controls.

In the manufacturing procedure, the carbon fiber-woven fabric was manually laid to form a preform and polylactic acid (PLA) threads were sewn into the preform at pre-determined intervals. After this, the vacuum-assisted resin infusion (VARI) process was used for resin impregnation and curing. Finally, the VaSC method was used when the PLA threads were evaporated in an oven, leaving hollow channels. In order to make their decomposition temperature much lower than the glass transition temperature of the composite matrix resin, some catalyst was added into the PLA threads.

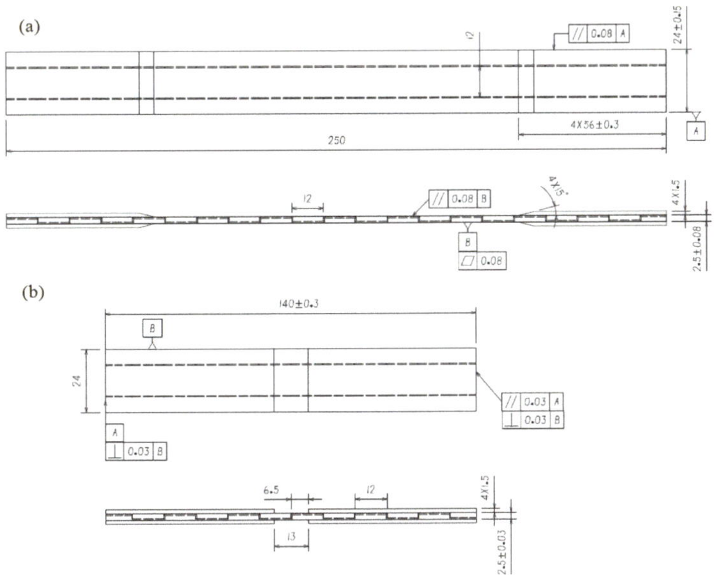

Figure 1. Configuration and geometric parameters (mm) of specimens: tensile specimen (**a**) and compression specimen (**b**).

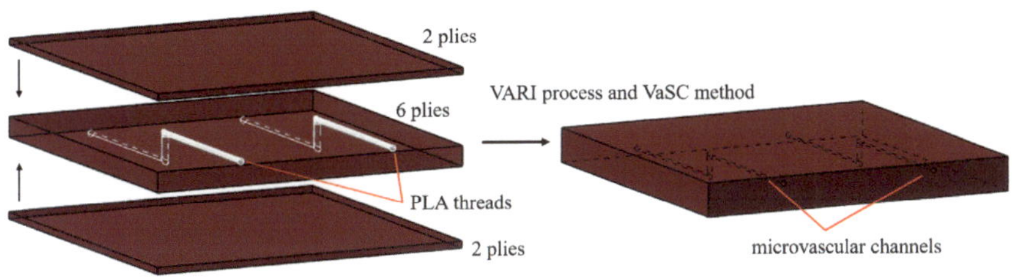

Figure 2. Specimen structure and manufacturing process schematic.

2.3. Mechanical Testing

Tensile and compression tests were conducted on an INSTRON-8801 testing machine (Norwood, MA, USA) in a standard laboratory environment (23 ± 2 °C, 50 ± 10% relative humidity), using displacement control loading at a rate of 2 mm/min. The width and thickness of the gage section of each specimen were measured three times before testing, and the average values were taken. Five microvascular specimens and five blank specimens were tested in both experiments. The microvascular specimens for tensile and compression test were numbered XT-P-1~5 and XC-P-1~5, while the blank specimens were numbered XT-C-1~5 and XC-C-1~5 accordingly.

The tensile test was conducted according to ASTM D3039; strain in the longitudinal and transverse directions of the specimen was measured using two extensometers. The stiffness of the specimen was calculated using data from the longitudinal extensometer within the range of 1000 με to 3000 με. Extensometers were removed when the longitudinal strain was 5000 με, then the specimen was stretched to failure. The failure load and mode were recorded. The setup of the test is shown in Figure 3.

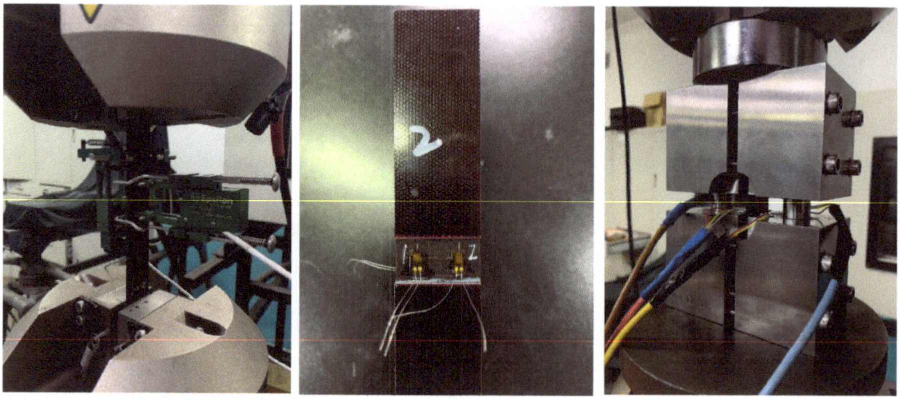

(**a**) (**b**) (**c**)

Figure 3. Test set-up and extensometer installation in tensile test (**a**). Strain gauges on compression specimen (**b**). Test set-up in compression test (**c**).

The compression test was conducted according to ASTM D6641. Four strain gauges were used on the gage section to measure the strain of the specimen. The stiffness of the specimen was calculated using the gauges' data within the range of 1000 με to 3000 με. The specimen was loaded until failure, and the failure load and mode were recorded. The location of the strain gauges and the setup of the test are shown in Figure 3.

All specimens were determined to have failed when visible fracture occurred and the load of the testing machine rapidly decreased by more than 30%.

2.4. Result and Analysis

Test results are shown in Table 2. The tensile stiffness and strength of the specimens with microvascular channels reduced by 6.1% and 11.9% compared with the control group, while the compressive strength reduced by 10.3% but the stiffness increased by 2.1%. The stress–displacement curves of the tensile specimens and stress–strain curves of the compression specimens are shown in Figure 4. It can be observed that the microvascular has an evident effect on the strength of the laminates, but a relatively smaller effect on the stiffness. When the strain is small, the curves of the two types of specimens are very close to each other. The increased stiffness of the compression specimens with microvascular channels may be due to the dispersion of the material properties, which has a greater effect than that of the microvascular channels.

Table 2. Test results.

Specimen Type	Specimens	Tensile Stiffness/GPa	Tensile Strength/MPa	Specimens	Compression Stiffness/GPa	Compression Strength/MPa
Blank	XT-C-1	43.51	419.75	XC-C-1	41.34	475.67
	XT-C-2	43.36	436.13	XC-C-2	42.22	455.59
	XT-C-3	43.85	430.25	XC-C-3	41.26	477.40
	XT-C-4	45.14	466.70	XC-C-4	39.30	467.92
	XT-C-5	44.05	454.45	XC-C-5	40.40	437.92
	Average	43.98	441.45	Average	40.90	462.9
Microvascular	XT-P-1	41.15	376.07	XC-P-1	41.48	416.11
	XT-P-2	41.46	400.55	XC-P-2	41.20	391.21
	XT-P-3	40.63	404.09	XC-P-3	42.38	438.65
	XT-P-4	41.44	386.48	XC-P-4	42.27	413.13
	XT-P-5	41.72	376.82	XC-P-5	41.45	416.27
	Average	41.30	388.80	Average	41.75	415.10
	Variations of AVG/%	−6.1	−11.9	Variations of AVG/%	2.1	−10.3

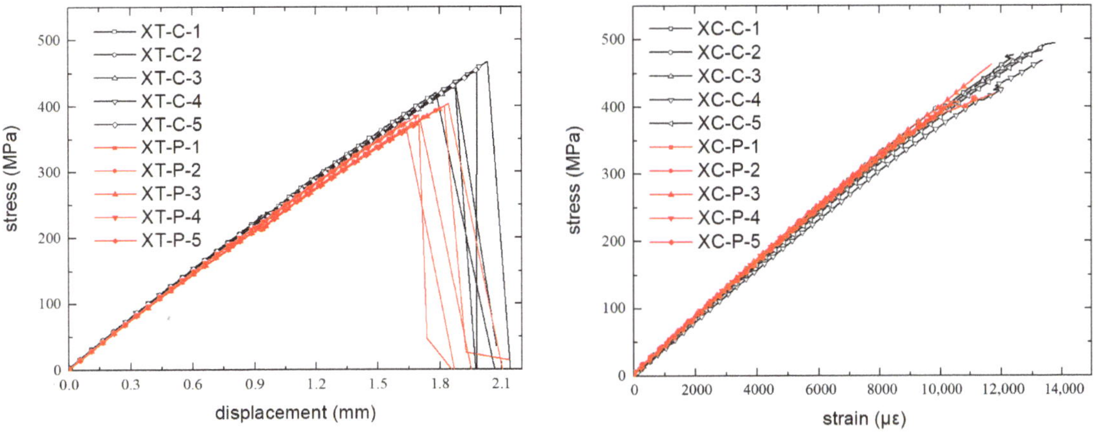

(**a**) (**b**)

Figure 4. Stress–displacement curves of tensile specimens (**a**) and stress–strain curves of compression specimens (**b**).

The failure modes of the tensile specimens are shown in Figure 5a. Three of the specimens failed at the middle of the gage section. Figure 5b shows the fracture section captured by optical microscopy and the z-direction channels are visible, suggesting that the z-direction microvascular channels are found at key sections of these specimens when carrying tensile load. Specimens XT-P-1 and XT-P-2 failed near the grip section where stress concentration existed, but their strengths were not significantly lower than others.

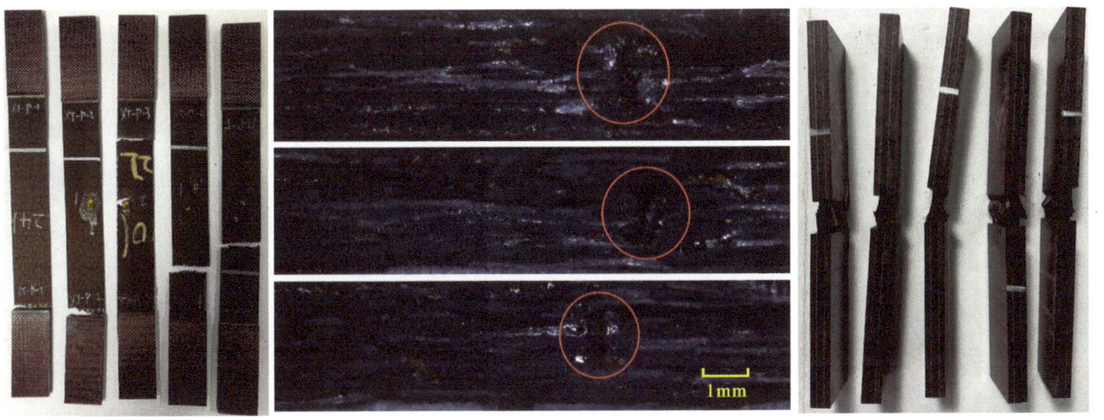

(**a**) (**b**) (**c**)

Figure 5. Failure modes of the specimens, lateral tensile failure at middle of gage section or top of grip section (**a**), enlarged view of the tensile fracture section (z-direction channels are circled in red) (**b**), and brooming compression failure at middle of gage section (**c**).

For compression specimens, the primary failure mode was brooming fracture in the middle of the gage section where the z-direction microvascular channel was located, as shown in Figure 5c; however, it was difficult to determine the z-direction channel due to the extensive damage in the matrix of the compression specimens.

When bearing loads, the z-direction microvascular channel causes stress concentration around it, thereby reducing the strength. However, since the microvascular channel only causes a slight bending of fibers in local areas, which is also a feature of the woven fabric material itself, the fibers are not cut and the total fiber volume does not change, so that the stiffness of laminates is barely affected.

3. Finite Element Model

To further investigate the failure mechanism of laminates containing microvascular channels and to conduct parametric studies, it is necessary to establish a detailed finite element model; however, modeling the whole structure is too hard to achieve. It is advisable to consider in-plane and z-direction microvascular channels separately. The mechanical properties of laminates containing in-plane microvascular channels have been the focus of some studies, in which it was observed that the mechanical properties of the laminates along the direction of the in-plane channel were less affected [13,14,16,18]. Also, the experimental results in Section 2 indicate that the z-direction of microvascular channels is a critical position. Therefore, the model will focus on the z-direction of microvascular channels and its influence on the mechanical properties of the laminate. Finite element analysis was performed on ABAQUS 6.14 software.

3.1. Model Generation

3.1.1. Resin-Rich Region

In laminates, the resin-rich region around microvascular channels can cause stress concentration or fiber bending, which are key factors affecting the mechanical performance of the laminate. It is important to consider the resin-rich region in the FE model. Ma et al. [27] developed a numerical method to predict the shape of the resin-rich region around microvascular channels, and this paper refers to this method to determine the length of the resin-rich region.

The z-direction microvascular channels in woven fabric composite will lead to the formation of two perpendicular resin-rich regions within the warp and weft fiber yarns in one ply. Since the forming process involves stitching sacrificial lines into the preform before resin impregnation, the fibers are still continuous. Typical intralaminar resin-rich region in the current specimens are shown in Figure 6.

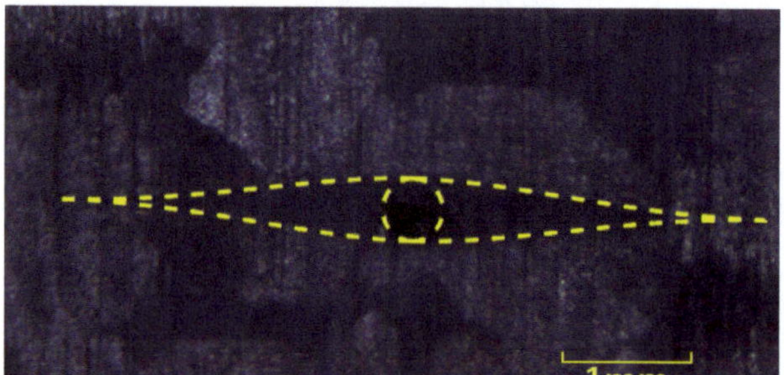

Figure 6. Intralaminar resin-rich region around the microvascular channel. (The boundaries are marked with dotted line).

Some specimens were cut to measure the length of the resin-rich region. The measurement results were compared with the numerical results as illustrated in Figure 7. The measurement results show obvious dispersion, which is primarily due to the bending of fiber in the fabric material, and so the complete resin-rich region might not be observable on certain thickness sections. By adjusting the parameter values in the numerical methods, the calculated results exceeded 95% of the measured values. The lengths of the resin-rich region used in the FE models are presented in Table 3.

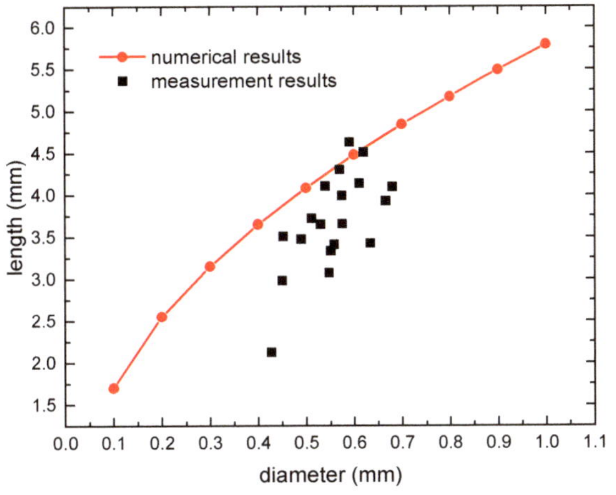

Figure 7. Comparison of numerical and measurement results of resin-rich region length.

Table 3. Resin-rich region length in FE model.

Microvascular Diameter/mm	Resin-Rich Region Length/mm	Microvascular Diameter/mm	Resin-Rich Region Length/mm
0.1	1.70	0.6	4.48
0.2	2.55	0.7	4.84
0.3	3.15	0.8	5.17
0.4	3.65	0.9	5.49
0.5	4.04	1.0	5.79

3.1.2. Variation of Fiber Volume Fraction

The fiber volume fraction around the microvascular channel will change as the fibers are pushed away, and the mechanical properties of the local material will be affected. The authors of [23] illustrated the necessity of considering this factor in in-plane microvascular models. This paper also considers this factor in the model of the z-direction microvascular channels first, and then compares different modeling approaches without it.

Since there are two resin-rich regions in the warp and weft yarns in one layer, it is necessary to model them separately. This paper assumed that the fabric layer was divided into warp and weft sublayers in the model, and each sublayer was treated as an equivalent unidirectional ply. By assigning appropriate material properties, the mechanical performances of the combination of two sublayers remained the same as those of the original fabric material, and the properties of the sublayers and fabric layers satisfy the following relationships:

$$E_L = E_T = \frac{1}{2}(E_{11} + E_{22}) \tag{1}$$

$$\nu_L = \nu_T = \frac{\nu_{21}E_{11} + \nu_{12}E_{22}}{E_{11} + E_{22}} \tag{2}$$

$$G_{LT} = G_{12} \tag{3}$$

$$X_{Lt} = X_{Tt} = \frac{1}{2}\left(X_t + \frac{X_t}{E_{11}}E_{22}\right) \tag{4}$$

$$X_{Lc} = X_{Tc} = \frac{1}{2}\left(X_c + \frac{X_c}{E_{11}}E_{22}\right) \tag{5}$$

where the subscripts L and T represent the longitudinal and transverse direction of the fabric. E_L, E_T, ν_L, ν_T, G_{LT}, X_{Lt}, X_{Tt}, and X_{Lc}, X_{Tc} are the elastic modulus, Poisson's ratio, shear modulus, tensile strength, and compression strength of the woven fabric. E_{11}, E_{22}, ν_{12}, ν_{21}, G_{12}, X_t, and X_c are the elastic modulus, Poisson's ratio, shear modulus, tensile strength and compression strength of the unidirectional ply. The out-of-plane properties of the two materials are considered the same.

In order to ensure that the simulation results of the fabric material are consistent with the nominal values, it is necessary to apply correction factors when calculating the modulus and strength parameters of the unidirectional ply based on the properties of the fibers and resin. Referring to the Chamis [28] model, the calculation formulas are as follows:

$$E_{11} = \alpha_1\left(V^f E^f_{11} + V^m E^m\right) \tag{6}$$

$$E_{22} = \alpha_2 \frac{E^f_{22} E^m}{E^f_{22} - \sqrt{V^f}\left(E^f_{22} - E^m\right)} \tag{7}$$

$$\nu_{12} = \alpha_3\left(V^f \nu^f_{11} + V^m \nu^m\right) \tag{8}$$

$$G_{12} = \alpha_4 \frac{G^f_{12} G^m}{G^f_{12} - \sqrt{V^f}\left(G^f_{12} - G^m\right)} \tag{9}$$

$$G_{23} = \alpha_5 \frac{G^f_{23} G^m}{G^f_{23} - \sqrt{V^f}\left(G^f_{23} - G^m\right)} \tag{10}$$

$$X_t = \beta_t\left(V^f \sigma_f + V^m \frac{\sigma_f}{E^f_{11}} E^m\right) \tag{11}$$

$$X_c = \beta_c\left(V^f \sigma_f + V^m \frac{\sigma_f}{E^f_{11}} E^m\right) \tag{12}$$

where the subscripts f and m represent the fiber and resin. V^f, V^m are the fiber and resin's volume fraction. E^m, G^m, and ν^m are the elastic modulus, Poisson's ratio, and shear modulus of the resin. E^f_{11}, E^f_{22} are the elastic moduli in the longitudinal and transverse directions of the fiber. ν^f_{12}, G^f_{12}, and G^f_{23} are the Poisson's ratio and shear modulus of the fiber. σ_f is the fiber's strength. α and β are correction factors.

The properties of the fibers and resin used in the research are shown in Table 4, and the fiber volume fraction is 55%. The correction factors must be adjusted to ensure that the failure strain in the fiber direction of the unidirectional ply matches that of the actual fabric material, as indicated in Table 5. The calculated properties of the equivalent unidirectional ply are presented in Table 6. It is assumed that the transverse tensile and compressive strengths of the unidirectional ply are same as the strengths of the matrix. The material properties of the unidirectional ply used in the following models are all calculated using the method described in this section.

Table 4. Mechanical properties of the fiber and resin.

E^f_{11}/GPa	E^f_{22}/GPa	G^f_{12}/GPa	G^f_{23}/GPa	ν^f_{12}	σ_f/MPa	E^m/GPa	G^m/GPa	ν^m	S^m_t/MPa	S^m_c/MPa
230	13.8	9	4.8	0.2	3530	3.2	1.13	0.42	80	180

Table 5. Correction factor value.

α_1	α_2	α_3	α_4	α_5	β_t	β_c
0.8118	0.8257	1.5279	1.1322	1.1538	0.5321	0.5735

Table 6. Equivalent unidirectional ply properties calculated by modified Chamis model.

E_{11}/GPa	E_{22}/GPa	ν_{12}	G_{12}/GPa	G_{12}/GPa	X_t/MPa	X_c/MPa
103.86	6.14	0.46	3.64	3.00	1038.6	1119.8

3.2. Failure Criteria and Material Property Degradation

3.2.1. CFRP Material

Three-dimensional Hashin failure criteria [29], Chang fiber-shear failure criteria [30], and Ye delamination failure criteria [31] were employed to predict the different damage modes in the laminate, which are explained in detail as follows:

Fiber failure:

$$\left(\frac{\sigma_{11}}{X_T}\right)^2+\left(\frac{\tau_{12}}{S_{12}}\right)^2+\left(\frac{\tau_{13}}{S_{13}}\right)^2\geq 1\ (\sigma_{11}\geq 0) \tag{13}$$

$$\left(\frac{\sigma_{11}}{X_C}\right)^2\geq 1\ (\sigma_{11}< 0) \tag{14}$$

Matrix failure:

$$\left(\frac{\sigma_{22}}{Y_T}\right)^2+\left(\frac{\tau_{12}}{S_{12}}\right)^2+\left(\frac{\tau_{23}}{S_{23}}\right)^2\geq 1\ (\sigma_{22}\geq 0) \tag{15}$$

$$\left(\frac{\sigma_{22}}{Y_C}\right)^2+\left(\frac{\tau_{12}}{S_{12}}\right)^2+\left(\frac{\tau_{23}}{S_{23}}\right)^2\geq 1\ (\sigma_{22}< 0) \tag{16}$$

Fiber–matrix shear failure:

$$\left(\frac{\sigma_{11}}{X_C}\right)^2+\left(\frac{\tau_{12}}{S_{12}}\right)^2+\left(\frac{\tau_{13}}{S_{13}}\right)^2\geq 1\ (\sigma_{11}< 0) \tag{17}$$

Delamination:

$$\left(\frac{\sigma_{33}}{Z_T}\right)^2+\left(\frac{\tau_{13}}{S_{13}}\right)^2+\left(\frac{\tau_{23}}{S_{23}}\right)^2\geq 1\ (\sigma_{33}\geq 0) \tag{18}$$

$$\left(\frac{\sigma_{33}}{Z_C}\right)^2+\left(\frac{\tau_{13}}{S_{13}}\right)^2+\left(\frac{\tau_{23}}{S_{23}}\right)^2\geq 1\ (\sigma_{33}< 0) \tag{19}$$

where σ_{11}, σ_{22}, and σ_{33} are normal stress components along the longitudinal, transverse, and thickness directions, respectively. τ_{12}, τ_{13}, and τ_{23} are shear stress components. X_T and X_C are tensile and compressive strengths along the longitudinal direction. Y_T and Y_C are tensile and compressive strengths along the transverse direction. Z_T and Z_C are tensile and compressive strengths along the thickness direction. S_{12}, S_{13}, and S_{23} are shear strengths. The properties of unidirectional CFRP were obtained by the method presented in Section 3.1.2 and listed in Tables 1 and 6.

Once the stress state of an element satisfies any of the above failure criteria, stiffness parameters of the element will be degraded to a certain value according to the degradation rules proposed by Camanho and Matthews [32] as listed in Table 7.

Table 7. Stiffness degradation rules of composite.

Failure Mode	Stiffness Degradation Rule
Fiber failure	0.07 × all parameters
Matrix failure	$0.2 \times E_{22}, G_{12}, G_{23}, \mu_{12}, \mu_{23}$
Fiber–matrix shear failure	$0.2 \times G_{12}, \mu_{12}$
Delamination	$0.2 \times E_{33}, G_{13}, G_{23}, \mu_{13}, \mu_{23}$

3.2.2. Resin

For the resin-rich region, the parabolic criterion [33] was used to determine the initiation of resin damage:

$$\frac{3J + I(S_{mc} - S_{mt})}{S_{mc}S_{mt}} = 1\ (I \geq 1) \tag{20}$$

$$-\frac{3J = I(S_{mc} - S_{mt})}{S_{mc}S_{mt}} = 1\ (I < 1) \tag{21}$$

where S_{mt} and S_{mc} are the unidirectional tensile and compressive strength of resin as listed in Table 4. I and J are defined as follows:

$$I = \widetilde{\sigma}_{m1} + \widetilde{\sigma}_{m2} + \widetilde{\sigma}_{m3} \tag{22}$$

$$J = \frac{1}{6}\left[(\widetilde{\sigma}_{m1} - \widetilde{\sigma}_{m2})^2 + (\widetilde{\sigma}_{m1} - \widetilde{\sigma}_{m3})^2 + (\widetilde{\sigma}_{m2} - \widetilde{\sigma}_{m3})^2\right] \tag{23}$$

where $\widetilde{\sigma}_{m1}$, $\widetilde{\sigma}_{m2}$, and $\widetilde{\sigma}_{m3}$ are principle stress components. The stiffness parameters of failure elements are degraded according to the relation of $E^m = 0.2E^m, \mu^m = 0.2\mu^m$. The failure criteria and material property degradation rules of CFRP and resin are defined in the VUMAT subroutine of ABAQUS.

3.3. Model Details

The model with z-direction microvascular channels that considers the resin-rich region, fiber bending, and variations in material properties is shown in Figure 8, referred to as model A. In this model, it is assumed that the fiber volume fraction linearly decreases from the edge of the channel to the surrounding area, while ensuring the conservation of the total fiber volume. Different colors in the elements represent different material properties. This was accomplished by using Python script. The coordinates of each node were read, and the fiber volume fraction was calculated based on the distance from the center of the element to the center of the channel. Then, the material parameters were calculated according to Formulas (6)~(12) and assigned to the corresponding elements.

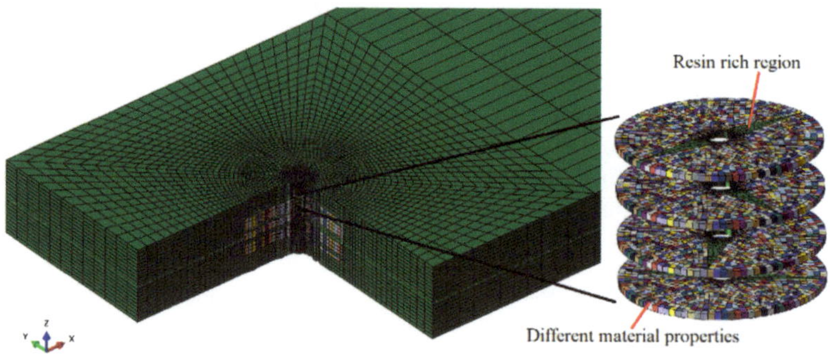

Figure 8. Z-direction microvascular channel in model A (Different colors of elements represent different material properties).

Symmetric models were established to save computational time. The tensile model only includes the gage section, while the compression model does not include reinforcement tabs. The z-direction microvascular structure was tied to the overall structure. One end of the model was fixed, the other was coupled with a reference point where the displacement load is applied, and the reaction force was extracted. In the tensile model, the displacements of the side nodes were extracted to calculate the stiffness, while in the compression model, the strain at the center of the surface was extracted for the same purpose, consistent with the experimental measurement method.

The fiber bending region and resin-rich region near the channel were meshed finely to investigate the damage propagation in more detail with a maximum element size of 130 μm in the x- and y-directions. The main element type was C3D8R, with a few C3D6 elements in the resin-rich region, and the total number of elements in tensile and compression models were 110,344 and 104,616, respectively. The models of the laminates, including boundary conditions, loading condition, and nodes for result output, are shown in Figure 9.

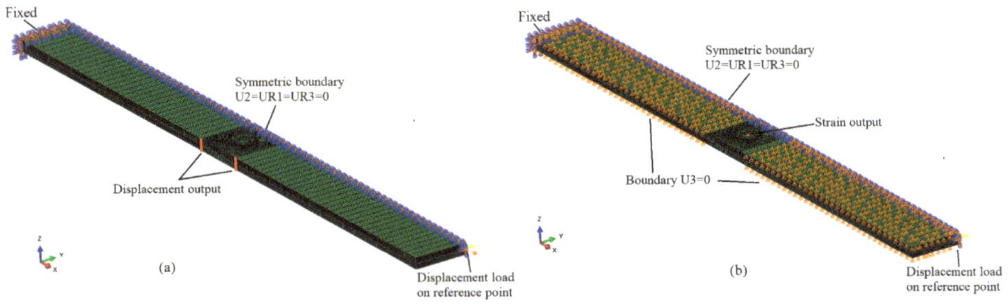

Figure 9. Boundary conditions, loading condition, and nodes for result output: tensile model (**a**) and compression model (**b**).

To investigate the necessity of this modeling approach, several other z-direction microvascular models were also established for comparison. Model B did not consider the variation in fiber volume fraction compared to model A. Model C did not divide the fabric layer into two unidirectional ply and applied fabric material directly compared to model B. These two models had the same mesh as model A. Model D did not consider the resin-rich regions, bending of fibers near the channel, and variation in fiber volume fraction. The model contained 45,292 elements and also applied fabric material properties. This modeling process is simple but equivalent to cutting the fibers and reducing the total fiber volume. The control models are shown in Figure 10.

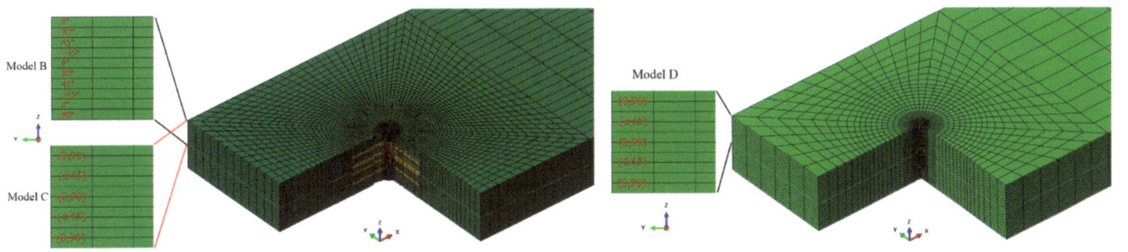

Figure 10. Schematic of the comparison model.

3.4. Validation and Comparison of the FE Models

The tensile and compressive performances of the microvascular laminates calculated by each model, along with their comparison with experimental results, are listed in Table 8. It shows that since model A considers the variation in the fiber volume fraction around the microvascular channel, which results in higher material strength near the channel, the strengths of the laminate are slightly greater than those of model B. However, the results of model B are still very close to the experimental data. The strength results of model C and D are significantly lower than the experimental results, indicating that these modeling approaches do not accurately reflect the actual performances of the structure. In FE models, the element layers with a longitudinal resin-rich region bear greater load because of higher stiffness and the total load reaches peak value when they fail. So, the strengths of the whole model mainly depend on the properties of these layers, which explains why model C has a larger error with a fabric material assigned. In model D, the fibers are cut and stress concentration is severer, which is inconsistent with reality. The details are discussed in the following section. As for the stiffness performances of the laminates, since the impact of the microvascular channels is minimal, the simulation results of all models are relatively close.

Table 8. Comparison of experiment and simulation results.

Properties	Experiment		Control Model		Model A		Model B		Model C		Model D	
	Control	Vascular	FEM	Error/%	FEM	Error/%	FEM	Error/%	FEM	Error/%	FEM	Error/%
Tensile stiffness/GPa	43.98	41.30	43.67	−0.7	42.95	4.0	42.98	4.1	43.05	4.2	43.29	4.8
Tensile strength/MPa	441.5	388.8	434.7	−1.5	386.6	−0.6	379.1	−2.5	315.1	−19.0	337.7	−13.1
Compression stiffness/GPa	40.90	41.75	43.49	6.3	41.69	−0.1	41.37	−0.9	42.87	2.7	42.86	2.7
Compression strength/MPa	462.9	415.1	469.5	1.4	424.2	2.2	413.1	−0.5	350.5	−15.6	360.4	−13.2

The damage configurations of test and simulation results are listed in Table 9. The tensile specimen was polished to observe the channels and the z-direction channel is circled in red. As can be seen, the numerical results all exhibit lateral tensile or compressive failures. The failure fractures of models A and B are relatively more jagged and closer to the test results, further demonstrating the better rationality of models A and B.

Table 9. Experimental and numerical damage configurations.

Load	Model A	Model B	Model C	Model D	Experiment
Tensile				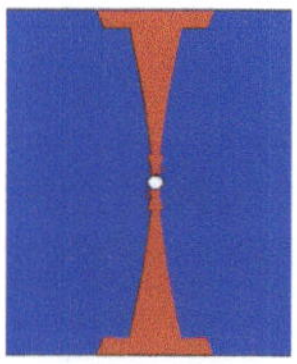	
Compression			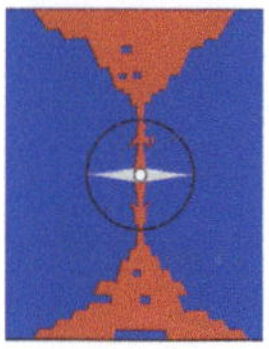	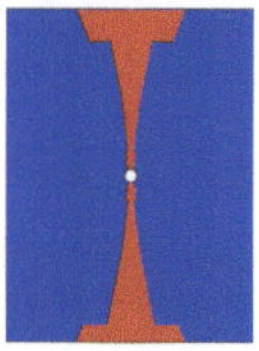	

Based on the above results, it can be concluded that when modeling the z-direction microvascular channel in woven fabric material, the resin-rich regions formed within the warp and weft fibers and the bending of fibers should be considered. Furthermore, the fabric layer should be equivalently modeled as two unidirectional plies. It is not necessary to consider the variation in the fiber volume fraction around the microvascular channel, which simplifies the modeling process appropriately.

3.5. Damage Mechanism

Figure 11 presents the load–displacement curves of the microvascular laminates numerically in model A, as well as the damage propagation process in different layers of the laminate.

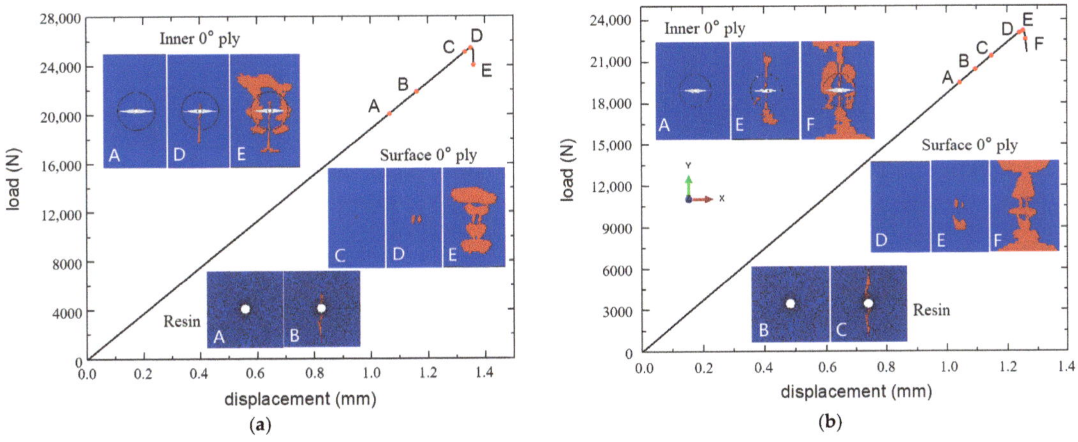

Figure 11. Load–displacement curves and damage propagation of specimens: tensile (**a**) and compression (**b**).

During the tensile process, fiber damage first appeared around the microvascular channel in the internal 0° layers of the laminate at point A (19.5 kN). At point B (20.4 kN), resin damage appeared around the microvascular channel in the internal 90° layers, and extended to the entire resin-rich region at point C (21.4 kN). Before reaching the peak load, the fiber damage in the internal 0° layers continued to extend towards the edges of the laminate, while the surface layers showed no significant damage. After reaching the peak load at point E (23.2 kN), extensive fiber damage rapidly occurred in all layers until the specimen ultimately failed.

During the compression process, fiber damage first appeared around the microvascular channel in the internal 0° layer at point A (20.1 kN), and damage in the resin-rich region around in the 90° layers also occurred. The damage had almost completely penetrated the resin-rich region by point B (21.9 kN). Before reaching the peak load, the fiber damage in the internal layers continued to extend from the edge of the channel towards the laminate edges, with no significant damage in the surface layers. After reaching the peak load at point D (25.4 kN), extensive fiber damage occurred near the channel and at the tips of the resin-rich region in the internal 0° layers, while significant damage also appeared in the surface layers, leading to the final failure of the specimen.

Figure 12 presents the S_{11} stress maps around the microvascular channel in the 0° layer during the loading process, with the laminate under tensile or compression load of 70 MPa and showing no damage. For comparison, the stress maps calculated by model D without resin-rich regions are also presented under the same load conditions. When the resin-rich regions are present, the stress concentration factor around the microvascular channel is approximately 1.44, whereas it is about 2.3 when not considering the resin-rich region. It can be seen that although the presence of resin-rich regions leads to stress concentration, their impact is significantly less than directly drilling holes in the laminate, resulting in a relatively smaller loss in the strength of the laminate.

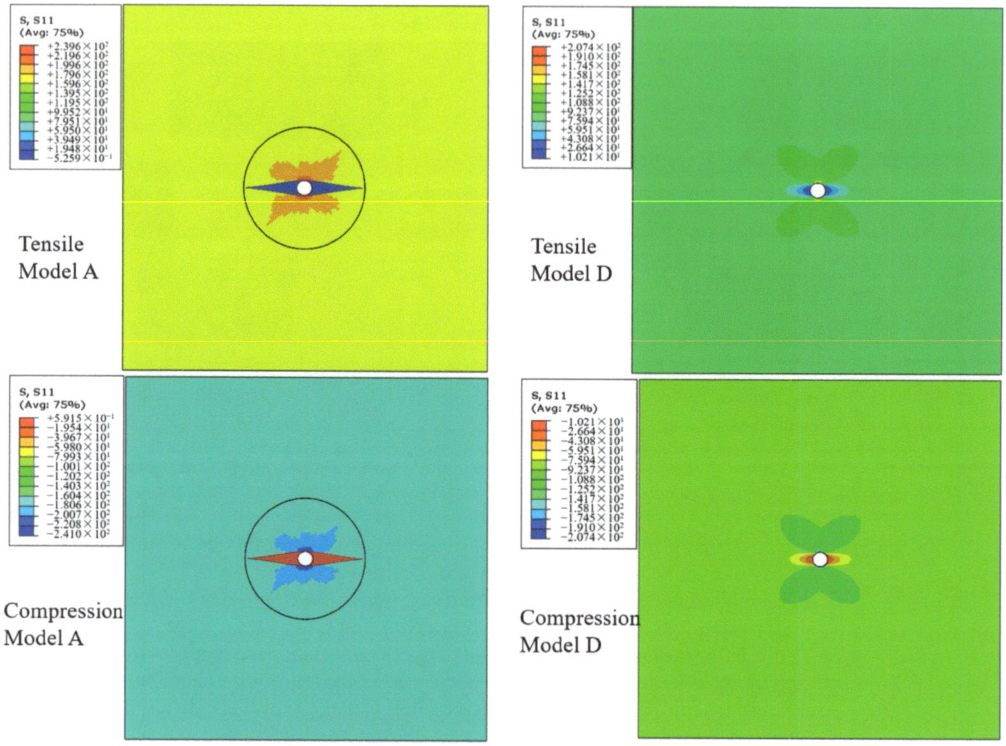

Figure 12. Stress distribution around the microvascular channel.

4. Parameter Study

From Section 3, it is known that model B can calculate the mechanical properties of the laminates accurately and the modeling approach is easier than with model A. Therefore, this modeling method was utilized to investigate the influence of microvascular parameters on the mechanical properties of the laminates.

4.1. Diameter

The materials and layup sequence are the same as those in Section 3, with channel diameters ranging from 0.1 mm to 1 mm. This range is commonly used in current research, for smaller diameters are not conducive to the flow of the repair agent while larger diameters affect the laminate's mechanical performances too much, which is impractical. The spacing of channels, *S*, was set to 8, 12, and 18 mm. The changes in the stiffness and tensile/compressive strengths of the laminates with varying channel diameters are shown in Figure 13. All calculation results presented as a percentage relative to the results of the control model.

It can be observed that the mechanical properties of the microvascular laminates are generally lower than those of the control laminates. However, within the diameter range of 0.1~1 mm, the decrease in laminate stiffness is less than 3%, with a slightly increasing rate of decrease, while the strength of the laminates decreases approximately linearly with channel diameters. Microvascular channels with a spacing of 8 mm and a diameter of 1 mm can reduce the tensile and compressive strength of the laminates by about 15%.

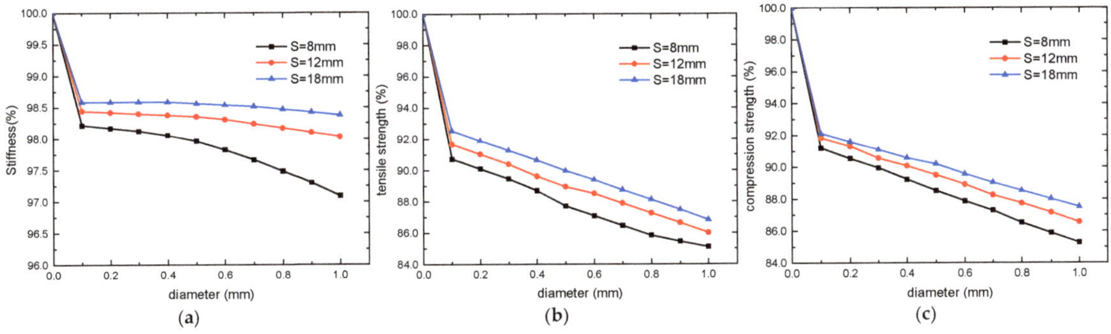

Figure 13. Variation of mechanical properties with diameter of microvascular channels: stiffness (**a**), tensile strength (**b**), and compression strength (**c**).

4.2. Spacing

With microvascular channel diameters *D* set to 0.1, 0.5, and 1 mm, the changes in the stiffness and tensile/compressive strengths of the laminates with varying spacing of channel are shown in Figure 14. It can also be seen that within the current range of design parameters, the stiffness of the laminates is minimally affected by the microvascular channel. When the spacing between channels exceeds 30 mm, there is almost no change in laminate stiffness, and the variations in tensile and compressive strengths are also relatively minor. However, when the spacing is less than 30 mm, the decrease in the mechanical properties of the laminates becomes significantly more pronounced.

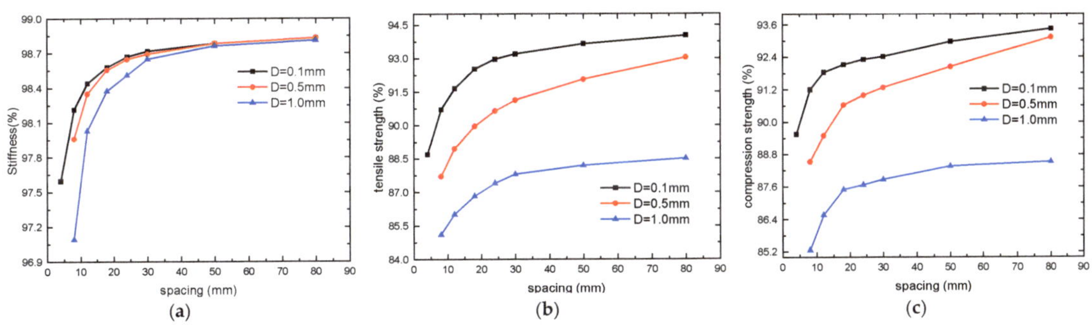

Figure 14. Variation of mechanical properties with spacing of microvascular channel: stiffness (**a**), tensile strength (**b**), and compression strength (**c**).

4.3. Volume Fraction

The volume fraction of the microvascular channel in the laminate is also an important design parameter in practical applications. Based on the microvascular configuration in this paper, the volume fractions of the microvascular channel corresponding to various channel diameters and spacing are calculated. Figure 15 presents the variation in the mechanical properties of the laminates under specified channel diameters. Since the volume fraction of the microvascular channel with a 0.1 mm diameter is extremely small, making the curve difficult to observe, only the curves with 0.5 mm and 1 mm diameters are presented. Furthermore, as the stiffness of the laminates is minimally affected, the focus is primarily on the variation in strength. It can be observed that the laminate with larger channel diameter has a lower slope on the strength curve. As the volume fraction of the microvascular channel increases, the slope of the strength curves decreases gradually.

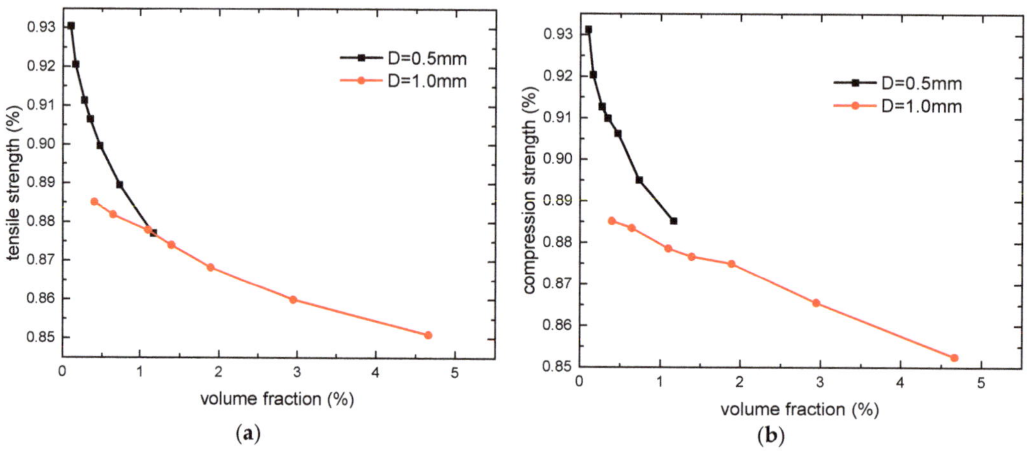

Figure 15. Variation in mechanical properties with volume fraction of microvascular channel: tensile strength (**a**) and compression strength (**b**).

4.4. Parameter Design Criteria

When designing microvascular self-healing composite structures, it is necessary to consider both the mechanical properties of the laminate and the damage repair capability of the microvascular channel. Generally, to ensure that self-healing is triggered in time before the damage propagation significantly affects the overall performance of the structure, the spacing of the microvascular channel must be less than a certain value. At the same time, there may also be requirements that the mechanical properties of the laminate should not fall below a certain level, or the volume fraction of the microvascular channel should not exceed a certain level when designing structures. So, the variation of the mechanical properties of the microvascular composites can be determined through experiments or finite element methods first, and then the range of microvascular design parameters can be determined based on these limitations. Larger diameter and spacing are better within the allowable design range.

5. Conclusions

Tensile and compressive performances of woven fabric CFRP laminates containing three-dimensional microvascular channels were investigated experimentally. Different finite element models with z-direction microvascular channels were established and verified by experimental results. The validated models were used to investigate the damage propagation process and failure mechanism of the laminates under tensile and compressive loads. Then, a parameter study was conducted. The following conclusions can be obtained:

(1) The Z-direction microvascular channel has a critical position, which is prone to damage under tensile and compressive loads. It has minimal effect on the stiffness of laminates, but a certain effect on the strength. With a channel diameter of 0.5 mm and a spacing of 12 mm, the tensile and compressive strengths decrease by approximately 10% to 12% compared with blank laminates.

(2) The FE models with z-direction microvascular channels which consider different orientations of the resin-rich region formed in the warp and weft fiber yarns agree well with the test results. For woven fabric CFRP composites with z-direction microvascular channels, equivalent unidirectional ply material properties should be assigned separately in an FE model for more accurate calculation, while the effect of variations in the fiber volume fraction around the microvascular channel can be disregarded.

(3) Within the common microvascular diameter ranging from 0.1 mm to 1 mm, stiffness variation of the laminates is small, while the laminate strength varies approximately linearly with the channel diameter.
(4) The combined investigation using experimental and numerical methods makes it possible to conveniently determine the mechanical properties of laminates with microvascular layers, which provides references for structural design. Moreover, it is also necessary to establish design methods for microvascular parameters in order to apply microvascular composites in engineering structures, which is a key direction for future study.

Author Contributions: Conceptualization, Z.A. and X.C.; Data curation, X.G. and Y.C.; Investigation, Z.A., D.Z., X.G. and Y.C.; Methodology, Z.A., X.C. and Y.M.; Project administration, X.C.; Resources, X.C.; Supervision, X.C.; Writing—original draft, Z.A.; Writing—review and editing, X.C. and Z.A. All authors have read and agreed to the published version of the manuscript.

Funding: This research received no external funding.

Institutional Review Board Statement: Not applicable.

Data Availability Statement: Data are contained within the article.

Conflicts of Interest: Dafang Zhao is an employee of AVIC Manufacturing Technology Institute and the specimens were manufactured there. Yihao Ma is an employee of China North Industries Group Research Institute of Navigation and Control Technology. The remaining authors have no conflict of interests to declare.

References

1. Liang, F.F.; Jin, D.; He, Y. Research on the maintenance of composite material structure. *China Adhes.* **2019**, *28*, 57–61. [CrossRef]
2. Wang, J.T.; Tang, J.; Chen, D.D.; Xing, S.L.; Liu, X.Y.; Hao, J.Y. Intrinsic and Extrinsic Self-healing Fiber-reinforced Polymer Composites: A Review. *Polym. Compos.* **2023**, *44*, 6304–6323. [CrossRef]
3. Willocq, B.; Odent, J.; Dubois, P.; Raquze, J. Advances in Intrinsic Self-Healing Polyurethanes and Related Composites. *RSC Adv.* **2020**, *10*, 13766–13782. [CrossRef] [PubMed]
4. Ma, Y.H.; Du, X.Y.; Hu, R.W.; Zhao, D.F.; Li, B.T.; Cheng, X.Q. Development of Self-Healing Composite Materials with Microvascular Networks. *Polym. Mater. Sci. Eng.* **2018**, *34*, 166–172. [CrossRef]
5. Pang, J.W.C.; Bond, I.P. A Hollow Fibre Reinforced Polymer Composite Encompassing Self-Healing and Enhanced Damage Visibility. *Compos. Sci. Technol.* **2005**, *65*, 1791–1799. [CrossRef]
6. Williams, G.; Trask, R.; Bond, I.P. A Self-Healing Carbon Fibre Reinforced Polymer for Aerospace Applications. *Compos. Part A Appl. Sci. Manuf.* **2007**, *38*, 1525–1532. [CrossRef]
7. Patrick, J.F.; Krull, B.P.; Garg, M.; Mangun, C.L.; Moore, J.S.; Sottos, N.R.; White, S.R. Robust Sacrificial Polymer Templates for 3D Interconnected Microvasculature in Fiber-Reinforced Composites. *Compos. Part A Appl. Sci. Manuf.* **2017**, *100*, 361–370. [CrossRef]
8. Norris, C.J.; White, J.A.P.; McCombe, G.; Chatterjee, P.; Bond, I.P.; Trask, R.S. Autonomous Stimulus Triggered Self-Healing in Smart Structural Composites. *Smart Mater. Struct.* **2012**, *21*, 094027. [CrossRef]
9. Patrick, J.F.; Hart, K.R.; Krull, B.P.; Diesendruck, C.E.; Moore, J.S.; White, S.R.; Sottos, N.R. Continuous Self-Healing Life Cycle in Vascularized Structural Composites. *Adv. Mater.* **2014**, *26*, 4302–4308. [CrossRef]
10. Luterbacher, R.; Coope, T.S.; Trask, R.S.; Bond, I.P. Vascular Self-Healing within Carbon Fibre Reinforced Polymer Stringer Run-out Configurations. *Compos. Sci. Technol.* **2016**, *136*, 67–75. [CrossRef]
11. Sakurayama, N.; Minakuchi, S.; Takeda, N. Sensing and Healing of Disbond in Composite Stiffened Panel Using Hierarchical System. *Compos. Struct.* **2015**, *132*, 833–841. [CrossRef]
12. Jabbar, M.; Nawab, Y.; Shaker, K. Self-Healing Composites: Healing Using Rapid Cure Resin Systems. In *Rapid Cure Composites*; Hameed, N., Peerzada, M., Salim, N., Parameswaranpillai, J., Eds.; Woodhead Publishing: Sawston, UK, 2023; pp. 107–120. ISBN 978-0-323-98337-2.
13. Kousourakis, A.; Bannister, M.K.; Mouritz, A.P. Tensile and Compressive Properties of Polymer Laminates Containing Internal Sensor Cavities. *Compos. Part A Appl. Sci. Manuf.* **2008**, *39*, 1394–1403. [CrossRef]
14. Devi, U.; Pejman, R.; Phillips, Z.J.; Zhang, P.; Soghrati, S.; Nakshatrala, K.B.; Najafi, A.R.; Schab, K.R.; Patrick, J.F. A Microvascular-Based Multifunctional and Reconfigurable Metamaterial. *Adv. Mater. Technol.* **2021**, *6*, 2100433. [CrossRef]
15. Saeed, M.U.; Li, B.; Chen, Z.; Cui, S. Fabrication of Microchanneled Composites by Novel Selective Polymer Degradation. *Mater. Manuf. Process.* **2016**, *31*, 2057–2063. [CrossRef]
16. Coppola, A.M.; Thakre, P.R.; Sottos, N.R.; White, S.R. Tensile Properties and Damage Evolution in Vascular 3D Woven Glass/Epoxy Composites. *Compos. Part A Appl. Sci. Manuf.* **2014**, *59*, 9–17. [CrossRef]

17. Norris, C.J.; Bond, I.P.; Trask, R.S. The Role of Embedded Bioinspired Vasculature on Damage Formation in Self-Healing Carbon Fibre Reinforced Composites. *Compos. Part A Appl. Sci. Manuf.* **2011**, *42*, 639–648. [CrossRef]
18. Nguyen, A.T.T.; Orifici, A.C. Structural Assessment of Microvascular Self-Healing Laminates Using Progressive Damage Finite Element Analysis. *Compos. Part A Appl. Sci. Manuf.* **2012**, *43*, 1886–1894. [CrossRef]
19. Huang, C.Y.; Trask, R.S.; Bond, I.P. Characterization and Analysis of Carbon Fibre-Reinforced Polymer Composite Laminates with Embedded Circular Vasculature. *J. R. Soc. Interface* **2010**, *7*, 1229–1241. [CrossRef]
20. Al-Shawk, A.; Tanabi, H.; Sabuncuoglu, B. Investigation of Stress Distributions in the Resin Rich Region and Failure Behavior in Glass Fiber Composites with Microvascular Channels under Tensile Loading. *Compos. Struct.* **2018**, *192*, 101–114. [CrossRef]
21. Demiral, M.; Tanabi, H.; Sabuncuoglu, B. Experimental and Numerical Investigation of Transverse Shear Behavior of Glass-Fibre Composites with Embedded Vascular Channel. *Compos. Struct.* **2020**, *252*, 112697. [CrossRef]
22. Zhao, Y.; Wang, Z.Q.; Jiang, P.; Wang, A.Y.; Chang, Z.P.; Kang, Y.G. Numerical Simulation on Mechanical Properties and Damage Behavior of CFRP with Self-Healing Microvascular Channels. *Mech. Adv. Mater. Struct.* **2022**, *31*, 297–313. [CrossRef]
23. Ran, G.L.; Ma, Y.H.; An, Z.Q.; Zhao, D.F.; Guo, X.; Cheng, X.Q. Study of tensile properties of laminates containing microvascular channels with different diameters. *J. Beijing Univ. Aeronaut. Astronaut.* **2022**, *48*, 1–15. [CrossRef]
24. Fu, H.M.; Zhang, Y.B. On the Distribution of Delamination in Composite Structures and Compressive Strength Prediction for Laminates with Embedded Delaminations. *Appl. Compos. Mater.* **2011**, *18*, 253–269. [CrossRef]
25. *ASTM D3039*; Standard Test Method for Tensile Properties of Polymer Matrix Composite Materials. American Society for Testing Materials: West Conshohocken, PA, USA, 2017.
26. *ASTM D6641*; Standard Test Method for Compressive Properties of Polymer Matrix Composite Materials Using a Combined Loading Compression (CLC) Test Fixture. American Society for Testing Materials: West Conshohocken, PA, USA, 2023.
27. Ma, Y.H.; Cheng, X.Q.; Zhang, J.K.; Zhao, D.F.; Huang, W.J. Prediction of Resin Pocket Geometry around Rigid Fiber Inclusion in Composite Laminate by Hot-Pressing of Prepregs. *J. Compos. Mater.* **2020**, *54*, 1987–1999. [CrossRef]
28. Chamis, C.C. Mechanics of Composite Materials: Past, Present, and Future. *J. Compos. Technol. Res.* **1989**, *11*, 3–14. [CrossRef]
29. Hashin, Z. Failure Criteria for Unidirectional Fiber Composites. *J. Appl. Mech.* **1980**, *47*, 329–334. [CrossRef]
30. Chang, F.K.; Chang, K.Y. A Progressive Damage Model for Laminated Composites Containing Stress Concentrations. *J. Compos. Mater.* **1987**, *21*, 834–855. [CrossRef]
31. Ye, L. Role of Matrix Resin in Delamination Onset and Growth in Composite Laminates. *Compos. Sci. Technol.* **1988**, *4*, 257–277. [CrossRef]
32. Camanho, P.P.; Matthews, F.L. A Progressive Damage Model for Mechanically Fastened Joints in Composite Laminates. *J. Compos. Mater.* **1999**, *33*, 2248–2280. [CrossRef]
33. Fiedler, B.; Hojo, M.; Ochiai, S.; Schulte, K.; Ando, M. Failure Behavior of an Epoxy Matrix under Different Kinds of Static Loading. *Compos. Sci. Technol.* **2001**, *61*, 1615–1624. [CrossRef]

polymers

Article

Online and Ex Situ Damage Characterization Techniques for Fiber-Reinforced Composites under Ultrasonic Cyclic Three-Point Bending

Aravind Premanand [1,*], Mario Prescher [2], Michael Rienks [1], Lutz Kirste [2] and Frank Balle [1,3,4]

1 Department for Sustainable Systems Engineering (INATECH), Faculty of Engineering, University of Freiburg, 79110 Freiburg, Germany; michael.rienks@inatech.uni-freiburg.de (M.R.); frank.balle@inatech.uni-freiburg.de (F.B.)
2 Fraunhofer Institute for Applied Solid State Physics (IAF), 79108 Freiburg, Germany; mario.prescher@iaf.fraunhofer.de (M.P.); lutz.kirste@iaf.fraunhofer.de (L.K.)
3 Freiburg Materials Research Center (FMF), 79104 Freiburg, Germany
4 Fraunhofer Institute for High Speed Dynamics, Ernst Mach Institute (EMI), 79104 Freiburg, Germany
* Correspondence: aravind.premanand@inatech.uni-freiburg.de

Abstract: With ultrasonic fatigue testing (UFT), it is possible to investigate the damage initiation and accumulation from the weakest link of the composite material in the very high cycle fatigue (VHCF) regime in a shorter time frame than conventional fatigue testing. However, the thermal influence on the mechanical fatigue of composites and the scatter in fatigue data for composites under ultrasonic cyclic three-point bending loading still need to be investigated. In this study, we conducted interrupted constant-amplitude fatigue experiments on a carbon-fiber satin-fabric reinforced in poly-ether-ketone-ketone (CF-PEKK) composite material. These experiments were carried out using a UFT system, which operates at a cyclic frequency of 20 kHz with a pulse–pause sequence. Various parameters, such as the CF-PEKK specimen's surface temperature, acoustic activity, and the ultrasonic generator's input resonance parameters, were measured during cyclic loading. During experiment interruption, stiffness measurement and volumetric damage characterization in the CF-PEKK specimens using 3D X-ray microscopy (XRM) were performed. The locations of damage initiation and accumulation and their influence on the changes in in situ parameters were characterized. Under fixed loading conditions, damage accumulation occurred at different locations, leading to scattering in fatigue life data. Further, the damage population decreased from the surface to the bulk of the composite material.

Keywords: ultrasonic fatigue testing; carbon fiber reinforced polymers; online monitoring; fatigue damage; X-ray microscopy

Citation: Premanand, A.; Prescher, M.; Rienks, M.; Kirste, L.; Balle, F. Online and Ex Situ Damage Characterization Techniques for Fiber-Reinforced Composites under Ultrasonic Cyclic Three-Point Bending. *Polymers* **2024**, *16*, 803. https://doi.org/10.3390/polym16060803

Academic Editors: Xiaoquan Cheng, Wenjun Huang and Qian Zhang

Received: 27 February 2024
Revised: 11 March 2024
Accepted: 11 March 2024
Published: 13 March 2024

1. Introduction

Continuous fiber-reinforced polymer matrix composites (PMCs) with textile reinforcements are considered an essential class of high-performance materials as they exhibit improved damage tolerance behavior compared to unidirectional (UD) cross-ply laminates [1]. Other advantages of woven-fabric composites include high toughness, dimensional stability for various temperatures, and simple manufacturing [2]. Reliable designs of structures made out of carbon fabric-reinforced polymers (CFRPs) require an accurate evaluation of their fatigue life under service loading conditions [3]. Two notable applications for textile composites are aerospace [4] and wind energy engineering [5].

During their service life of up to 30 years, aerospace and wind energy industry structures are often loaded with more than 10^8 cycles [6]. Many material configurations resulting from the fibers, fiber architectures, matrices, manufacturing methods, as well as stacking sequences make the generic fatigue characterization difficult [7]. The loading directions in the structures also have a strong influence on the fatigue of composites. When the stress state is not simply tensile, uni-axial, and aligned with the fibers, severe loads are placed on the matrix, reducing the composite's fatigue resistance [8]. To fully utilize the mechanical properties of CFRPs for lightweight applications, the fatigue behavior and underlying failure mechanisms must be well understood.

1.1. Fatigue Damage in Woven Fabric-Reinforced Composites

The main fatigue damage mechanisms observed during fatigue loading of woven-fabric composites can be classified into micro-structural damage within the strand of fibers and macroscopic damage across the different layers. The micro-structural damage mechanisms include matrix micro-cracks, fiber-matrix debonding, fiber breakage, and crack coupling, synonymous with laminated composites built up of UD layers [2]. The macroscopic damage mechanisms include transverse cracks in the weft fiber bundles, shear failure in the wrap fiber bundles, cracks in matrix regions, delamination between weft and warp fiber bundles, delamination between adjacent fabric layers, tensile failure of warp bundles, and final fracture [2]. Due to the interlacing fiber architecture in woven-fabric composites, the damage accumulation is gradual until the start of the final phase of fatigue life. Further, the textile reinforcements in woven composites limit the growth of cracks and delamination compared to UD reinforcements [9]. Compared to UD laminates, the main damage mechanisms due to fatigue of woven laminates are transverse cracks, delaminations, and fiber failure [10–14]. Lamon et al. (2023) [15] observed that one of the main damage mechanisms in woven fiber architecture before final failure is the initiation and propagation of off-axis cracks (transverse cracks) in the fiber bundles. During the final phase of damage growth, all the damage modes can interact and grow rapidly. The final failure is usually marked by the fracture of warp bundles, leading to the ultimate failure of the laminate.

Most of the work on fatigue damage analysis of woven composites has been performed in low- and high-cycle fatigue regimes of composites, and the limited knowledge of very high cycle fatigue (VHCF) is compensated by conservative designs [16]. The limited works in the VHCF regime can be explained by the very long test periods to realize experiments beyond 10^7 cycles. Accelerated fatigue testing methods, where the cyclic frequency is increased, offer a potential solution to reduce the experimental time to reach the VHCF regime. Various authors have studied different CFRP composites since the first cyclic three-point bending test setup was developed for conducting VHCF experiments in an ultrasonic fatigue test system for composites [17]. A carbon fiber-reinforced in polyphenylene sulfide (CF-PPS) composite system stacked with 2/2 twill fabric in an orthotropic layup exhibited a five-stage damage accumulation starting from fiber matrix debonding coalescing to transverse cracks and micro-delaminations between 0° and 90° fiber rovings. This damage mechanism was followed by the development of meta- and macro-delaminations, which led to the final failure of the short beam specimens [18,19]. In contrast, during another study on a CF–epoxy composite system stacked with 4-Harness satin fabric in a quasi-isotropic layup, sudden failure by fracture at the bottom of the specimens was observed [20]. A CF–epoxy composite system with plain weave fabric stacked with alternating 0° and 90° layers was reported to undergo pitting corrosion with damages in the epoxy matrix [21].

1.2. Damage Monitoring Techniques

Different non-destructive techniques have been applied to detect and evaluate fatigue damage in composites. This includes ultrasonic C-scans for evaluation of delamination onset and growth [22–24], acoustic emission technique to capture material degradation and fatigue damage evolution [25–28], and thermography [29–32]. The use of X-ray computed tomography (X-CT) as a non-destructive inspection technique for assessing damage in composites has increased in the last decade [33]. This is mainly due to the possibility of volumetric damage characterization with a sufficient resolution (up to 1 μm) [34]. Yu et al. (2016) [35] performed X-ray tomography at different stages of fatigue life for

3D woven glass fiber-reinforced composite specimens by performing interrupted axial fatigue experiments. Jespersen et al. (2016) [36] studied the progression of fatigue damage in a uni-directional, non-crimp glass fabric-reinforced composite under tension–tension loading. Here, the damage progression as a function of stiffness degradation was examined by CT scans on four different samples, each at different phases: one during the initial drop, two during stable stiffness degradation, and one close to final failure. Later, Djabali et al. (2019) [37] investigated the fatigue damage mechanisms in thick CF–epoxy laminate subjected to bending loads by combining in situ monitoring techniques such as acoustic emission (AE) and digital imaging correlation (DIC) and X-CT. This combination enabled the identification and quantification of fatigue damage.

1.3. Motivation for Current Work

The knowledge and experience on fatigue of composites in the VHCF regime conducted using a UFT system under three-point bending by the authors [38] as well as others [18,20,21] are limited to the surface of the composite specimens. Under a pure bending load case, maximum cyclic compressive stresses occur at the top layer of the composite specimen, and tensile stresses occur at the bottom layer. However, since the resonance condition for a three-point bending load case leads to short specimens, typically in the order of 30–35 mm for carbon fiber-reinforced polymer composites, cyclic shear stresses occur in the regions between the loading nose and support units. Light optical or scanning electron microscopy (used in [18,20,21,38]) can provide excellent results on the state of damage anywhere in the specimen but is limited to only one plane. Specimens have to be physically sliced to acquire information across multiple planes. The goal of this investigation was to determine the volumetric damage state of the composite specimens due to ultrasonic cyclic loading and relate it with the existing damage monitoring techniques that are already used for UFT.

2. Material and Specimen Geometry

Recently, the applicability of the CF-PEKK composite material was demonstrated by the Thermoplastic Affordable Primary Aircraft Structure (TAPAS) Consortium for small and large aircraft structures [39–42]. PEKK is a semi-crystalline thermoplastic polymer with a glass transition temperature (T_g) of ≈160 °C [43] and a processing temperature slightly lower than that of PEEK [44]. Thus, the carbon fiber five-harness satin fabric-reinforced poly-ether-ketone-ketone (CF-PEKK) composite material was chosen for the current study. The composite laminates were manufactured by Toray Advanced Composites (Nijverdal, The Netherlands). The physical properties of the CF-PEKK composite laminate provided in Table 1 were already published in [38,45].

Table 1. Properties of CF-PEKK composite laminate previously published in [45].

Elastic Properties	Values
Fiber architecture	5H satin weave fabric
Fiber volume fraction [%]	50
Consolidated ply thickness [mm]	0.31 ± 0.01
Number of plies [-]	13
Composite layup [-]	$[0/90/0/90/0/90/\underline{0}]_s$
Glass transition temperature [°C]	* 160
Melting temperature [°C]	* 337
Tensile modulus 0° [GPa]	58.0
Ultimate tensile strength 0° [MPa]	** 776
Ultimate compressive strength 0° [MPa]	** 585

Values obtained from material data sheets provided by the manufacturer, [43] * and [44] **.

The dimensions of the composite specimens have to be evaluated using modal analysis to conduct cyclic loading experiments in ultrasonic fatigue testing systems. The evaluated dimension enables the composite specimens to be stimulated with resonance vibrations in the desired eigenmode at a frequency of 20 kHz [17]. The dimensions of the cyclic three-point bending specimen for testing the CF-PEKK material at a cyclic frequency of 20 kHz were previously determined during an existing study [46]. The specimen dimensions were 34 × 15 × 4.1 mm^3 [46].

For determining the flexural properties of fiber-reinforced polymers under three-point bending, ASTM D7264/D7264M-15 [47] recommends a minimum span-to-thickness ratio of 16:1. However, the obtained specimen geometry results in a span-to-thickness ratio of 4.5:1. This produces flexural as well as shear stresses during cyclic loading. However, in a previous investigation by the authors on the CF-PEKK specimens [46], monotonic or quasi-static experiments failed in the middle of the specimen due to normal stresses. Thus, normal stresses were evaluated for CF-PEKK specimens for the current span-to-thickness ratio [48].

The lateral movement during resonance oscillations can be avoided by accurately positioning the specimens with the support pins at the nodal points where displacements are zero during transverse bending eigenmode [17]. For the CF-PEKK specimen, the span length or the distance between the support pins was 18.6 mm [46]. Another study experimentally validated the desired oscillation mode during ultrasonic cyclic loading for the CF-PEKK specimens using 3D Scanning Laser Doppler Vibrometry (3D-SLDV) from Polytec GmbH (Waldbronn, Germany) [48].

3. Methodology

3.1. Description of the UFT System

An in-house UFT system was developed using the ultrasonic load train manufactured by Herrmann Ultraschalltechnik GmbH (Karlsbad, Germany) at INATECH, University of Freiburg. A schematic of the loading system and the online monitoring units, along with their model names, are provided in Figure 1. A mechanical oscillation amplitude up to 57 µm can be achieved with this loading setup and the resonance unit. This is shown in Figure 2. Consequently, the fatigue experiments are displacement-controlled. A control unit is realized using LabVIEW 2018 software in a computer with the Windows operating system. Further information about the resonance train can be found in [48]. The developed LabVIEW interface allows for the user to

- control and record the ultrasonic generator's input resonance parameters at a sampling rate of 1 kHz [38,49],
- send the trigger signals to the optical laser microphone to record the acoustic activity, and
- capture the maximum surface temperature on the CF-PEKK specimen at a sampling rate of 32 Hz [38,45,49].

Two data acquisition systems (DAQ) from National Instruments were integrated with the control unit to realize high-speed data transfer between the ultrasonic generator, the online monitoring units, and the computer with the Windows operating system. Compressed air cooling was realized during the entire duration of the fatigue experiments, with four nozzles blowing air on all four sides of the specimen.

3.2. Online Monitoring Techniques

During the fatigue experiments, the resonance data for every ultrasonic pulse can be fetched at the end of each ultrasonic pulse from the ultrasonic generator. These data include input displacement, generator power, and resonance frequency. The evolution of the maximum surface temperature of the specimen due to ultrasonic oscillations was monitored using an infrared (IR) camera of type TIM640 from Micro-Epsilon Messtechnik GmbH (Ortenburg, Germany) at a frame rate of 32 Hz. Up to 4 areas can be defined on the surface of the specimens, and the maximum temperature as a function of time is recorded at the sampling frequency of the IR camera for the defined areas. The setup of the IR camera and the typical resonance data due to undamaged and damaged conditions were already investigated in a previous study by the authors [38,49]. An optical laser microphone was placed at a distance less than 50 mm perpendicular to the CF-PEKK specimens (see Figure 2).

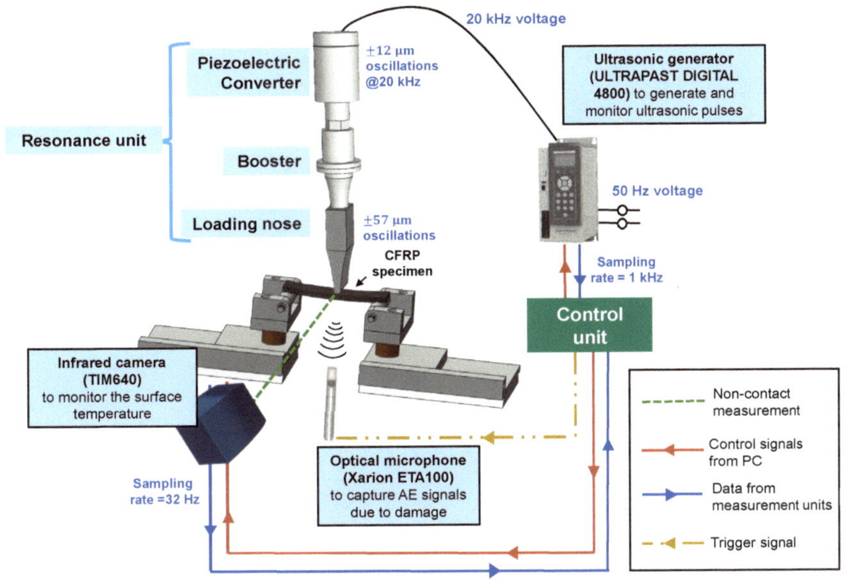

Figure 1. A schematic of the experimental setup showing ultrasonic fatigue testing and integrated and online monitoring systems for performing cyclic three-point bending loads on CFRP specimens.

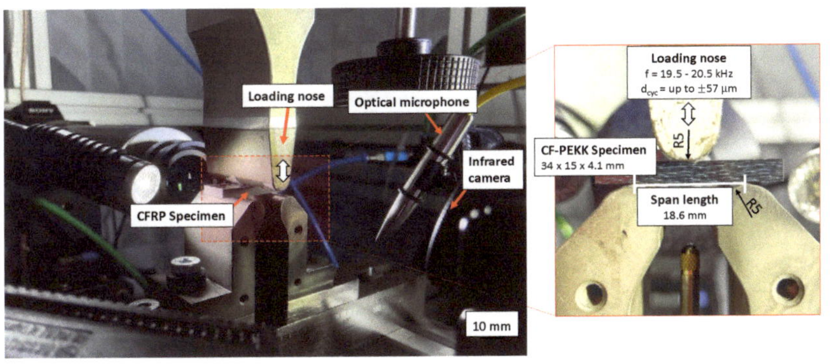

Figure 2. The ultrasonic fatigue testing system for three-point bending loading of CF-PEKK specimens.

3.3. Working Principle of the Optical Laser Microphone

An optical laser microphone of type ETA100Ultra from Xarion Laser Acoustics GmbH (Vienna, Austria) was used for measuring the acoustic activity due to damage initiation and accumulation during fatigue of CF-PEKK specimens under ultrasonic cyclic three-point bending loading conditions. The membrane-less optical laser microphone operates on the principle of interferometry with a high-frequency bandwidth between 10 Hz and 2 MHz. This measurement system consists of a sensor head, a controller, an amplifier, and a measurement computer [50]. The control unit sends laser light through an optical fiber to the sensor head. The sensor head consists of a pair of parallel, semi-reflective mirrors. During the cyclic loading of the CF-PEKK specimen, the sound pressure around the specimen is influenced by the presence of damage. This change in sound pressure affects the wavelength of light, which is detected optically by changing the refractive index within a Fabry–Pérot etalon. This is shown in Figure 3.

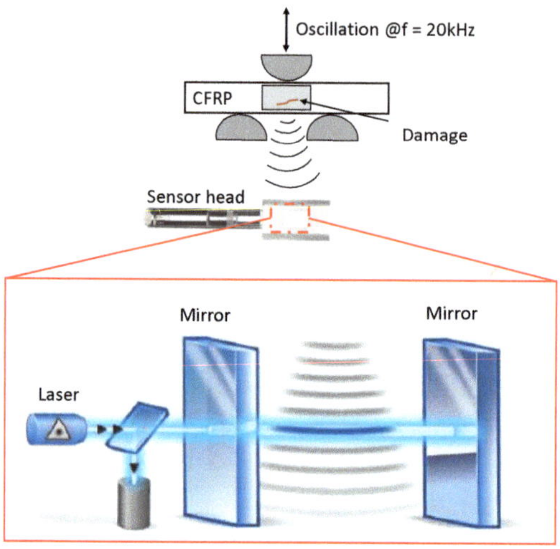

Figure 3. Damage detection technique using optical laser microphone [51].

To conduct online measurements, the analog output of the XARION sensor was sampled by a QASS Optimizer4D system from QASS GmbH (Wetter, Germany). It is a measuring system that performs real-time Fast Fourier Transform of the signals using a 24 bit A/D converter and operates up to a sampling rate of 4 MHz.

3.4. Ex Situ Damage Characterization Using 3D X-ray Microscopy

Undamaged and damaged CF-PEKK samples at different phases of fatigue life were characterized using X-ray microscopy (XRM) imaging. A Zeiss Xradia Versa 520 system from the Fraunhofer Institute for Applied Solid State Physics (Freiburg, Germany) was used for damage investigation. The XRM consists of

- a sealed transmission X-ray source (30–160 kV),
- a 2 k × 2 k pixel noise-suppressed charge-coupled detector in
- a tunable dual-stage detector system with multiple objectives with different magnifications.

In the dual-stage magnification system, the recorded images are enlarged by geometric magnification in the first step. In the second step, a scintillator converts the X-rays into visible light, which is then optically magnified by an optical lens. Combining two stages allows for image capture at a submicron resolution, even for large working distances. A spatial resolution of 1 μm is achievable using the Zeiss Xradia Versa 520 system. The CF-PEKK specimens were mounted as shown in Figure 4.

A scanning resolution of 10 μm on a volume of $18.6 \times 15 \times 4.1$ mm^3 (see Figure 5a) took ≈72 h for each specimen scan. All the chosen specimens were scanned once before applying cyclic loading, after experiment interruption, and after failure, thus three times in total. Figure 5b shows the top view of the scanned volume with the area covered by the loading nose to visualize the damaged area from the XRM scan results presented in Section 4.4.

The Zeiss "Scout-and-scan" software (version number: 16.1.14271.44713) was used to control the X-ray measurements. Since the specimens for three-point bending experiments have a flat geometry, the high-aspect-ratio tomography (HART) mode was used. The HART mode allows variable projection distances such that fewer projections are collected along the width and more along the thickness of the specimens.

After performing the X-ray scans, a virtual volume for each specimen was computationally reconstructed from the recorded projection images using the Zeiss XRM reconstruction software (version number: 16.1.14271.44713). A virtual volume is a 3D matrix that segments the sample into voxels. Each matrix element is assigned a grayscale value proportional to the material's atomic weight in the specimen's corresponding voxel. For further analysis of X-ray data, ORS Dragonfly Pro software (version number: 2022.2.0.1361) was used [52]. This software was used to visualize the slices of the reconstructed specimen through the reconstructed virtual volume.

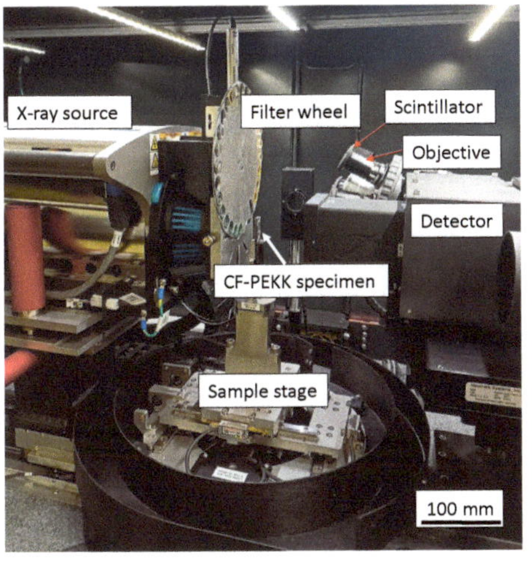

Figure 4. A CF-PEKK specimen mounted inside the Zeiss Xradia Versa 520 3D X-ray microscope (XRM).

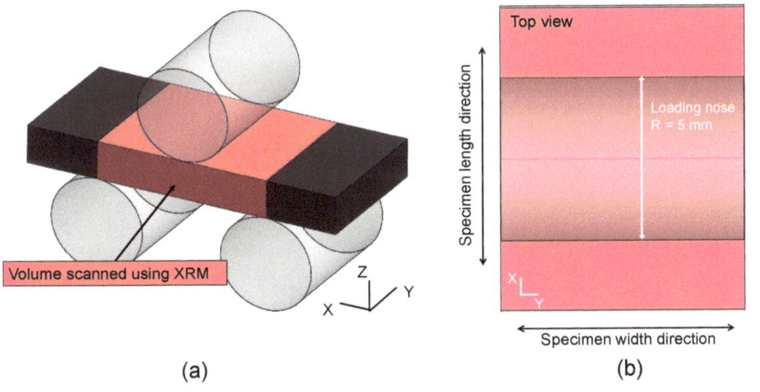

Figure 5. (**a**) A schematic of the CF-PEKK specimens to show the volume scanned using the XRM and (**b**) the top view of the scanned volume of the CF-PEKK specimens representing the length and width directions, as well as the width of loading nose.

3.5. Abort Criteria for VHCF Testing at 20 kHz

For displacement-controlled experiments performed using the UFT system, the definition of failure defined in ISO 13003:2003-12 [53] (a standard for determining fatigue properties under cyclic loading) is not applicable as the cyclic loads decrease due to the increase in the compliance (or decrease in stiffness) of the specimens in the presence of damage. Thus, additional failure criteria are required for the UFT systems [54]. The specimen failure or abort (termination) criteria of all the fatigue experiments are given as follows:

- when the maximum surface temperature in any of the defined areas exceeds 80 °C (0.5 × T_g of PEKK polymer),

- when the power required by the ultrasonic generator is greater than 700 W as this would mean that there is a significant damping between the resonance unit and specimen, leading to a surface temperature greater than 80 °C, or
- when the resonance frequency drops below 19.5 kHz.

3.6. Interrupted Constant Amplitude Fatigue Experiments at 20 kHz

Four interrupted constant amplitude fatigue (CAF) experiments were performed at a normal stress amplitude of 45.1 MPa. These experiments had pulse and pause durations of 50 ms and 300 ms, respectively. This pulse–pause sequence results in an effective frequency = 2885 Hz. Here, the specimens were interrupted one time, each at different stages of fatigue lives as shown in Figure 6, to measure the change in static stiffness and volumetric damage. At this stress amplitude (45.1 MPa), the maximum time for an experiment surviving 5 × 10^8 cycles (95% confidence limit) is 2 days. Thus, an experimental campaign at a higher stress amplitude was chosen to minimize the overall test time.

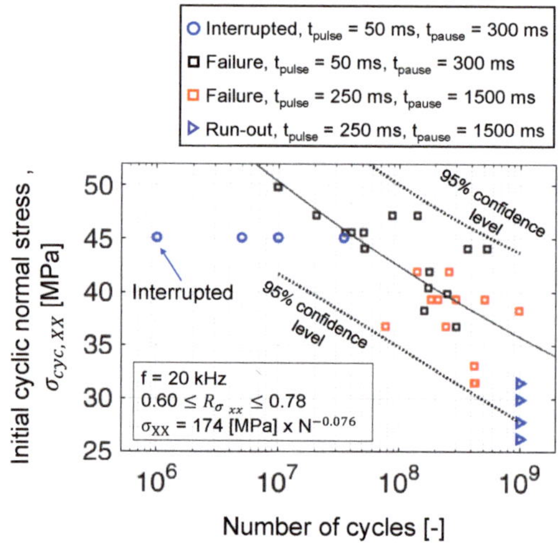

Figure 6. Interrupted CAF experiments denoted with blue circular markers overlaid in the SN diagram previously published in [38].

The change in static bending stiffness was measured using a table-top static three-point bending test system from ZwickRoell (Ulm, Germany) with a load capacity of 2.5 kN (model name: ZHU/zwickiLine+). The same radius of the loading nose and support pins (radius R = 5 mm) were used with the same span length defined for ultrasonic fatigue experiments. The normalized bending stiffness was obtained by calculating the slope of the force–displacement curves up to a maximum static displacement of 150 µm. The static experiments for stiffness measurements were carried out at a cross-head speed of 1 mm/min as per the ASTM D2344/D2344M-16 standard [55], which is a standard for applying three-point bending loads on short beam composites.

4. Results and Discussion

The four CAF experiments at a cyclic stress amplitude of 45.1 MPa and at a frequency of 20 kHz resulted in fatigue lives still within the 95% confidence intervals (see Figure 7). The number of cycles to failure (N_{dam}) defined by abort criteria mentioned in Section 3.5 for the four specimens is given in Table 2. Here, N_{int} corresponds to the number of load cycles after which the CAF experiments were interrupted, and N_{dam} refers to the number of load cycles to damage. The fraction between N_{int} and N_{dam} corresponds to the percentage of fatigue life at which the specimens were taken out of the UFT system for residual stiffness and XRM measurements.

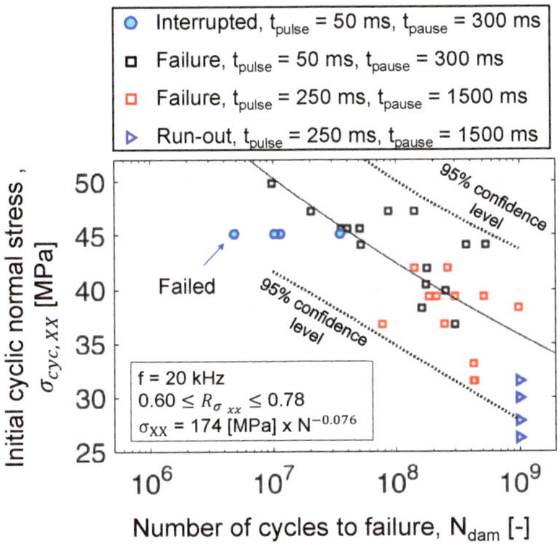

Figure 7. SN diagram with failure of interrupted CAF experiments along with results previously published in [38].

Table 2. Load cycles to interruption and damage for CAF experiments conducted at $\sigma_{cyc,XX}$ = 45.1 MPa.

Specimen ID	N_{int} [-]	N_{dam} [-]	N_{int}/N_{dam} %
CF-PEKK01	10^6	4.82×10^6	21
CF-PEKK02	5×10^6	1.15×10^7	44
CF-PEKK03	10^7	1.03×10^7	97
CF-PEKK04	3.5×10^7	3.52×10^7	99

4.1. Temperature and Resonance Data

The surface temperature of the CF-PEKK specimens and input resonance signals from the ultrasonic generator were recorded during ultrasonic cyclic loading for all the experiments. The input resonance signals include power expended by the generator, resonance frequency of ultrasonic oscillations, and cyclic displacement amplitude. The temperature and resonance signals were in good agreement during cyclic loading of CF-PEKK specimens in a previous investigation by the authors [49]. From the power expended during every ultrasonic pulse (duration of 50 ms), the energy dissipation can be calculated by determining the area of the power–time curve. Likewise, the peak temperature at the end of each pulse can be calculated from the time–temperature signal. The change in dissipated energy is the difference between the energy dissipated at the end of each pulse and the undamaged condition. The specimens resonate at the desired frequency at undamaged conditions, leading to lesser energy loss. This energy loss increases with increased damping in the presence of damage. In the damaged condition, the energy dissipation and the surface temperature of the specimen increase. The increase in maximum surface temperature at the specimen's central region since the experiment's start is denoted as relative temperature. The peak temperature and the energy dissipated at the end of each ultrasonic pulse can be correlated to the fatigue cycles by multiplying the effective frequency (≈1010 Hz) by the number of ultrasonic pulses. These calculations were performed using a Matlab routine, and the results are shown in Figure 8.

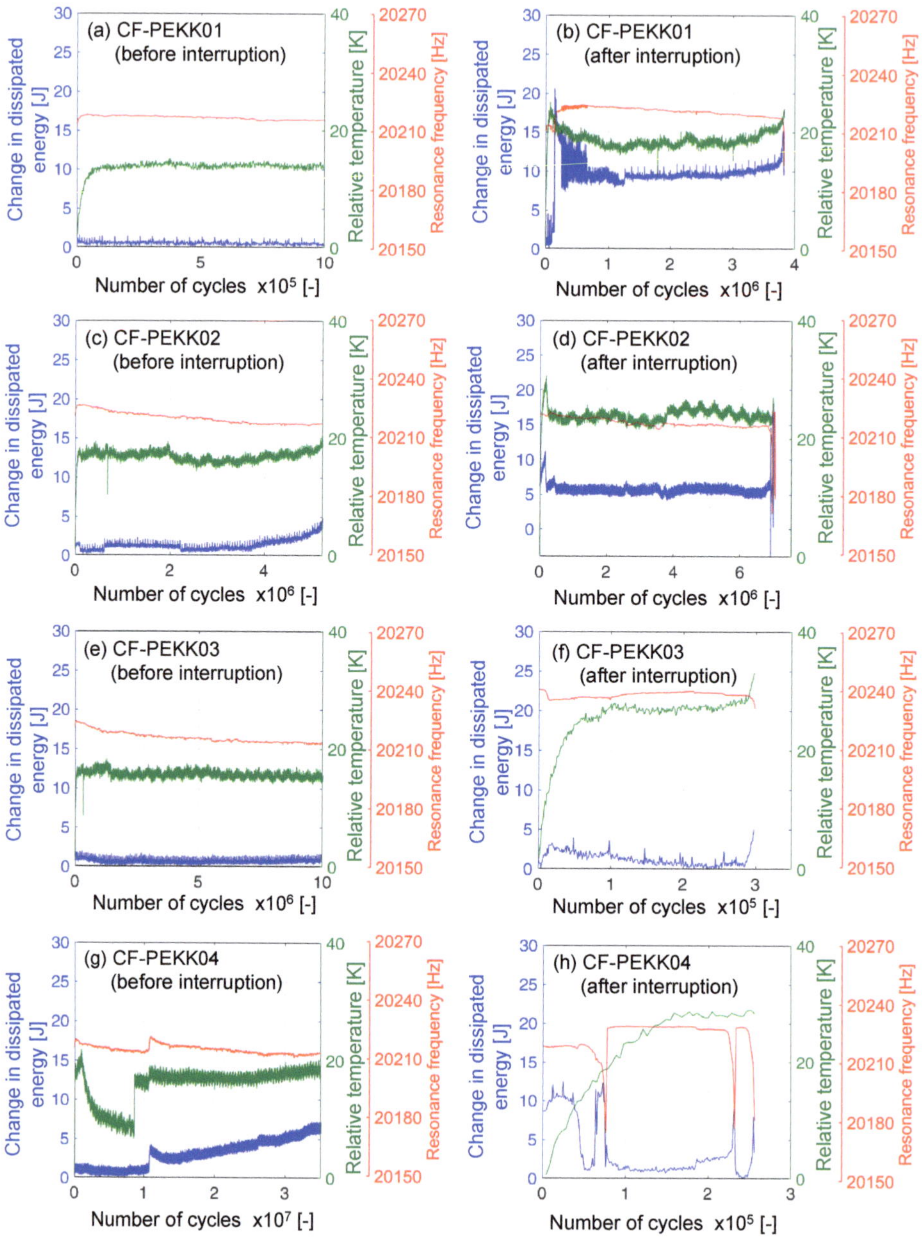

Figure 8. Change in dissipated energy per ultrasonic pulse, relative temperature evolution and drop in resonance frequencies calculated using time–temperature and input resonance signals for the four CF-PEKK specimens tested at $\sigma_{cyc,XX}$ = 45.1 MPa before and after interruption for ex situ damage characterization.

The subfigures on the left side (Figure 8a,c,e,g) were recorded before the interruption of the fatigue experiments, and the subfigures on the right side (Figure 8b,d,f,h) were recorded after the interruption of the fatigue experiments until damage. In all these figures, disturbances in the energy loss, resonance frequencies, and relative temperature curves leading to local impulsive peaks could be due to the initiation of any damage, leading to deviation from the ideal resonance conditions. Of all four specimens, CF-PEKK02 and CF-PEKK04 showed an evident increase in energy loss after 4×10^6 and 1×10^7 cycles, respectively. In contrast, CF-PEKK03 only showed a small gradual drop in resonance frequency before the interruption of the experiment after 10^7 cycles and was found to be damaged after 3×10^5 cycles after reload. The relative temperatures on all specimens without damage were less than 20 K for the cyclic normal stress amplitude of 45.1 MPa and increased beyond 20 K due to damage accumulation. Therefore, the end of the ultrasonic fatigue experiment due to failure can be confirmed by a sudden increase in energy loss, surface temperature, or a drop in resonance frequency.

4.2. Normalized Drop in Bending Stiffness

In the literature, the stiffness drop due to fatigue damage in composites is attributed to the initiation, accumulation, and propagation of multiple transverse cracks [2,11,30,56–58]. The normalized drop in monotonic (static) bending stiffness is plotted against normalized cycles to fatigue failure in Figure 9. Here, the CF-PEKK01 specimen, taken out of the UFT system after 21% of fatigue life, shows the maximum drop at the 20% mark (about 5% stiffness drop), leading to the shortest fatigue life to failure. Similarly, the CF-PEKK04 specimen taken out after 99% of fatigue life shows the least drop in stiffness (about 3%), leading to the longest fatigue life. CF-PEKK02 and CF-PEKK03 specimens had an equal stiffness drop of about 7% after 50% of fatigue life, leading to closer fatigue lives. Since the abort criteria also include a temperature limit of 80 °C, which can also be localized heating in the vicinity of damage, the interruption of the fatigue experiment depends on the side of the specimen where damage accumulated. This means damage can accumulate near the surface monitored by the IR camera or near the opposite surface, leading to a slightly different severity in damage accumulation until the experiment is aborted. This difference in the accumulation of damage leads to different stiffness drops ranging from 10 to 20%. An experiment leading to up to 20% stiffness drop occurs when the damage occurs on the opposite side of the specimen, which is not monitored by the IR camera.

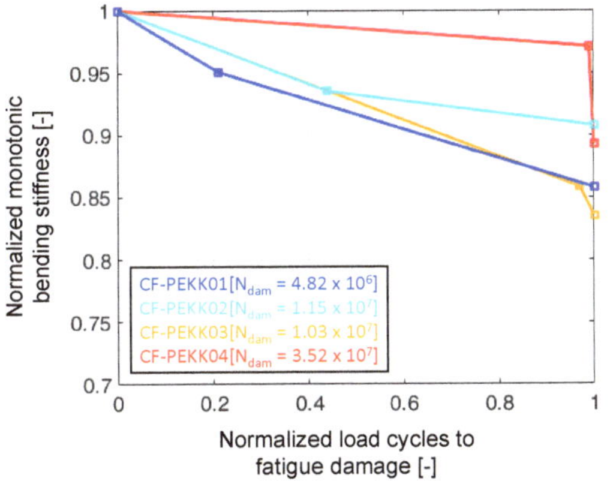

Figure 9. Normalized monotonic (static) bending stiffness drop for the four CF-PEKK specimens due to ultrasonic cyclic three-point bending at $\sigma_{cyc,XX}$ = 45.1 MPa.

Looking back at Figure 8a, the relative temperature, energy loss, and resonance frequency signals do not have any disturbance up to 10^5 cycles. However, the stiffness drop is the maximum of all experiments at their 20% fatigue life. One reason for this could be the development of multiple microscopic damage mechanisms that only reduce the global stiffness of the specimen but are not captured by the temperature and input resonance signals. In other words, the infrared thermography and resonance data are inadequate to capture the microscopic damage observed in CF-PEKK01 before interruption. After reloading CF-PEKK01 with the same loading parameters, the temperature and input resonance signals detect further development of microscopic damage (see Figure 8b). This can be seen from a strong signal disturbance within the first 10^6 cycles. A similar trend can also be seen in the case of the CF-PEKK03 specimen, where the resonance frequency drops by $\approx$10 Hz, corresponding to a stiffness drop of 15% until test interruption (at 97% fatigue life). After reloading the specimen, the relative temperature increased from 15 K to 27 K (see Figure 8e,f). Thus, the relative temperature and input resonance signals present an accurate damage condition in the composite specimens when examined together with the residual stiffness.

If these four specimens can be collectively considered as representative of the residual stiffness change during ultrasonic fatigue testing, one could say that the stiffness drops steeply within 20% of fatigue life, after which a gradual degradation is seen up to 97%. Between 97 and 100% of fatigue life, the residual stiffness drops strongly, leaving the specimen's state to satisfy one of the abort criteria described for UFT.

4.3. *Observations from Results Obtained Using the Optical Laser Microphone*

Figure 10a–d shows the Fast Fourier Transform (FFT) spectra of the four CF-PEKK specimens obtained using the optical laser microphone at different stages of fatigue for a normal stress amplitude of 45.1 MPa. The first tallest peak in Figure 10a–d corresponds to the loading or the excitation frequency of the UFT system. The tiny peak before 20 kHz is recorded due to the sound of the compressed air cooling setup. Several smaller amplitude (higher harmonics) peaks follow the highest peak at 20 kHz. About 7–9 peaks up to frequencies less than 200 kHz can be seen for undamaged CF-PEKK specimens (green curves). The peak count is about 22–24 and can even be observed up to 500 kHz for damaged specimen conditions (red curves). This increase in the number of peaks from undamaged to damaged conditions could be caused due to friction between the damaged surfaces. In other words, the higher harmonics could indicate the presence of the so-called breathing cracks in the oscillating structure [59]. These cracks open and close during cyclic oscillations [60]. Such a scenario in woven composites could occur with transverse cracks before growing into delamination [10,57,61] along the weft fiber bundle direction.

For the CF-PEKK specimen interrupted after 21% of its fatigue life (Figure 10a), the peaks in the interrupted FFT signal almost overlap with the FFT signals of the undamaged state. This marks the absence of macroscopic damage accumulation at about 20% fatigue life. The peaks in the FFT signal of the CF-PEKK specimen interrupted after 44% of fatigue life (Figure 10b) overlap mostly with the peaks in the FFT signals of the damaged state. In contrast, for the CF-PEKK specimens interrupted after 97 and 99% of fatigue lives, the peaks of the FFT signals from the damaged state are more pronounced than the interrupted states. The peaks in the interrupted state for CF-PEKK03 and CF-PEKK04 do not align completely with the undamaged state, suggesting the potential presence of damage, which can be described using the XRM observations.

4.4. *X-ray Microscopy (XRM) Observations*

The XRM analysis of CF-PEKK01 showed that damage initiated at the top side of the specimen due to local compressive stresses after 21% fatigue life, which failed after 4.82×10^8 cycles at the same location. Figure 11a shows the only locations where damage was found across the specimen volume. This damage was found on the top layer of the CF-PEKK specimen, which is also represented as a schematic in Figure 5b. This damage location is typical for short beam shear specimens [38,45]. Since this damage initiation location is away from the specimen surface monitored using the IR camera, this damage is not detected by the temperature signals in Figure 8a. The final damage after specimen failure was an accumulation of longitudinal and transverse cracks across the top layer of the CF-PEKK01 specimen (see Figure 11b). A transverse crack can be seen to propagate to the specimen edge monitored using thermography and thus captured in the relative temperature signal in Figure 8b.

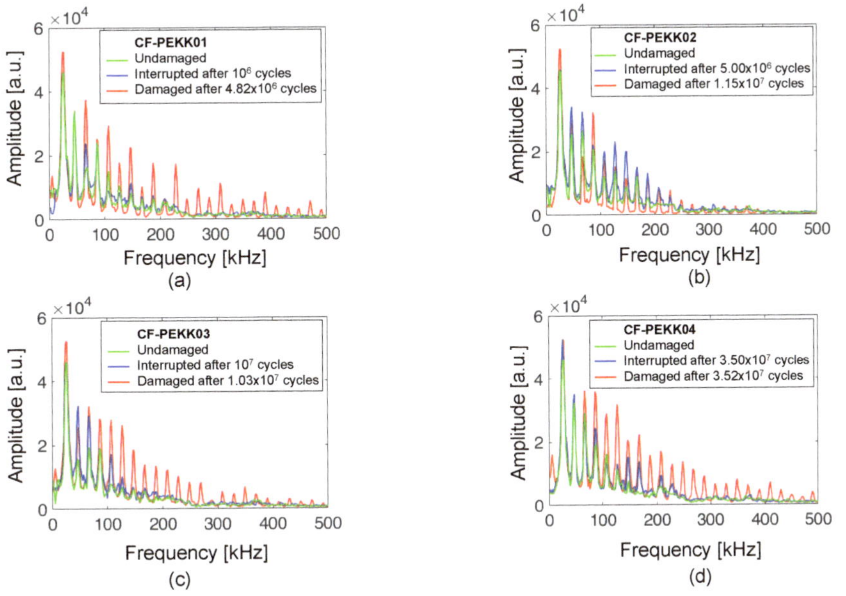

Figure 10. FFT spectra of 4 CAF experiments interrupted at (**a**) 10^6 cycles, (**b**) 5×10^6 cycles, (**c**) 10^7 cycles, and (**d**) 3.5×10^7 cycles.

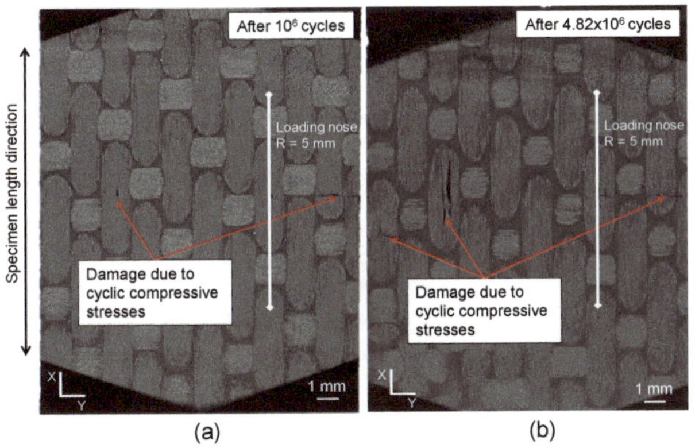

Figure 11. XRM images of CF-PEKK01 specimen showing the state of damage from the top view after (**a**) 21% and (**b**) end of fatigue life.

Figure 12 shows the results of XRM scans after interruption and damage for the CF-PEKK02 specimen. Figure 12a corresponds to the top view of the CF-PEKK specimen after 5×10^6 cycles. Figure 12b is obtained by slicing the specimen along its width, and Figure 12c is a zoomed-in view of this sliced plane showing damage initiation near the edge of the CF-PEKK specimen. Figure 12d is obtained by slicing the specimen along its length as shown by slicing lines in Figure 12a,e depicting the zoomed-in view showing longitudinal crack initiation at the matrix-warp fiber bundle interface. At the interrupted state, damage due to compressive stresses directly under the loading nose can be seen in Figure 12a,d,e. A slice along the width of the specimen shows damage initiation at the edge of the specimen due to cyclic shear stresses with fiber bundle/matrix debonding (Figure 12b,d). The XRM scan performed after the end of the fatigue experiment (after 1.15×10^7 load cycles) on the same specimen (CF-PEKK02) showed severe damage in the same locations spotted during the interrupted scan. This severity is marked with gray shaded regions in Figure 12f–j. Due to cyclic shear stresses, warp fiber bundle kinking was observed in the region between the support unit and loading nose (Figure 12f,i,j). The damage initiation observed near the edge of the CF-PEKK specimen in Figure 12b,c was found to propagate into the bulk of the material (see Figure 12g,h). The damage propagated to about two-third of the width of the specimen from the edge to the bulk. This can be seen in Figure 12h with damage features such as transverse interlaminar delamination.

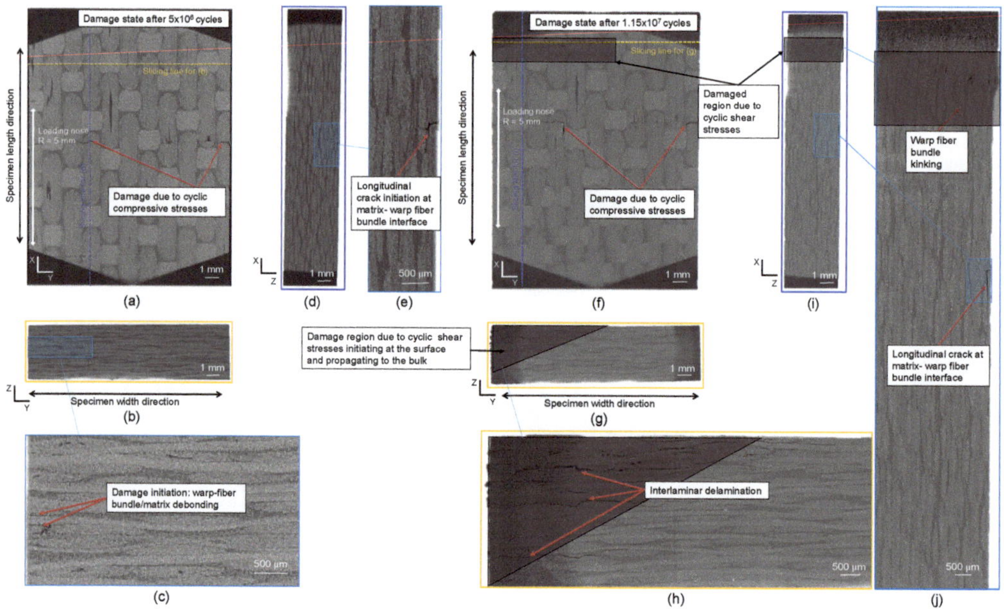

Figure 12. XRM images of the CF-PEKK02 specimen showing the state of damage from the (**a**) top view, (**b**) a slice across the YZ plane, (**c**) the zoomed-in view showing damage initiation in the YZ plane, (**d**) a slice across the XZ plane, (**e**) the zoomed-in view showing damage initiation in the XZ plane after 44% fatigue life, (**f**) the top view, (**g**) a slice across the YZ plane, (**h**) the zoomed-in view showing damage propagation in the YZ plane, (**i**) a slice across the XZ plane and (**j**) the zoomed-in view showing damage propagation in the XZ plane of the specimen after failure.

For the CF-PEKK03 specimen, the damage was observed in the top layer (see Figure 13a). Here, the warp fiber bundles were found to debond from the PEKK matrix (see Figure 13b,c), leading to transverse cracks after 10^7 that grew longer, leading to failure after 1.03×10^7 cycles (see Figure 13d,e). Here, the damage initiated at the top layer and propagated within the same layer as warp fiber bundle debonding. Such a damage mechanism could have resulted in a small resonance frequency drop in Figure 8e. However, since this damage occurred away from the specimen surface monitored by the infrared camera, it was not captured by the temperature signals. Only the final failure leading to a transverse crack up to the edge of the specimen was captured by the IR camera in Figure 8f. A longitudinal crack was found (see Figure 13g,h) when the specimen was sliced along the XZ plane near the specimen center. Until this crack development, no change in energy loss could be observed in Figure 8f, meaning the failure was sudden.

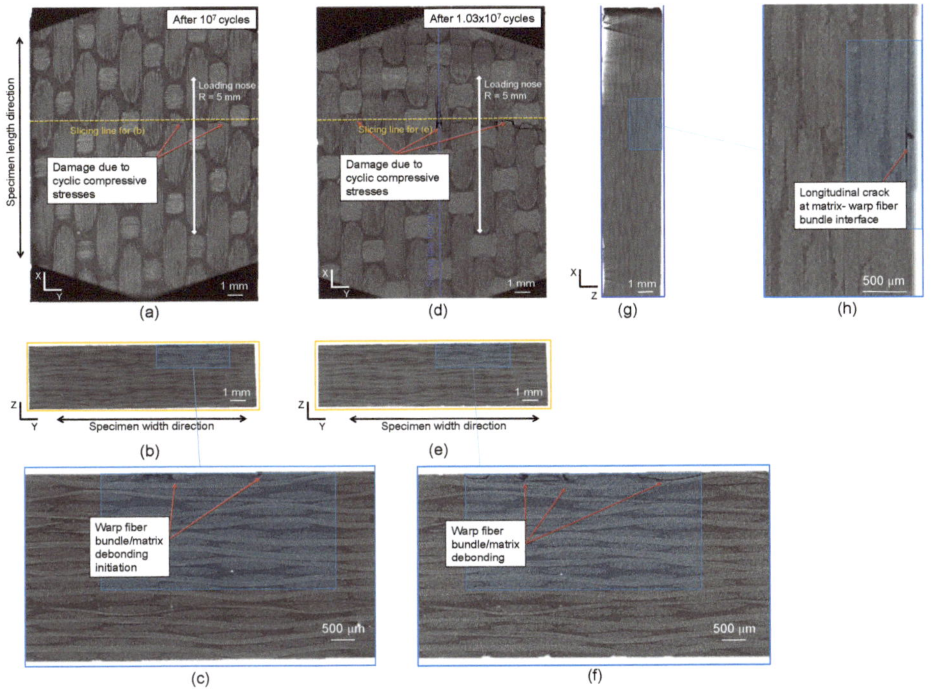

Figure 13. XRM images of the CF-PEKK03 specimen showing the state of damage from the (**a**) top view, (**b**) a slice across YZ plane, (**c**) the zoomed-in view to show damage initiation in the YZ plane after 97% fatigue life, (**d**) the top view, (**e**) a slice across the YZ plane, (**f**) the zoomed-in view to show damage propagation in the YZ plane, (**g**) a slice across the XZ plane and (**h**) the zoomed-in view to show damage propagation in the XZ plane of the specimen after failure.

The XRM scans of the CF-PEKK04 specimen showed the absence of any macroscopic damage due to shear stresses up to 99 % of fatigue life, with only a single crack observed on the top of the specimen due to local compressive stresses from the loading nose (see Figure 14a). The XRM scans after the end of the fatigue experiment (Figure 14b–f) showed damage morphologies similar to the damaged state of CF-PEKK02, meaning the damage severely accumulated in the last percent of the fatigue life. These results align with those obtained from the FFT spectra for interrupted and damaged conditions of the CF-PEKK04 specimen. Here, warp fiber bundle kinking was observed due to cyclic shear stresses and longitudinal cracks below the top layer due to local compressive stresses.

4.5. Discussion on the Results Obtained Using Optical Laser Microphone and XRM

The damage initiates at the surface due to the free edge effect [62,63] and accumulates across the width of the specimen in the form of transverse cracks [64]. In five-harness satin fabric architecture, transverse cracks are off-axis cracks that initiate in the matrix and propagate along the weft bundle direction, leading to weft fiber bundle delamination (see Figure 12c,h) [30]. Now, the definition of crack initiation in composite materials is even harder to define than in metals and other homogenous materials. Reifsneider (1980) described fatigue in composites as a sequential damage development with crack nucleation in off-axis plies. This could be a transverse crack. However, damage initiation can be identified at micro- and macroscopic levels [65]. Caiulo and Kachanov (2010) [66] reported that there is no correlation between the clustering of microcracks and stiffness reduction at the structural level. This is why the definition of Salkind (1972) [67] for damage initiation as the crack of detectable size seems more practical but dependent on the damage monitoring system.

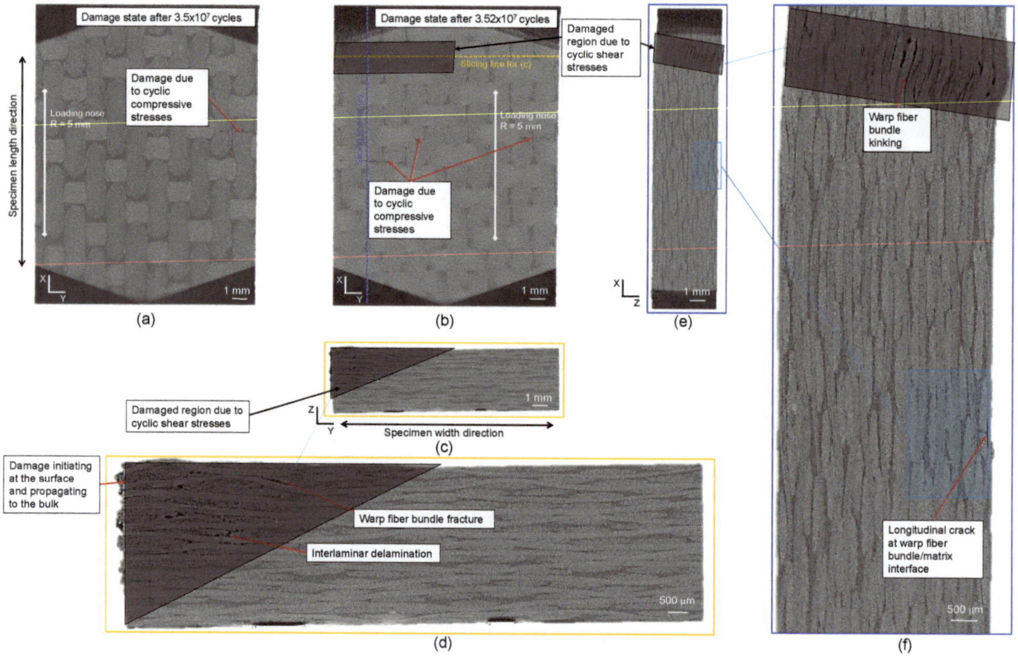

Figure 14. XRM images of the CF-PEKK04 specimen showing the state of damage from the (**a**) top view after 99% fatigue life, (**b**) the top view, (**c**) a slice across the YZ plane, (**d**) the zoomed-in view showing damage propagation in the YZ plane, (**e**) a slice across the XZ plane and (**f**) the zoomed-in view showing damage propagation in the XZ plane of the specimen after failure.

The higher-order harmonics due to damage during ultrasonic fatigue testing were already investigated using signals from a single-point Laser Doppler Vibrometer and optical laser microphone [68]. In another study, acoustic signals from a contactless optical laser microphone were used for non-linear analysis of damage evolution in CFRP during VHCF [69]. Studies on material non-linearity parameters are not new. A substantial monotonic increase in material non-linearity of Ti-6-Al-4V due to fatigue loading was reported in [70,71]. Later, this approach was used to detect early damage detection in CFRP [69,72]. However, the damage cannot be quantified when the optical laser microphone is the only in situ damage characterization technique. These results are much more useful when correlated with another technique that quantifies the shape and size of damage. Furthermore, each higher harmonic peak shown in Figure 10 should be matched to a specific damage feature to predict the damage state inside the composite specimen by just observing the FFT spectra during ultrasonic loading. Ex situ XRM scans are useful to compare with the results obtained using the optical laser microphone. Only macroscopic damage could be captured in this investigation since the CF-PEKK specimens are scanned using XRM without static load. This challenge could be overcome by performing in situ XRM measurements or applying static loading through a small fixture during XRM scans. Neither of these options was immediately possible with the current 3D XRM system at the Fraunhofer IAF facility. Furthermore, a higher resolution (in the order of 1 µm) scan could provide more information on the microscopic damage features. However, since damage can occur anywhere in the volume between the support pins ($18.6 \times 15 \times 4.1$ mm^3), the scan time would also increase significantly ($\times 100$). A higher resolution scan in selected areas where damage can occur is proposed for future investigations. Nevertheless, combining these two measurement techniques provides two important observations.

- The damage accumulation rate can be different even if the loading parameters are the same depending on the interaction between the local stress state at the weakest links in the composite material.
- Damage due to cyclic compressive stresses and the stress concentration at the contact between the loading nose and specimen occurs at the top layer of the CF-PEKK specimens. A longitudinal crack results from the accumulation of weft fiber bundle-matrix debonding.
- The damage due to cyclic shear stresses initiates at the surface due to the free-edge effect and progresses to the bulk of the specimen.

The different damage accumulation rates result in different fatigue lives for specimens tested at the same loading conditions. Due to cyclic compressive stresses and stress concentration between the loading nose and specimens, the damage accumulation occurs in the form of longitudinal cracks in the XZ plane (see Figure 15b) and transverse cracks in the YZ plane, both debonding the warp fibers from the PEKK matrix. The damage due to cyclic shear stresses was found to occur in one of the four regions, between the support pins and loading nose, where high positive and negative shear stresses are witnessed. Figure 15a shows a representative schematic of the damaged volume due to cyclic shear stresses. The damage initiates at the surface and progresses towards the bulk of the material.

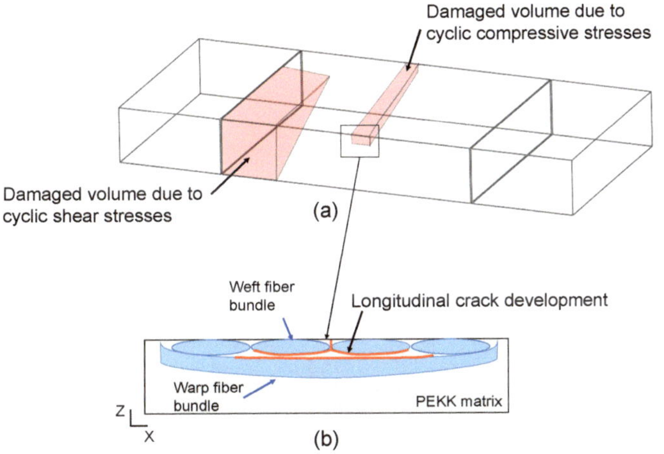

Figure 15. (**a**) A schematic of the CF-PEKK specimen showing damage sites observed using XRM and (**b**) a sketch of the longitudinal crack.

The final damaged state results from accelerated structural degradation due to damage accumulation resulting in localized surface temperature reaching 80 °C or when the specimen loses its ability to resonate. However, for CF-PEKK specimens, damage occurred at different locations on the surface due to high tensile, compressive, or shear stresses and their interaction with the weakest links [38] since they are short beams loaded with shear and bending forces. This means the damage accumulation rate depends on the location of damage initiation and the local stresses around it.

The final failure morphologies in Figures 12h and 14d show the saturation of cracks near the surface and propagation into interlaminar (or inter-ply) delamination. This observation is essential for ultrasonic fatigue testing of composites. If the temperature in the bulk of the specimen is much higher than the temperature at the surface, the damage will also be adverse due to the softening of the polymer material. In other words, if a surface temperature of 80 °C were measured using an IR camera, a higher temperature in the bulk of the material would mean increased damage accumulation in the bulk. In contrast, Figures 12h and 14d show that the damage due to shear stresses initiates at the edge and accumulates only up to about two-third of the width of the specimens, and the damage population decreases from the edge to the bulk. This means that the temperatures at the bulk of the CF-PEKK specimens cannot be much higher than at the edges of the specimens. Therefore, assessing damage on the surface of the composite specimens loaded under ultrasonic cyclic three-point bending using thermography and digital light optical microscopy [49] would provide valid results.

5. Summary and Conclusions

Thermography provides essential insights into damage accumulation that occurs at the surface of composite specimens and is extensively used in the fatigue characterization of composites. An important question to be addressed for ultrasonic fatigue testing of composites is the state of damage in the bulk of the material and the possibility of temperature influence due to friction between the damaged surfaces. Furthermore, only one side of the specimen is usually monitored using the infrared (IR) camera. This prompts another question: how can thermography be used to reliably capture the fatigued condition of the composite specimens if damage occurs either in the bulk or on the opposite side of the specimen, which is not covered by thermography?

Macroscopic damage anywhere in the composite changes the resonance characteristics (energy loss by ultrasonic generator and resonance frequency), which was reflected in the thermal response of the specimen with increased heating. This increased heating of the specimens occurs due to the loss of the CF-PEKK specimen's capacity to oscillate at ideal resonance conditions due to damage accumulation. Therefore, the surface characterization (thermography and light optical microscopy) could be related to the global properties (dynamics response and residual stiffness) of the CF-PEKK specimens. However, the damage state in the bulk of the material is still unknown.

This investigation mapped the damage initiation and accumulation locations using 3D X-ray microscopy (XRM). By looking into the volume of the damaged specimen, it was possible to relate the physical damage to the observed thermal and dynamic response of the CF-PEKK material. Furthermore, XRM provides strong evidence that the thermal influence on fatigue damage in composites due to ultrasonic oscillations can be minimized or even avoided with suitable ultrasonic test parameters (pulse duration, pulse–pause ratio) that lead to a surface temperature below $0.5 \times T_g$ [°C] of the polymer matrix. This can be confirmed with damage population near the surface rather than in the bulk of the CF-PEKK specimens.

Finally, with the use of an optical laser microphone, it was observed that damage in the composite specimen affects not only the eigenmode frequency but also the higher harmonics of the eigenmode frequency. However, more explanation for the change in amplitude and peak count of these higher harmonics needs to be further investigated to use such a technology for online damage assessment during ultrasonic fatigue testing of fiber-reinforced composites.

From this investigation, the following conclusions were drawn:

- The very high cycle fatigue damage in a CFRP specimen under ultrasonic cyclic three-point bending, due to shear stresses, initiates and populates at the surface.
- The temperature in the bulk of the specimen cannot be much higher than the temperature measured at the surface using infrared thermography.
- The scatter in the fatigue life of composites under cyclic three-point bending is due to the different damage initiation sites and the local stress state in the vicinity of initiated damage, leading to different damage accumulation rates.
- A combination of different measurement techniques is required to adequately capture the damage accumulation in fiber-reinforced polymers under ultrasonic cyclic three-point bending.

Author Contributions: Conceptualization, A.P.; methodology, A.P., M.P. and M.R.; software, M.R.; formal analysis, A.P. and M.P.; investigation, A.P. and M.P.; resources, L.K.; data curation, A.P. and M.P.; writing—original draft preparation, A.P.; writing—review and editing, M.P., L.K. and F.B.; supervision, L.K. and F.B.; project administration, F.B.; funding acquisition, F.B. All authors have read and agreed to the published version of the manuscript.

Funding: This work was funded by the Carl Zeiss Foundation (Stuttgart, Germany) under the framework of "Cross-scale characterization of robust functional material systems (SCHARF)".

Institutional Review Board Statement: Not applicable.

Data Availability Statement: Data are contained within the article.

Acknowledgments: The authors thank the Carl Zeiss Foundation for supporting the SCHARF research project at INATECH, University of Freiburg. The authors acknowledge the support from the Open Access Publication Fund of the University of Freiburg. Finally, the Walter-and-Ingeborg-Herrmann Foundation is highly acknowledged for its support of all our ultrasonic-related activities.

Conflicts of Interest: The authors declare no conflicts of interest. The funders had no role in the study's design, data collection, analysis, interpretation, manuscript writing, or decision to publish the results.

References

1. Gorbatikh, L.; Lomov, S.V. Damage accumulation in textile composites. In *Modeling Damage, Fatigue and Failure of Composite Materials*; Elsevier: Amsterdam, The Netherlands, 2016; pp. 41–59. [CrossRef]
2. Naik, N.K. Woven-fibre thermoset composites. In *Fatigue in Composites*; Elsevier: Amsterdam, The Netherlands, 2003; pp. 296–313. [CrossRef]
3. Kawai, M. 2D woven fabric composites under fatigue loading of different types and in different environmental conditions. In *Fatigue of Textile Composites*; Elsevier: Amsterdam, The Netherlands, 2015; pp. 149–191. [CrossRef]
4. Paepegem, W.V. Fatigue testing and online inspection of carbon textile composites for aeronautical applications. In *Fatigue of Textile Composites*; Elsevier: Amsterdam, The Netherlands, 2015; pp. 353–382. [CrossRef]
5. Zangenberg, J.; Brøndsted, P. Fatigue life in textile composites used for wind energy engineering. In *Fatigue of Textile Composites*; Elsevier: Amsterdam, The Netherlands, 2015; pp. 403–440. [CrossRef]
6. Bathias, C.; Paris, P.C. *Gigacycle Fatigue in Mechanical Practice*; CRC Press: Boca Raton, FL, USA, 2004. [CrossRef]
7. Vassilopoulos, A.P.; Keller, T. Introduction to the Fatigue of Fiber-Reinforced Polymer Composites. In *Engineering Materials and Processes*; Springer: London, UK, 2011; pp. 1–23. [CrossRef]
8. Harris, B. A historical review of the fatigue behaviour of fibre-reinforced plastics. In *Fatigue in Composites*; Elsevier: Amsterdam, The Netherlands, 2003; pp. 3–35. [CrossRef]
9. Quaresimin, M.; Ricotta, M., Fatigue response and damage evolution in 2D textile composites. In *Fatigue of Textile Composites*; Elsevier: Amsterdam, The Netherlands, 2015; pp. 193–221. [CrossRef]
10. Fujii, T.; Amijima, S.; Okubo, K. Microscopic fatigue processes in a plain-weave glass-fibre composite. *Compos. Sci. Technol.* **1993**, *49*, 327–333. [CrossRef]
11. Naik, R.A.; Patel, S.R.; Case, S.W. Fatigue Damage Mechanism Characterization and Modeling of a Woven Graphite/Epoxy Composite. *J. Thermoplast. Compos. Mater.* **2001**, *14*, 404–420. [CrossRef]
12. Nishikawa, Y.; Okubo, K.; Fujii, T.; Kawabe, K. Fatigue crack constraint in plain-woven CFRP using newly-developed spread tows. *Int. J. Fatigue* **2006**, *28*, 1248–1253. [CrossRef]
13. Fruehmann, R.; Dulieu-Barton, J.; Quinn, S. Assessment of fatigue damage evolution in woven composite materials using infra-red techniques. *Compos. Sci. Technol.* **2010**, *70*, 937–946. [CrossRef]
14. Wu, T.; Yao, W.; Xu, C. A VHCF life prediction method based on surface crack density for FRP. *Int. J. Fatigue* **2018**, *114*, 51–56. [CrossRef]
15. Lamon, F.; Maragoni, L.; Carraro, P.A.; Quaresimin, M. Fatigue damage evolution in woven composites with different architectures. *Int. J. Fatigue* **2023**, *167*, 107365. [CrossRef]
16. Shabani, P.; Taheri-Behrooz, F.; Samareh-Mousavi, S.S.; Shokrieh, M.M. Very high cycle and gigacycle fatigue of fiber-reinforced composites: A review on experimental approaches and fatigue damage mechanisms. *Prog. Mater. Sci.* **2021**, *118*, 100762. [CrossRef]
17. Backe, D.; Balle, F.; Eifler, D. Fatigue testing of CFRP in the Very High Cycle Fatigue (VHCF) regime at ultrasonic frequencies. *Compos. Sci. Technol.* **2015**, *106*, 93–99. [CrossRef]
18. Backe, D.; Balle, F. Ultrasonic fatigue and microstructural characterization of carbon fiber fabric reinforced polyphenylene sulfide in the very high cycle fatigue regime. *Compos. Sci. Technol.* **2016**, *126*, 115–121. [CrossRef]

19. Balle, F.; Backe, D. Very high cycle fatigue of carbon fiber reinforced polyphenylene sulfide at ultrasonic frequencies. In *Fatigue of Materials at Very High Numbers of Loading Cycles*; Springer Fachmedien Wiesbaden: Wiesbaden, Germany, 2018; pp. 441–461. [CrossRef]
20. Weibel, D.; Balle, F.; Backe, D. Ultrasonic Fatigue of CFRP—Experimental Principle, Damage Analysis and Very High Cycle Fatigue Properties. *Key Eng. Mater.* **2017**, *742*, 621–628. [CrossRef]
21. Cui, W.; Chen, X.; Chen, C.; Cheng, L.; Ding, J.; Zhang, H. Very High Cycle Fatigue (VHCF) Characteristics of Carbon Fiber Reinforced Plastics (CFRP) under Ultrasonic Loading. *Materials* **2020**, *13*, 908. [CrossRef] [PubMed]
22. Beghini, M.; Bertini, L.; Vitale, E. Analysis of fatigue delamination growth in carboresin specimens with central hole. *Compos. Struct.* **1991**, *17*, 257–274. [CrossRef]
23. Scarponi, C.; Briotti, G. Ultrasonic technique for the evaluation of delaminations on CFRP, GFRP, KFRP composite materials. *Compos. Part B Eng.* **2000**, *31*, 237–243. [CrossRef]
24. Mouritz, A.P.; Townsend, C.; Khan, M.S. Non-destructive detection of fatigue damage in thick composites by pulse-echo ultrasonics. *Compos. Sci. Technol.* **2000**, *60*, 23–32. [CrossRef]
25. Dzenis, Y.A. Cycle-based analysis of damage and failure in advanced composites under fatigue 1. Experimental observation of damage development within loading cycles. *Int. J. Fatigue* **2003**, *25*, 499–510. [CrossRef]
26. Gutkin, R.; Green, C.; Vangrattanachai, S.; Pinho, S.T.; Robinson, P.; Curtis, P.T. On acoustic emission for failure investigation in CFRP: Pattern recognition and peak frequency analyses. *Mech. Syst. Signal Process.* **2011**, *25*, 1393–1407. [CrossRef]
27. Liu, P.; Chu, J.; Liu, Y.; Zheng, J. A study on the failure mechanisms of carbon fiber/epoxy composite laminates using acoustic emission. *Mater. Des.* **2012**, *37*, 228–235. [CrossRef]
28. Katunin, A. Characterization of damage evolution during fatigue of composite structures accompanied with self-heating effect by means of acoustic emission. *J. Vibroeng.* **2018**, *20*, 954–962. [CrossRef]
29. Toubal, L.; Karama, M.; Lorrain, B. Damage evolution and infrared thermography in woven composite laminates under fatigue loading. *Int. J. Fatigue* **2006**, *28*, 1867–1872. [CrossRef]
30. Montesano, J.; Fawaz, Z.; Bougherara, H. Non-destructive assessment of the fatigue strength and damage progression of satin woven fiber reinforced polymer matrix composites. *Compos. Part B Eng.* **2015**, *71*, 122–130. [CrossRef]
31. Palumbo, D.; Finis, R.D.; Demelio, P.G.; Galietti, U. A new rapid thermographic method to assess the fatigue limit in GFRP composites. *Compos. Part B Eng.* **2016**, *103*, 60–67. [CrossRef]
32. Huang, J.; Pastor, M.L.; Garnier, C.; Gong, X. A new model for fatigue life prediction based on infrared thermography and degradation process for CFRP composite laminates. *Int. J. Fatigue* **2019**, *120*, 87–95. [CrossRef]
33. Garcea, S.C.; Wang, Y.; Withers, P.J. X-ray computed tomography of polymer composites. *Compos. Sci. Technol.* **2018**, *156*, 305–319. [CrossRef]
34. Böhm, R.; Stiller, J.; Behnisch, T.; Zscheyge, M.; Protz, R.; Radloff, S.; Gude, M.; Hufenbach, W.A. A quantitative comparison of the capabilities of in situ computed tomography and conventional computed tomography for damage analysis of composites. *Compos. Sci. Technol.* **2015**, *110*, 62–68. [CrossRef]
35. Yu, B.; Blanc, R.; Soutis, C.; Withers, P.J. Evolution of damage during the fatigue of 3D woven glass-fibre reinforced composites subjected to tension–tension loading observed by time-lapse X-ray tomography. *Compos. Part A Appl. Sci. Manuf.* **2016**, *82*, 279–290. [CrossRef]
36. Jespersen, K.M.; Zangenberg, J.; Lowe, T.; Withers, P.J.; Mikkelsen, L.P. Fatigue damage assessment of uni-directional non-crimp fabric reinforced polyester composite using X-ray computed tomography. *Compos. Sci. Technol.* **2016**, *136*, 94–103. [CrossRef]
37. Djabali, A.; Toubal, L.; Zitoune, R.; Rechak, S. Fatigue damage evolution in thick composite laminates: Combination of X-ray tomography, acoustic emission and digital image correlation. *Compos. Sci. Technol.* **2019**, *183*, 107815. [CrossRef]
38. Premanand, A.; Balle, F. Influence of pulse duration on the fatigue behavior of a carbon-fiber-reinforced composite under cyclic 3-point bending at 20 kHz. *Fatigue Fract. Eng. Mater. Struct.* **2024**. [CrossRef]
39. Dubé, M.; Chazerain, A.; Hubert, P.; Yousefpour, A.; Bersee, H.E. Characterization of resistance-welded thermoplastic composite double-lap joints under static and fatigue loading. *J. Thermoplast. Compos. Mater.* **2013**, *28*, 762–776. [CrossRef]
40. van den Brink, W.W.; Bruins, R.; Lantermans, P.; Groenendijk, C. Analysis and testing of a thermoplastic composite stiffened panel under compression. In Proceedings of the 5th Aircraft Structural Design Conference, Manchester, UK, 4–6 October 2016.
41. van Dooren, K.; Labans, E.; Tijs, B.; Bisagni, C.; Waleson, J. Analysis and testing of a thermoplastic composite stiffened panel under compression. In Proceedings of the 22nd International Conference on Composite Materials (ICCM22), Melbourne, Australia, 11–16 August 2019; pp. 2336–2342.
42. Ochôa, P.; Groves, R.M.; Benedictus, R. Systematic multiparameter design methodology for an ultrasonic health monitoring system for full-scale composite aircraft primary structures. *Struct. Control Health Monit.* **2019**, *26*, e2340. [CrossRef]
43. Toray Cetex TC1320 PEKK Product Data Sheet. Available online: https://www.toraytac.com/media/409dcc72-6aff-4643-86b8-20b5f464f038/lNO6TA/TAC/Documents/Data_sheets/Thermoplastic/UD%20tapes,%20prepregs%20and%20laminates/Toray-Cetex-TC1320_PEKK_PDS.pdf (accessed on 29 November 2022).
44. Toray Cetex TC1200 PEEK Product Data Sheet. Available online: https://www.toraytac.com/media/7765d981-1f9f-472d-bf24-69a647412e38/Pr7gdw/TAC/Documents/Data_sheets/Thermoplastic/UD%20tapes,%20prepregs%20and%20laminates/Toray-Cetex-TC1200_PEEK_PDS.pdf (accessed on 29 November 2022).

45. Premanand, A.; Rogala, T.; Wachla, D.; Amraei, J.; Katunin, A.; Khatri, B.; Rienks, M.; Balle, F. Fatigue strength estimation of a CF/PEKK composite through self-heating temperature analysis using cyclic bending tests at 20 kHz. *Compos. Sci. Technol.* **2023**, *243*, 110218. [CrossRef]
46. Premanand, A.; Balle, F. Influence of Pulse-Pause Sequences on the Self-Heating Behavior in Continuous Carbon Fiber-Reinforced Composites under Ultrasonic Cyclic Three-Point Bending Loads. *Materials* **2022**, *15*, 3527. [CrossRef] [PubMed]
47. *ASTM D7264*; Standard Test Method for Flexural Properties of Polymer Matrix Composite Materials. ASTM: West Conshohocken, PA, USA, 2021. [CrossRef]
48. Premanand, A.; Balle, F. Stress and strain calculation method for orthotropic polymer composites under axial and bending ultrasonic fatigue loads. *Ultrasonics* **2023**, *135*, 107130. [CrossRef] [PubMed]
49. Premanand, A.; Rienks, M.; Balle, F. Damage assessment during ultrasonic fatigue testing of a CF-PEKK composite using self-heating phenomenon. *Int. J. Fatigue* **2024**, *180*, 108084. [CrossRef]
50. Fischer, B. Optical microphone hears ultrasound. *Nat. Photon.* **2016**, *10*, 356–358. [CrossRef]
51. Preißer, S.; Fischer, B.; Panzer, N. Listening to Ultrasound with a Laser. *Opt. Photonik* **2017**, *12*, 22–25. [CrossRef]
52. Dragonfly 2022.2 [Computer Software]. Available online: https://www.theobjects.com/dragonfly (accessed on 15 December 2023).
53. *ISO13003:2003*; Fibre-Reinforced Plastics-Determination of Fatigue Properties under Cyclic Loading Conditions. ISO: Geneva, Switzerland, 2003.
54. Fitzka, M.; Karr, U.; Granzner, M.; Melichar, T.; Rödhammer, M.; Strauss, A.; Mayer, H. Ultrasonic fatigue testing of concrete. *Ultrasonics* **2021**, *116*, 106521. [CrossRef]
55. *ASTM D2344*; ASTM Standard Test Method for Short-Beam Strength of Polymer Matrix Composite Materials and Their Laminates. ASTM: West Conshohocken, PA, USA, 2022. [CrossRef]
56. Gao, F.; Boniface, L.; Ogin, S.; Smith, P.; Greaves, R. Damage accumulation in woven-fabric CFRP laminates under tensile loading: Part 1. Observations of damage accumulation. *Compos. Sci. Technol.* **1999**, *59*, 123–136. [CrossRef]
57. Kumar, R.; Talreja, R. Fatigue damage evolution in woven fabric composites. In Proceedings of the 41st Structures, Structural Dynamics, and Materials Conference and Exhibit, Atlanta, GA, USA, 3–6 April 2000; American Institute of Aeronautics and Astronautics: Reston, VA, USA, 2000. [CrossRef]
58. Vallons, K.A.; Drozdzak, R.; Charret, M.; Lomov, S.V.; Verpoest, I. Assessment of the mechanical behaviour of glass fibre composites with a tough polydicyclopentadiene (PDCPD) matrix. *Compos. Part A Appl. Sci. Manuf.* **2015**, *78*, 191–200. [CrossRef]
59. Al-hababi, T.; Cao, M.; Alkayem, N.F.; Shi, B.; Wei, Q.; Cui, L.; Šumarac, D.; Ragulskis, M. The dual Fourier transform spectra (DFTS): A new nonlinear damage indicator for identification of breathing cracks in beam-like structures. *Nonlinear Dyn.* **2022**, *110*, 2611–2633. [CrossRef]
60. Kharazan, M.; Irani, S.; Noorian, M.A.; Salimi, M.R. Nonlinear vibration analysis of a cantilever beam with multiple breathing edge cracks. *Int. J. Non-Linear Mech.* **2021**, *136*, 103774. [CrossRef]
61. Patel, S. Durability of a graphite/epoxy woven composite under combined hygrothermal conditions. *Int. J. Fatigue* **2000**, *22*, 809–820. [CrossRef]
62. Lecomte-Grosbras, P.; Paluch, B.; Brieu, M. Characterization of free edge effects: Influence of mechanical properties, microstructure and structure effects. *J. Compos. Mater.* **2013**, *47*, 2823–2834. [CrossRef]
63. Zhang, C.; Binienda, W.K. A meso-scale finite element model for simulating free-edge effect in carbon/epoxy textile composite. *Mech. Mater.* **2014**, *76*, 1–19. [CrossRef]
64. Hosoi, A.; Arao, Y.; Kawada, H. Transverse crack growth behavior considering free-edge effect in quasi-isotropic CFRP laminates under high-cycle fatigue loading. *Compos. Sci. Technol.* **2009**, *69*, 1388–1393. [CrossRef]
65. Magi, F.; Maio, D.D.; Sever, I. Damage initiation and structural degradation through resonance vibration: Application to composite laminates in fatigue. *Compos. Sci. Technol.* **2016**, *132*, 47–56. [CrossRef]
66. Caiulo, A.; Kachanov, M. On Absence of Quantitative Correlations between Strength and Stiffness in Microcracking Materials. *Int. J. Fract.* **2010**, *164*, 155–158. [CrossRef]
67. Salkind, M. *Fatigue of Composites*; ASTM International: West Conshohocken, PA, USA, 1972; pp. 143–169. [CrossRef]
68. Rabe, U.; Helfen, T.; Weikert, M.; Hirsekorn, S.; Herrmann, H.G.; Boller, C.; Backe, D.; Balle, F.; Eifler, D. Nonlinear ultrasonic testing of carbon fibre reinforced plastics in the very high cycle fatigue regime. In Proceedings of the Meetings on Acoustics, Cefalu, Sicily, Italy, 1–7 July 2012. [CrossRef]
69. Venkat, R.S.; Starke, P.; Boller, C. Acoustics based assessment of a composite material under very high cycle fatigue loading. In *Fatigue of Materials at Very High Numbers of Loading Cycles*; Springer Fachmedien Wiesbaden: Wiesbaden, Germany, 2018; pp. 463–485. [CrossRef]
70. Frouin, J.; Matikas, T.E.; Na, J.K.; Sathish, S. In-situ monitoring of acoustic linear and nonlinear behavior of titanium alloys during cycling loading. In Proceedings of the Nondestructive Evaluation of Aging Materials and Composites III, Newport Beach, CA, USA, 3–5 March 1999. [CrossRef]

71. Cantrell, J.H.; Yost, W.T. Nonlinear ultrasonic characterization of fatigue microstructures. *Int. J. Fatigue* **2001**, *23*, 487–490. [CrossRef]
72. Bagchi, S.; Venkat, R.S.; Boller, C.; Starke, P.; Mitra, M. Monitoring Early Damage Initiation of Very High Cycle Fatigued Composite Material Using a Nonlinearities Based Inverse Approach. In *Structural Health Monitoring 2015*; Destech Publications: Lancaster, PA, USA, 2015. [CrossRef]

polymers

MDPI

Article

Mechanical Behaviors of Polymer-Based Composite Reinforcements within High-Field Pulsed Magnets

Siyuan Chen [1,2], Tao Peng [1,2], Xiaotao Han [1,2], Quanliang Cao [1,2], Houxiu Xiao [1,2] and Liang Li [1,2,*]

[1] Wuhan National High Magnetic Field Center, Huazhong University of Science and Technology, Wuhan 430074, China; chen_siyuan@hust.edu.cn (S.C.); pengtao@mail.hust.edu.cn (T.P.); xthan@hust.edu.cn (X.H.); quanliangcao@hust.edu.cn (Q.C.); xiaohouxiu@mail.hust.edu.cn (H.X.)

[2] State Key Laboratory of Advanced Electromagnetic Engineering and Technology, Huazhong University of Science and Technology, Wuhan 430074, China

* Correspondence: liangli44@hust.edu.cn; Tel.: +86-27-87792334

Abstract: The development of pulsed magnets capable of generating magnetic fields exceeding 100 Tesla has been recognized as a crucial pursuit for advancing the scientific research on high magnetic fields. However, the operation of magnets at ultra-high magnetic fields often leads to accidental failures at their ends, necessitating a comprehensive exploration of the underlying mechanisms. To this end, this study investigates, for the first time, the mechanical behaviors of Zylon fiber-reinforced polymers (ZFRPs) within pulsed magnets from a composite perspective. The study begins with mechanical testing of ZFRPs, followed by the development of its constitutive model, which incorporates the plasticity and progressive damage. Subsequently, in-depth analyses are performed on a 95-T double-coil prototype that experienced a failure. The outcomes reveal a notable reduction of approximately 45% in both the radial and axial stiffness of ZFRPs, and the primary reason for the failure is traced to the damage incurred by the end ZFRPs of the inner magnet. The projected failure field closely aligns with the experiment. Additionally, two other magnet systems, achieving 90.6 T and 94.88 T, are analyzed. Finally, the discussion delves into the impact of transverse mechanical strength of the reinforcement and axial Lorentz forces on the structural performance of magnets.

Keywords: pulsed magnet; polymer-based composites; structural performances; damage and failure mechanism; Zylon fiber

Citation: Chen, S.; Peng, T.; Han, X.; Cao, Q.; Xiao, H.; Li, L. Mechanical Behaviors of Polymer-Based Composite Reinforcements within High-Field Pulsed Magnets. *Polymers* **2024**, *16*, 722. https://doi.org/10.3390/polym16050722

Academic Editors: Xiaoquan Cheng, Wenjun Huang and Qian Zhang

Received: 29 January 2024
Revised: 28 February 2024
Accepted: 5 March 2024
Published: 6 March 2024

1. Introduction

Pulsed magnets are the exclusive devices capable of generating repeated magnetic field intensity exceeding 50 T within milliseconds. They play an essential role as fundamental scientific research tools in the fields of condensed matter physics and magnetics [1–3]. The ability of magnets to manipulate the electronic states of matter is significantly enhanced with increasing magnetic field intensity. Consequently, research laboratories, such as the National High Magnetic Field Laboratory (NHMFL), Dresden High Magnetic Field Laboratory, and Wuhan National High Magnetic Field Center (WHMFC), have dedicated the past four decades to achieving magnetic field intensities of 100 T or higher [4–6]. The current world record stands at 100.75 T, achieved by NHMFL [7], while WHMFC attained 94.88 T in 2021 [6]. Both of them have set their sights on reaching 110 T in the future.

Nevertheless, the upgrading of pulsed magnets poses considerable challenges due to the substantial Lorentz forces involved. Nowadays, pulsed magnets employ an inner reinforcing structure, where each conductor layer is reinforced with fiber composites, as depicted in Figure 1a,b. Since the radial component predominantly contributes to the Lorentz forces (especially on the mid-plane of magnets), composites are designed with a winding angle of 90° and manufactured by customized winding machines (Figure 1c). Although this approach results in magnets with low axial stiffness and strength, it remains substantial for low-field operations. Nonetheless, as the field intensity increases, the axial forces induced by the radial stray field become significant (especially on the magnet ends), frequently leading to inadvertent damage. The composites at the magnet ends are found to undergo complete destruction, accompanied by dislocation of the wires [3,8–10]. The underlying failure mechanism remains unclear, and effective countermeasures have yet to be identified.

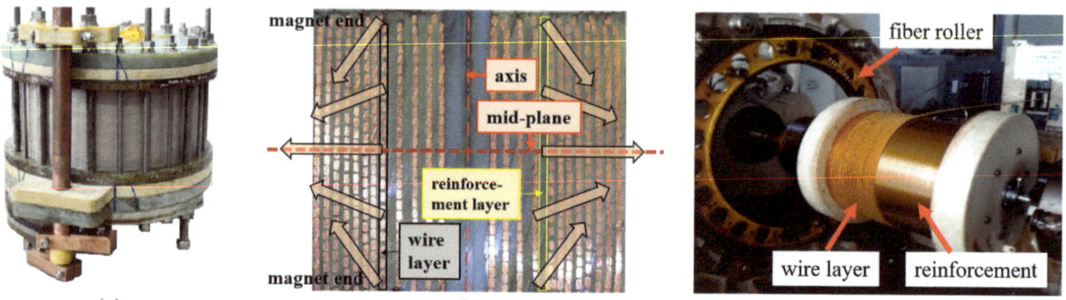

(a) (b) (c)

Figure 1. Structure of pulsed magnets and relating winding machine. (**a**) Mono-coil pulsed magnet of WHMFC; (**b**) magnet cross-section (arrow inside indicates the Lorentz forces subjected by conductors); (**c**) customized winding machine of WHMFC.

The unclear failure mechanism primarily arises from the inadequate analyses of composites within pulsed magnets. Traditionally, the composite reinforcements are considered as ideal anisotropic elastic materials, employing the von Mises failure criterion, and a design is deemed reasonable if the von Mises stress of the reinforcement remains below 80% of the fiber-direction ultimate tensile strength (UTS) [11–13]. However, this treatment overlooks the transverse and shear failure modes, as well as the nonlinear behaviors exhibited by composite materials. It should be noted that, during high-field operations, the reinforcement encounters radial and axial compression stresses exceeding 200 MPa, thereby increasing the likelihood of transverse damage [14–16]; however, the von Mises criteria would obscure this underlying damage. Additionally, laboratories currently adopt Zylon fibers as reinforcement fibers due to their exceptional strength and excellent electrical insulation. Nevertheless, Zylon fibers display poor impregnation with epoxy, resulting in a notable plastic behavior [17–19]. Hence, it is insufficient to solely consider the ideal anisotropic elasticity.

Moreover, the lack of systematical testing on Zylon fiber-reinforced polymers (ZFRPs) poses another problem. Significant research efforts have been dedicated to investigation of the longitudinal ultimate tensile strength of ZFRPs [20–23]. Only Huang et al. conducted measurements regarding the transverse compression strength of ZFRPs with cylindrical and bar-shaped specimens, but these results exhibited notable dispersion due to stress concentration effects [23]. As a result, the transverse and in-plane shear properties of ZFRPs remain insufficiently understood, impeding efforts to elucidate the underlying failure mechanism.

Taking all these factors into account, this paper presents, for the first time, a comprehensive analysis on the mechanical behavior of ZFRP reinforcements from a composite perspective. The study begins with comprehensive mechanical testing of ZFRPs, followed by the development of a plasticity-damage model. Subsequently, the simulation technique for pulsed magnets is proposed, and the mechanical behaviors of ZFRP reinforcements inside three double-coil magnet systems are thoroughly analyzed. These systems include a 95-T prototype that experienced failure, as well as two magnets achieving magnetic fields of 94.88 T and 90.6 T, respectively. Finally, the influence of the transverse mechanical properties of reinforcement and axial Lorentz forces on the structural failure are discussed.

2. Materials and Methods

In general, the ZFRP reinforcements within pulsed magnets are wound under high pre-stress, achieving a high fiber filling factor (V_f) of 0.8. However, due to the challenges encountered in manufacturing a standard specimen with such a high V_f, specimens with a moderate V_f of 0.53 were initially tested. Subsequently, the mechanical properties of ZFRPs with V_f of 0.8 were deduced based on a bridging model. The mechanical tests have been conducted in our recent work [24] but will still be outlined in this section for clarity.

2.1. Specimen Preparation

To begin with, a Zylon prepreg was manufactured using a one-step hot melt method. A thermoset epoxy with a moderate characteristic curing temperature was selected as the resin matrix. The epoxy exhibited a tensile elastic modulus of 3.2 GPa and UTS of 70 MPa, and its flexural strength and modulus were measured as 130 MPa and 1.07 GPa, respectively.

Subsequently, ZFRP laminates were produced using the autoclave-forming method. The temperature was gradually raised to 350 K at a rate of 2 K/min and held for 30 min. Next, the temperature was further increased to 400 K and maintained for 90 min, followed by a natural cooling process to room temperature. The entire process was conducted under a pressure of 0.8 MPa. Eight layers of prepreg were employed, resulting in a laminate with a thickness of 2 mm, while the V_f was determined to be 0.53.

Furthermore, an E-glass end tab was attached to the laminate using a low-temperature structural adhesive. The adhesion areas were polished and cleaned with acetone. To mitigate the stress concentration during transverse tension testing, wedge-shaped tabs were incorporated (according to ASTM D3039 [25]). Finally, laminates were machined into the desired shape using a diamond band saw system (EXKAT300CL). To mitigate the fluffing of Zylon fibers, the ZFRP laminate was sandwiched between two carbon/epoxy laminates during the cutting process.

2.2. Mechanical Tests

The tensile tests of ZFRPs were conducted under ASTM D3039, while the compression and in-plane shear tests were based on ASTM D6641 [26] and ASTM D3518 [27], respectively. The testing results are presented in Table 1. For clarity, the notation used is as follows: '1' denotes the fiber direction; '2' signifies the transverse direction; 'c' denotes compression properties while 't' signifies tension properties; 'E' denotes elastic modulus ('G' denotes the shear modulus in special); 'X' signifies strength; and '0' denotes the elastic limit strength. It should be noted that, due to the limitations of the environmental chamber, the results of 77 K were obtained through linear fitting or extrapolation.

Table 1. Mechanical properties of ZFRPs with V_f of 0.53.

Temperature	E_{11}^t (GPa)	υ_{12}	X_{11}^c (MPa)	E_{11}^c (GPa)	X_{22}^c (MPa)	E_{22}^c (GPa)	$X_{22}^{t,0}$ (MPa)	X_{22}^t (MPa)	E_{22}^t (GPa)	X_{12}^0 (MPa)	X_{12}^u (MPa)	G_{12} (GPa)
296 K	142	0.36	170	106	81.8	3.66	11.8	18.0	3.12	30.0	49.6	1.49
243 K	/	/	173	110	85.9	3.65	13.8	19.2	3.51	31.2	46.5	1.64
213 K	/	/	/	/	100	4.27	18.5	18.5	3.60	30.6	46.7	1.75
77 K	173.3 [20]	0.36 [20]	170 **	110 **	124.5 *	4.99 *	18.5 **	18.5 **	4.45 *	30.6 **	46.7 **	2.16 *

* Deduced by linear fitting. ** Deduced by extrapolation.

The stress value $X_{22}^{t,0}$ in Table 1 was defined as the stress at which the deviation from the linear zone reached 0.5 MPa, while the stress value $X_{12}^{t,0}$ was defined as the stress at which the attenuation of the secant in-plane shear modulus reached 7%. In addition, tension tests on symmetric laminates $(\pm 55°)_{2s}$ and $(\pm 67.5°)_{2s}$ were performed. Their tension strengths at 293 K were determined to be 32.6 MPa and 22.4 MPa, respectively. The tested stress–strain curves will be employed to verify the validity of material model in Section 3.3.

2.3. Properties with High V_f

Based on the testing results, the properties of Zylon fiber at 77 K were deduced as E_{11}^f = 327 GPa, E_{22}^f = 2.7 GPa, G_{12}^f = 1.9 GPa, and G_{23}^f = 1.3 GPa, according to the bridging model [26]. The bridging model utilizes the Mori–Tanaka approach to establish a relationship between the stresses in the fiber and the matrix, enabling the calculation of the strength and modulus of composites with a high degree of accuracy [28]. Subsequently, the bridging model was employed to predict the mechanical properties of ZFRPs with a V_f of 0.8. The predicted results are listed in Table 2 and will be used in the subsequent sections.

Table 2. Mechanical properties of ZFRPs with V_f of 0.8 at 77 K.

E_{11} (GPa)	E_{22} (GPa)	G (GPa)	ν_{12}	ν_{23}	X_{11}^t (GPa)	X_{11}^c (MPa)	$X_{22}^{t,0}$ (MPa)	$X_{22}^{t,u}$ (MPa)	X_{22}^c (MPa)	X_{12}^0 (MPa)	X_{12}^u (MPa)
263	3.36	2.06	0.0047	0.6	4.64	170	12	18.3	124	27.3	45.2

3. Constitutive Modeling of ZFRPs

To characterize the nonlinearity of ZFRPs caused by the internal damages and inherent plasticity, an anisotropic model of FRPs constructed in our previous work was employed [29]. The effectiveness of the model was demonstrated through biaxial-tension and open-hole-tension tests of glass and carbon FRP laminates. This section will outline the constructed model and verify its applicability to ZFRPs.

3.1. Plasiticity Evolution

An anisotropic elastoplastic model was built to capture the plasticity evolution. The effective engineering stress and elastic strain are represented in vector form as follows:

$$\tilde{\boldsymbol{\sigma}} = [\tilde{\sigma}_{11}, \tilde{\sigma}_{22}, \tilde{\sigma}_{33}, \tilde{\sigma}_{12}, \tilde{\sigma}_{23}, \tilde{\sigma}_{13}]^T \tag{1}$$

$$\boldsymbol{\varepsilon}^e = [\varepsilon_{11}^e, \varepsilon_{22}^e, \varepsilon_{33}^e, \varepsilon_{12}^e, \varepsilon_{23}^e, \varepsilon_{13}^e]^T \tag{2}$$

Based on the assumption of strain equivalence, these two quantities follow the relationship of

$$\dot{\tilde{\boldsymbol{\sigma}}} = \mathbf{M} \cdot \dot{\boldsymbol{\sigma}} = diag\left(\frac{1}{1-d_1}, \frac{1}{\eta_E(1-d_2)}, \frac{1}{\eta_E(1-d_2)}, \frac{1}{\eta_G(1-d_6)}, \frac{1}{\eta_G(1-d_6)}, \frac{1}{\eta_G(1-d_6)}\right) \cdot \dot{\boldsymbol{\sigma}} = \mathbf{M} \cdot \mathbf{D^e} \cdot \dot{\boldsymbol{\varepsilon}}^{\mathbf{e}} \tag{3}$$

where $\mathbf{D^e}$ denotes the Voigt form of the four-order stiffness tensor; $\mathbf{M}$ signifies the damage matrix; d_1, d_2, and d_6 correspond to the coefficients reflecting the fiber-direction, transverse, and shear micro-damages, respectively; and η_E and η_G correspond to the coefficients reflecting the transverse and shear macro-damages, respectively. The evolution of d and η will be introduced in Section 3.2.

The equivalent stress is formulated as follows [30]:

$$\overline{\tilde{\sigma}}_y = \sqrt{\frac{3}{2}\left[(\tilde{\sigma}_{22})^2 + (\tilde{\sigma}_{33})^2 + 2a(\tilde{\sigma}_{12})^2 + 2a(\tilde{\sigma}_{13})^2 + 2a(\tilde{\sigma}_{23})^2\right]} \tag{4}$$

where a represents the coupling between transverse plasticity and in-plane shear plasticity. The associated yield function and equivalent plastic strain are given by:

$$\Phi = \frac{1}{2}\tilde{\boldsymbol{\sigma}}^T \mathbf{P} \tilde{\boldsymbol{\sigma}} - \overline{\tilde{\sigma}}_y^2(\bar{\varepsilon}^p) \tag{5}$$

and

$$\overline{\tilde{\sigma}}_y(\bar{\varepsilon}^p) = \beta(\bar{\varepsilon}^p)^\alpha \tag{6}$$

where **P** is defined as diag(0,3,3,6a,6a,6a). The material-related coefficients a, α, and β are determined as 2.0, 0.3, and 900 for ZFRPs through in-plane shear tests. The initial yield stress is set as zero considering that 'α' is tiny. Details of the determination process of these coefficients can be found in [29].

The flowing vector **N** of the equivalent strain is

$$\mathbf{N} = \frac{\partial \Phi}{\partial \tilde{\sigma}} = \mathbf{P} \cdot \tilde{\sigma} \tag{7}$$

Thus far, the equivalent plastic strain can be expressed as

$$\dot{\bar{\varepsilon}}^{p} = 2\dot{\gamma}\bar{\bar{\sigma}}_y = \sqrt{\frac{2}{3}\left(\dot{\varepsilon}_{22}^{p}\right)^2 + \frac{2}{3}\left(\dot{\varepsilon}_{22}^{p}\right)^2 + \frac{1}{3a}\left(\dot{\varepsilon}_{12}^{p}\right)^2 + \frac{1}{3a}\left(\dot{\varepsilon}_{23}^{p}\right)^2 + \frac{1}{3a}\left(\dot{\varepsilon}_{13}^{p}\right)^2} \tag{8}$$

where γ is the Lagrange's plastic multiplier. Accordingly, the hardening law is

$$\dot{\bar{\varepsilon}}^{p} = \dot{\gamma}H = \dot{\gamma}\sqrt{\mathbf{N}^T \cdot \mathbf{Z} \cdot \mathbf{N}} \tag{9}$$

where H is the hardening modulus and **Z** is defined as diag(0,2/3,2/3,1/3a,1/3a,1/3a).

3.2. Damage Evolution

Puck's criteria were employed to determine the material damage. Considering that the ZFRP reinforcements of magnets are subjected to radial, axial, and hoop stresses, the three-dimensional (3D) form of Puck's criteria were utilized. Details of Puck's criteria can be found in [31]. Necessary coefficients of Puck's criteria were listed in Table 2.

The Cachan continuum damage model was employed to describe the evolution of micro-damages [32–34]. The damage strain energy density W_D is redefined in terms of the action-plane stresses as

$$W_D = \frac{1}{2}\left[\frac{\sigma_{11}^2}{E_{11}} - \frac{2v_{11}}{E_{11}}\sigma_{11}\sigma_n + \frac{\langle\sigma_n\rangle_+^2}{E_{22}(1-d_2)} + \frac{\langle\sigma_n\rangle_-^2}{E_{22}} + \frac{\tau_{nt}^2 + \tau_{n1}^2}{2G_{12}(1-d_6)}\right] \tag{10}$$

The corresponding damage-development equations are [30,31]:

$$\begin{cases} d_2 = \frac{\langle \sup\left(\sqrt{Y_{d'}(t) + bY_d(t)}\right) - Y_0\rangle_+}{Y_c} \\ d_6 = \frac{\langle \sup\left(\sqrt{Y_{d'}(t) + bY_d(t)}\right) - Y_0'\rangle_+}{Y_c'} \end{cases} \tag{11}$$

with

$$\begin{cases} Y_d = \frac{\partial W_D}{\partial d_2} = \frac{1}{2}\frac{\langle\sigma_n\rangle_+^2}{E_{11}(1-d_2)^2} \\ Y_{d'} = \frac{\partial W_D}{\partial d_6} = \frac{1}{2}\frac{\tau_{nt}^2 + \tau_{n1}^2}{G_{12}(1-d_6)^2} \end{cases} \tag{12}$$

The values of b, Y_0, Y_c, Y_0', and Y_c' were determined as 12.7, 0.21, 6.16, 0.3, and 4.16 for ZFRPs, through cyclic tensile tests of $(\pm 45^\circ)_{2s}$ and $(\pm 67.5^\circ)_{2s}$ laminates. The determination process of these coefficients can be found in [30].

Furthermore, the macro-damages were characterized using the macro-phenomenon method [35], which is given as Equation (13). In this equation, '0' signifies the tangent modulus at the onset of macro-damages and '$f_{\mathrm{E,IFF}}$' indicates Puck's stress exposure for inter-fiber fracture. Since the stiffness of ZFRPs degrades rapidly after the emergence of macro-damages, the parameters c, η, and ξ were set as 100, 0.01, and 0.8, respectively.

$$\begin{bmatrix} E_{22} \\ G_{12} \end{bmatrix} = \begin{bmatrix} E_{33} \\ G_{23} \end{bmatrix} = \begin{bmatrix} \eta_E \cdot E_{22}^0 \\ \eta_G \cdot G_{12}^0 \end{bmatrix} = \begin{bmatrix} \left(\frac{1-\eta_{rE}}{1+c_E(f_{E,IFF}-1)^{\xi_E}} + \eta_{rE}\right) \cdot E_{22}^0 \\ \left(\frac{1-\eta_{rG}}{1+c_G(f_{E,IFF}-1)^{\xi_G}} + \eta_{rG}\right) \cdot G_{12}^0 \end{bmatrix} \tag{13}$$

3.3. Numerical Implementation

The constitutive model was implemented in Fortran and integrated into the ANSYS 19.0 software by the UserMat feature. The role of UserMat is to update the Cauchy stress and consistent tangent stiffness by utilizing the received stress and strain at every material integration point during the solution phase.

For the Cauchy stress, the return mapping algorithm was employed. In a plastic loading condition, the elastoplastic Equations (1)–(9) are simplified to

$$\Phi(\Delta\gamma) = \frac{1}{2}\left[(\mathbf{I}+\Delta\gamma\mathbf{D^eP})^{-1}\cdot\tilde{\boldsymbol{\sigma}}^{\mathbf{trial}}_{\mathbf{n+1}}\right]^T\cdot\mathbf{P}\cdot\left[(\mathbf{I}+\Delta\gamma\mathbf{D^eP})^{-1}\cdot\tilde{\boldsymbol{\sigma}}^{\mathbf{trial}}_{\mathbf{n+1}}\right] - \bar{\bar{\sigma}}_y^2\left(\bar{\varepsilon}^{p,trial}_{n+1}+\Delta\gamma H\left(\tilde{\boldsymbol{\sigma}}_{\mathbf{n+1}}\right)\right) \tag{14}$$

The only unknown variable $\Delta\gamma$ can be calculated by Newton–Raphson iteration and the stress and damage state can be evaluated. For the consistent tangent stiffness $\mathbf{D^{ep}}$, a chain derivation was employed and the stiffness can be deduced according to Equation (15).

$$\mathbf{D^{ep}} = \mathbf{M}^{-1}\cdot\left[(\mathbf{D^e})^{-1}+\Delta\gamma\mathbf{P}-\frac{\Delta\gamma\cdot(\mathbf{NN^T})\cdot(\mathbf{ZP})}{H^2}+\frac{\mathbf{NN^T}}{2H\bar{\bar{\sigma}}_y\alpha\beta(\bar{\varepsilon}^p)^{\alpha-1}}\right]^{-1} \tag{15}$$

The built constitutive model was conducted on ZFRP laminates. A quarter finite element model built in [29], which is simply and reveals minimal mesh dependency, was employed. The comparison results were shown in Figure 2. The deviation of the predicted UTS was less than 3%. The simulated failure modes, either matrix tensile failure or matrix shear failure, were also in good agreement with the experimental ones. These verified the applicability of the constructed constitutive model to ZFRPs.

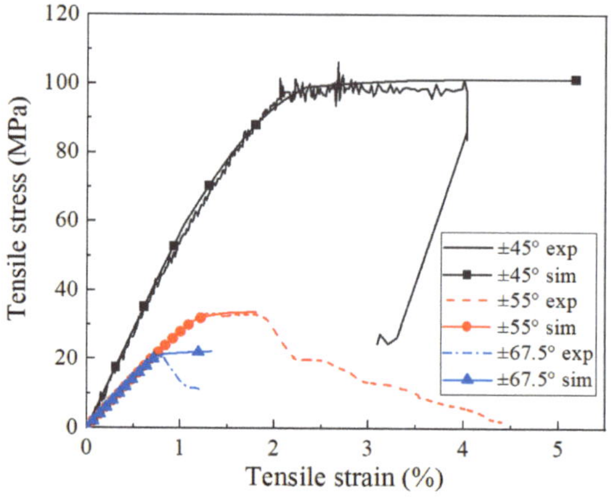

Figure 2. Comparisons between the tensile stress–strain curves of ZFRP symmetric laminates obtained experimentally and by simulation.

4. Simulation Technique for Magnets

4.1. Simulation Strategy

The simulation strategy for pulsed magnets was illustrated in Figure 3. Initially, the coupling field of the electrical circuit, magnetic field, and thermal field were solved using a sequential coupling method with a two-dimensional (2D) axisymmetric formulation. The circuit current was firstly solved; then, the calculated current was used to excite the magnet. The induced voltage of the magnet was refreshed by the magnetic penetration method [36]. Next, the Joule heat from the magnetic analysis was loaded on the thermal analysis module and refreshed the coil resistance accordingly [37].

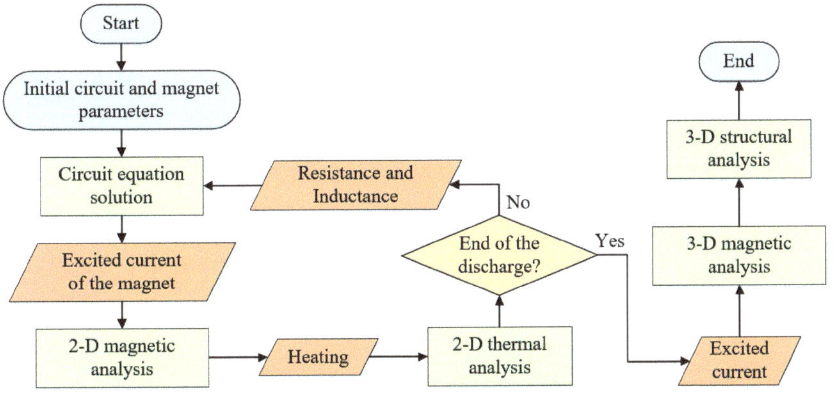

Figure 3. Flowchart of the simulation strategy for pulsed magnets.

Subsequently, the obtained excited current was applied to perform a 3D magnetic field analysis, utilizing the scalar magnetic potential method. Finally, the 3D Lorentz forces obtained from the 3D magnetic analysis were applied to the structural model, which are represented as

$$\begin{cases} F_r = \int\limits_{V_{node}} \left(\vec{J} \times \vec{B} \right) \cdot \mathbf{e_r} \cdot dV_{node} \\ F_z = \int\limits_{V_{node}} \left(\vec{J} \times \vec{B} \right) \cdot \mathbf{e_z} \cdot dV_{node} \end{cases} \quad (16)$$

where V_{node} is the effective volume of one single node, while $\mathbf{e_r}$ and $\mathbf{e_z}$ are the unit direction vector of radial direction and axial direction, respectively. It should be emphasized that the employment of a 2D form for electromagnetic and thermal analysis is used to reduce the computation time, considering the symmetry structure of magnets. The structural analysis was conducted in a 3D form to efficiently carry out a dynamic mechanical examination.

4.2. Modeling Method

For the finite-element modeling, the wires and reinforcements were built separately and share an overlapping line on the boundary. Both of them were meshed separately, while contact elements were inserted on the interfaces. Figure 4 shows the mesh of one single conductor layer and corresponding reinforcement. As for the magnetic and thermal analysis, the contact pairs were set as perfect heat dissipation and perfect magnetic contact in the thermal and magnetic analysis. As for the structural analysis, the Coulomb friction model was employed to characterize the interface behavior between the conductor layer and reinforcement, with a friction coefficient of 0.2 [38].

The 2D model present in [36] was employed for the electromagnetic and thermal analysis. A 3D semi-model was used for the structural analysis of the inner magnet to reduce the computing time and enhance convergence, while a 1/2 3D semi-model was used for the outer magnet. The difference is due to the discrepancy ratio of the inside diameter to the height of magnets. The boundary limits in the finite-element modeling included the axial displacement being fixed on the axial cross-section (mid-plane) and the symmetric boundary (U_{hoop} = 0) being applied on the hoop cross-section (for the outer magnet). Additionally, since the edges of the magnet are fixed by flanges, the radial and hoop displacement of the end wires were constrained, while the wires could deform freely axially.

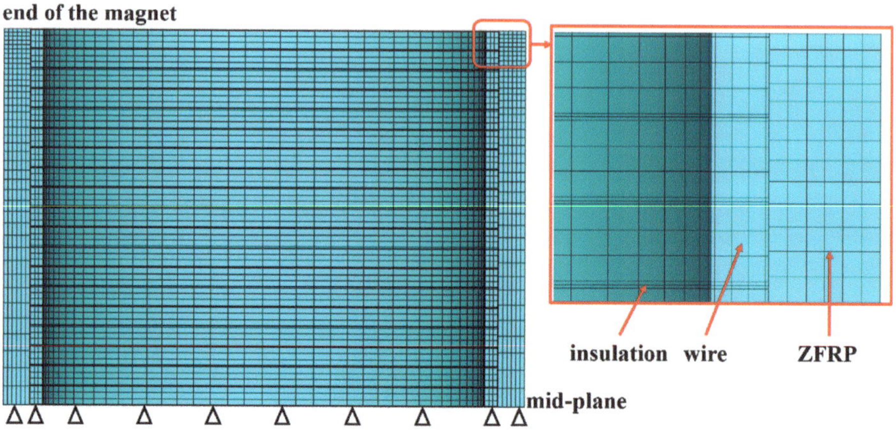

Figure 4. The 3D finite-element meshes of pulsed magnets (only shows one conductor layer and corresponding reinforcement).

The constitutive model constructed in Section 3 was employed to represent ZFRP reinforcements. In regard to wires, the Hill anisotropic plasticity model with kinematic hardening was utilized for Copper–Niobium. The longitudinal and transverse elastic moduli were determined as 78.7 GPa and 11.97 GPa, respectively, while the corresponding yield stresses were 370 MPa and 310 MPa. As for the hard copper wires, an isotropic plasticity model was adopted. The elastic moduli were determined as 42.9 GPa, while the corresponding yield stress was 277 MPa.

4.3. Electromagnetic Model Verification

The Lorentz forces are contingent on the magnet's current and hold a pivotal role in determining the effectiveness of the structural analysis. As a result, a two-layer coil was wound by CuNb and Glass fiber/epoxy to verify the built model, as shown in Figure 5a. This coil consists of 36 turns of wire winding, and the inner bore measures 44 mm. The thickness of the reinforcements are 6 mm and 6.3 mm, respectively. To safeguard against stray magnetic field interference, 30 mm G-10 flanges are employed to secure the inlet and outlet copper bars. The resistance and inductance of the coil stand at 32 mΩ and 80.26 μH, respectively, as measured by a RLC-bridge.

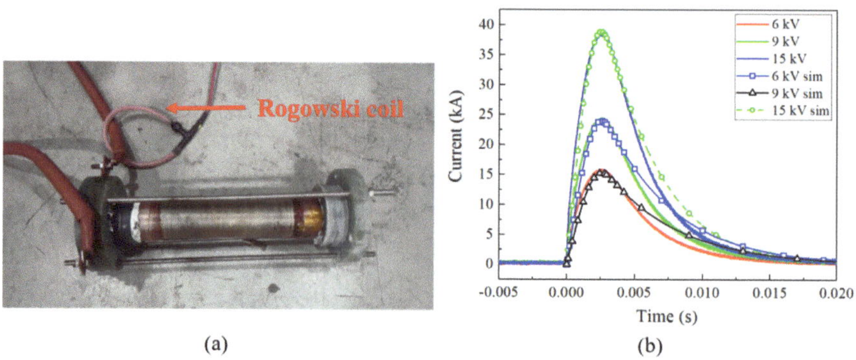

Figure 5. Electromagnetic model verification: (**a**) Experimental configuration; (**b**) Discharging current.

A 1.2 MJ capacitor bank with a capacitance of 3.84 mF was used. The line resistance and inductance were measured at 20 mΩ and 0.6 mH, respectively. During the discharge process, the voltage gradually increased from 6 kV to 9 kV to 15 kV, while the current was monitored using a Rogowski coil. The simulated discharging current closely matched the experimental data, as illustrated in Figure 5b. The primary deviations were observed in the declining portion of the current waveform due to the omission of the line resistance rise. This would not influence the peak current (or the calculated Lorentz forces) and is acceptable.

5. Structural Analysis of ZFRPs within Magnets

In this section, analyses were performed on a double-coil magnet system of WHMFC at first. It was originally designed with the von Mises stress failure criterion, and the designed peak magnetic field was 95 T [39]. Unfortunately, this prototype was broken at 83 T, where the stress level is much less than the design limit. Both the inner and outer magnet were destructed, and serious scorch marks were found at the end flanges. Therefore, this prototype was analyzed in detail to determine the reason for the breakage. Subsequently, two successful magnet systems, i.e., the 90.6-T and 94.88-T double-coil magnets of WHMFC, were analyzed in addition. The diameter and thickness of reinforcements within those three magnet were summarized in Table 3.

Table 3. Geometry of analyzed reinforcements.

Parameters	95-T Prototype	90.6-T Magnet	94.88-T Magnet
Number of inner-coil reinforcement layers	8	8	8
Diameter of inner-coil reinforcements (mm)	18.4; 30.2; 42.6; 57.4; 74.2; 93.0; 112.8; 132.6	18.4; 28.8; 40.6; 54.0; 68.8; 84.6; 101.0; 118.4	
Thickness of inner-coil reinforcements (mm)	2.7; 3.0; 4.2; 5.2; 6.2; 6.7; 6.7; 9.0	2.0; 2.7; 3.5; 4.2; 4.7; 5.0; 5.5; 6.0	3.0; 3.0; 5.0; 5.5; 6.5; 6.7; 6.7; 8.0
Number of outer-coil reinforcement layers	14	12	14
Diameter of outer-coil reinforcements (mm)	171.8; 196.6; 221.4; 246.2; 270.0; 289.8; 309.6; 326.4; 343.2; 360.0; 376.8; 393.6; 410.4; 427.2	140.3; 159.1; 177.9; 195.7; 212.5; 227.3; 239.1; 250.9; 262.7; 274.5; 286.3; 298.1	182.8; 206.6; 230.4; 254.2; 277.0; 295.8; 314.6; 330.4; 346.2; 362.0; 377.8; 393.6; 409.4; 425.2
Thickness of outer-coil reinforcements (mm)	6.5; 6.5; 6.5; 6.0; 4.0; 4.0; 2.5; 2.5; 2.5; 2.5; 2.5; 2.5; 2.5; 20.0	5.0; 5.0; 4.5; 4.0; 3.0; 1.5; 1.5; 1.5; 1.5; 1.5; 1.5; 20.0	6.5; 6.5; 6.5; 6.0; 4.0; 4.0; 2.5; 2.5; 2.5; 2.5; 2.5; 2.5; 2.5; 20.0

5.1. Inner Magnet Analysis

Analyses of the inner magnet were performed first and incorporated the small deformation formula. Figure 6 presents the stress distribution on the mid-plane at the designed peak magnetic field. ZFRPs in particular exhibited a maximum hoop stress of 2.93 GPa. The maximum axial compression stress in the wire reached 440 MPa, while that in ZFRPs reached 150 MPa. The maximum radial compression stress reached 280 MPa on the interfaces between the wires and ZFRPs, gradually decreasing towards zero in the radial direction. Furthermore, the absence of progressive failure consideration resulted in an approximate 18% increase in the axial compression stress on ZFRPs and a decrease in the hoop stress by 540 MPa. These changes are attributed to the significant variations of the anisotropy of ZFRPs, since the hoop stress increases as the anisotropy of the reinforcements [12].

Although no fiber fracturing was observed, macro-damages occurred on the outer surface of the end of the 2nd–8th reinforcements, about 17.5 turns away from the mid-plane (Figure 7a). It firstly occurred on the outer surface of the 7th ZFRPs when the magnetic field reached a value of 67 T. In addition, the most significant radial stiffness attenuation was observed on the internal surface of ZFRPs (Figure 7b). This can be attributed to the fact that the highest radial compression stress was exerted on the internal surface. The average radial stiffness on the mid-plane measured 2.9 GN/m, with a minimum value of 1.17 GN/m, representing a reduction compared to the original transverse stiffness of 5.29 GN/m. Similarly, the most substantial attenuation in the axial stiffness was observed on the mid-plane (Figure 7c). The average axial stiffness on the mid-plane was 2.81 GN/m, with a minimum value of 2.1 GN/m. In regard to the shear stiffness attenuation (Figure 7d), it was primarily located at the macro-damage position.

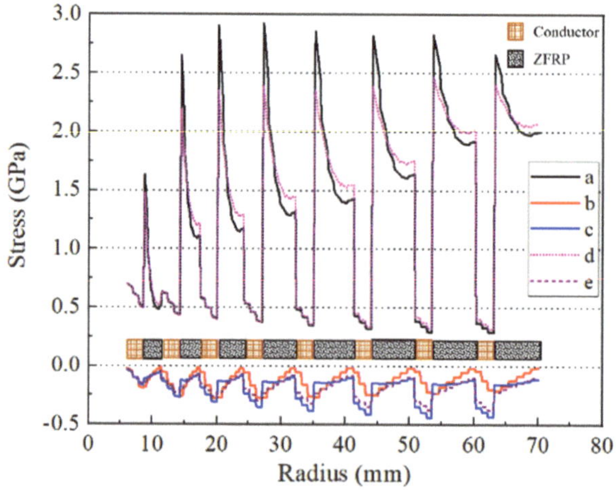

Figure 6. Stress distribution on the mid−plane of the inner magnet. Curves a–c correspond to the hoop, radial, and axial stresses of the proposed model, while curves d and e correspond to the hoop and axial stresses obtained using the ideal elastic model of ZFRPs.

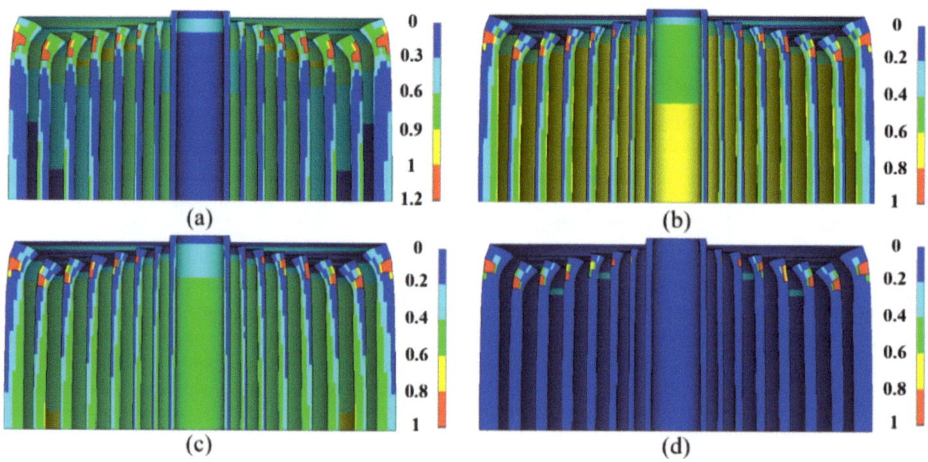

Figure 7. Stiffness attenuation of ZFRP reinforcements of the inner magnet at the designed peak magnetic field. The displacements scale factor is set as 10: (**a**) stress exposure (stress exposure greater than 1 indicated an emergence of macro-damages); (**b**) radial stiffness; (**c**) axial stiffness; (**d**) shear stiffness.

Figure 8 serves as a valuable reference to gain insight into the macro-damages. It elucidates that the primary factor contributing to macro-damages on the outer surface of the reinforcement was that the axial tensile stress exceeded the transverse tensile strength of ZFRPs. This axial tensile stress is linked to the forces of the end wires, the direction of which is $\arctan B_r/B_z$ against e_r (suppose the counterclockwise as the positive direction). Furthermore, Figure 8 reveals that both the inner and outer surfaces on the mid-plane of ZFRPs are susceptible to macroscopic failure, as the stress state closely approaches the failure envelope.

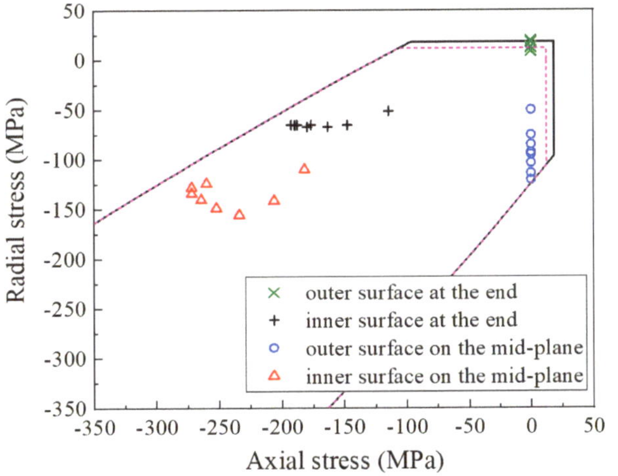

Figure 8. Axial vs. radial stress on the inner and outer surfaces of inner magnet reinforcements. The solid line indicates the macroscopic failure envelope, while the dashed line indicates the micro-failure envelop.

Subsequently, a dynamic structural analysis incorporating the large deformations formula was performed. The rise-time of the pulse was 3.54 ms, and each conductor layer and its corresponding reinforcement (abbreviate this entirety as CR) was analyzed separately. The analysis revealed that the first CR did not undergo any structural damage, while local buckling was observed in the 2nd to 7th CR. Once macro-damages occurred, the failure zone evolved much more rapidly than the statics (Figure 9a–c), ultimately leading to the complete destruction of the entire end-structure (Figure 9d). The end reinforcement could no longer constrain the wires.

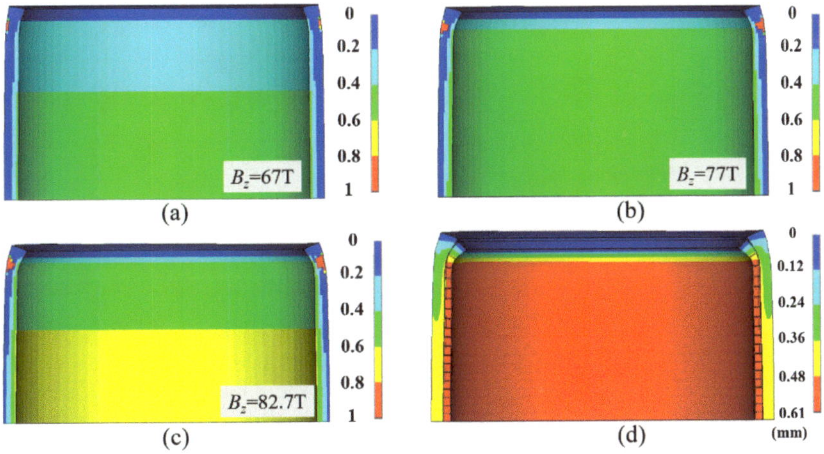

Figure 9. Radial stiffness attenuation of the ZFRP reinforcement of the 7th CR. The displacements scale factor is set as 10: (**a**) radial stiffness attenuation at 67 T; (**b**) radial stiffness attenuation at 77 T; (**c**) radial stiffness attenuation at 82.7 T; (**d**) radial displacement at 82.7 T.

Table 4 provides the buckling loads for each CR component. The 7th CR is the most vulnerable part. The predicted failure field exhibited good coincident to the actual numbers. From the aforementioned analyses, it becomes evident that the low transverse strength of ZFRPs is the critical factor leading to macro-damages, posing a significant risk to the magnet system.

Table 4. Buckling field of each CR component of the inner magnet.

	#1	#2	#3	#4	#5	#6	#7	#8
Buckling magnetic field (T)	-	93.05	93.87	94.86	94.83	92.51	82.76	-

5.2. Outer Magnet Analysis

The stress distribution of the outer magnet on the mid-plane, calculated under the designed peak magnetic field, is depicted in Figure 10. The maximum axial compression stress in the wires reached 569 MPa, while that in ZFRPs reached 249 MPa. Moreover, the maximum radial compression and hoop stress experienced by ZFRPs were 349 MPa and 2.95 GPa, respectively. Notably, the hoop stress decreased by 470 MPa with the absence of progressive failure, while the axial stress remained relatively unchanged. This negligible change in the axial stress can be attributed to the compression of the CR components within the outer magnet, which could effectively suppress the effect of stiffness attenuation. Furthermore, the maximum axial displacement was about 22% lower compared to the maximum displacement of the proposed model (8 mm). This 8 mm displacement could have severe implications in regard to the insulation.

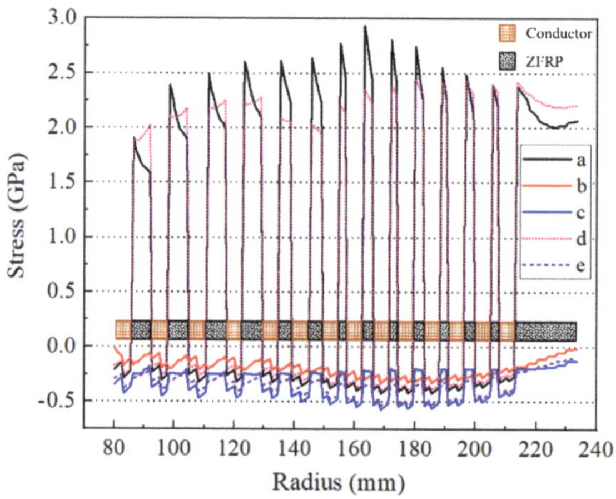

Figure 10. Stress distribution on the mid−plane of the outer magnet. Curves a–c correspond to the hoop, radial, and axial stresses of the proposed model, while curves d and e correspond to the hoop and axial stresses obtained using the ideal elastic model of ZFRPs.

Figure 11a illustrates the macroscopic failure and stiffness attenuation of ZFRPs at the designated peak magnetic field. No fiber fracture was detected. Macro-damages were observed on the outer surface of the 1st–4th ZFRP layers and the inner surface of the 9th and 10th ZFRP layers. This attributed to the imbalance in the transverse stress, which introduced significant shear stress. The attenuation of the stiffness is depicted in Figure 11b–d. On the mid-plane, the average radial, axial, and shear stiffness measured 1.62 GN/m, 1.35 GN/m, and 1.59 GN/m, respectively.

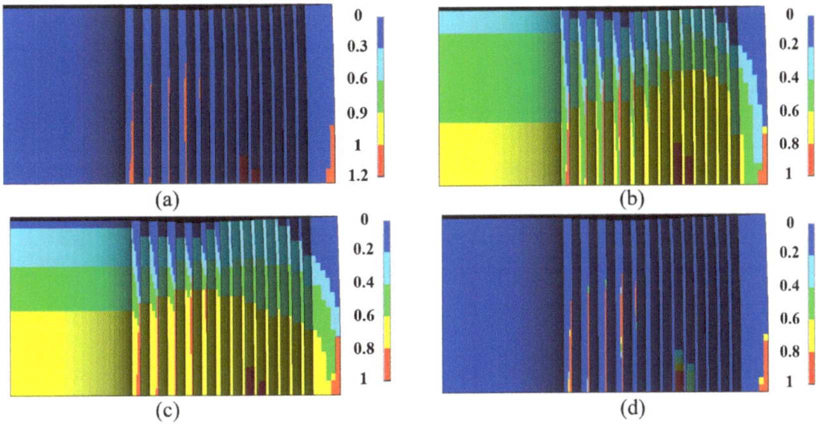

Figure 11. Stiffness attenuation of ZFRP reinforcements of the outer magnet: (**a**) stress exposure; (**b**) radial stiffness; (**c**) axial stiffness; (**d**) shear stiffness.

Finally, a dynamic analysis was conducted. Since the CR components of the outer coil were compressed against each other, no local buckling was observed. The stress distribution resembled closely that of the static analysis with a small deformation.

5.3. Other Examples

Two additional magnets, 94.88-T and 90.6-T magnet system of WHMFC, were analyzed in addition. The 94.88-T magnet system maintained the same configurations as the previous 95-T prototype, including the number of conductor layers, the wires in each conductor layer, and the inner bore dimensions. The pulse rise-time of the inner magnet is 3.72 ms and the ZFRP thickness is 3, 3, 5, 5.5, 6.5, 6.7, 7.2, 8 mm. Furthermore, the inner magnet was designed to contribute 55 T at the designed peak magnetic field of 96 T. The accumulated radial and axial Lorentz forces of the 7th CR were 7790 kN and 830 kN at designed peak field, which increased by 5.8% and 19% against the 7th CR of the 95-T prototype. The calculated damaged mode was similar but the calculated failure field increased to 92.9 T. This is mainly because that the ZFRP layers of the 94.88-T magnet system were much thicker.

For the 90.6-T magnet system, main parameters can be found in reference [6]. The pulse rise-time of the inner magnet was 4.8 ms, and the inner magnet was designed to contribute 48.4 T under the peak magnetic field of 90.6 T. The stiffness attenuation of the inner magnet at the designed peak magnetic field is shown in Figure 12. Macro-damages occurred on the outer surfaces of the end of the 4nd–8th reinforcements and the outer surfaces of the 5th–8th reinforcements on the mid-plane. The end damage was due to the axial tensile stress, and the damage on the mid-plane was attributed to the axial compression stress. The axial compression stresses on the mid-plane for the 5th–8th reinforcements measured 145 MPa, 154 MPa, 147 MPa and 149 MPa, respectively, while the radial stresses were zero. This imbalance of the transverse stress would lead to the shear failure mode. The calculated failure field is 89.2 T.

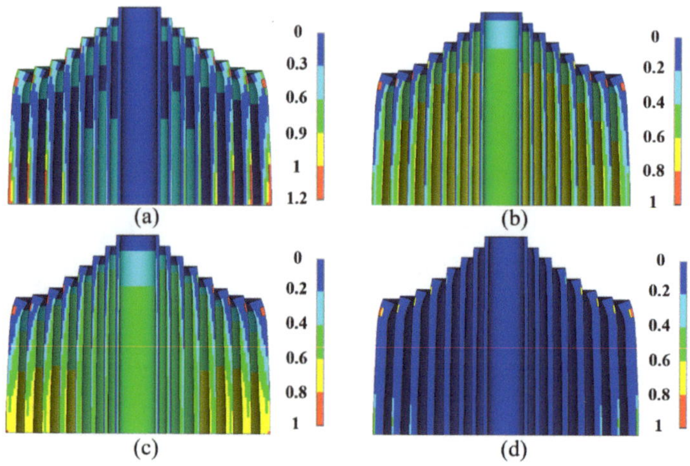

Figure 12. Stiffness attenuation of ZFRP reinforcements of the 90.6-T inner magnet. The displacements scale factor is set as 10: (**a**) stress exposure; (**b**) radial stiffness; (**c**) axial stiffness; (**d**) shear stiffness.

6. Discussions

6.1. Influence of the Transverse Strength of the Reinforcement

The transverse mechanical properties, denoted as X^0_{12}, X^u_{12}, $X^{t,0}_{22}$, $X^{t,u}_{22}$ and X^c_{22}, were artificially altered to discuss its impact on the magnets' performance. Both sets of the properties were increased to the same proportion, with the properties of T300/1034-C [29] serving as the target (when the increasing factor reached 1). The corresponding failure magnetic field of the 95-T prototype is shown in Figure 13. Since the Lorentz forces increased quadratically as the magnetic field grew, as depicted in Equation (16) (where J is linear to B), the failure magnetic field exhibited rapid escalation initially, followed by a deceleration in its growth once the increasing factor reached 0.4. Notably, when the increasing factor reached 0.9, the magnet remained intact. In a word, it is evident that the failure magnetic field increased with the enhancement of transverse properties. This observation underscores the importance of reinforcement materials possessing not only high longitudinal strength but also effective impregnation with epoxy to ensure robust transverse strength.

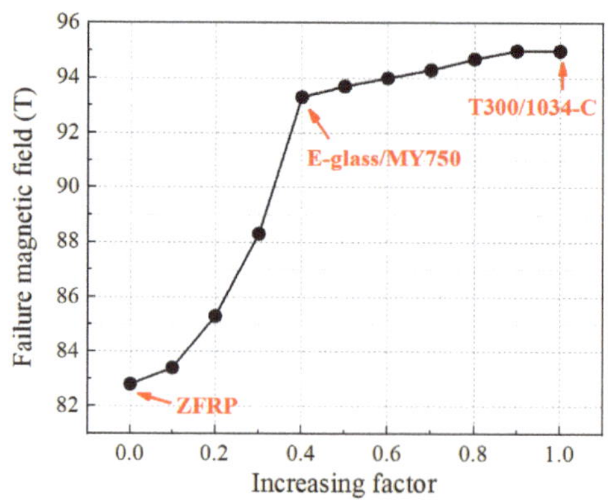

Figure 13. Influence of the transverse properties of reinforcements on the performance of magnets.

6.2. Influence of the Axial Lorentz Forces

Further to the transverse strength of reinforcements, another focus was the axial Lorentz forces within the magnet, which traditionally have received less attention in design considerations. The axial Lorentz forces were systematically scaled from 0.3 to 1.5 while keeping the radial Lorentz forces unchanged. The resulting angle at which the end wire forces were applied was changed from 5.6° to 26°. The corresponding failure magnetic field is depicted in Figure 14.

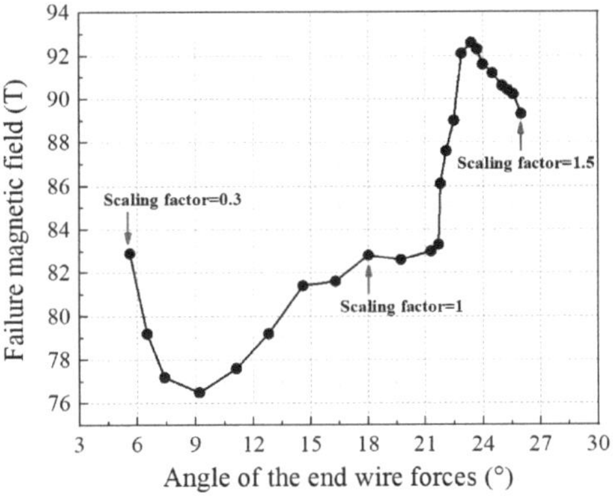

Figure 14. Influence of the axial Lorentz forces on the performance of magnets.

The general trend in the failure magnetic field variations can be roughly interpreted as follows: Under a moderate force angle of approximately 23°, the wires predominantly bear most of the axial forces and ZFRPs remain intact. As the axial forces increase further (with the axial force scaling factor exceeding 1.5), the wires approach their plastic limit, and ZFRPs gradually assume a greater share of the axial forces, leading to a reduction in the failure magnetic field. Conversely, when the end force angle is approximately 9°, ZFRPs take on the majority of axial forces, resulting in the lowest failure field. If the axial forces are further reduced, the failure field would increase. It should be emphasized that it is asymmetric to change the axial Lorentz forces in a physical sense; therefore, the trend of the curve in Figure 14 is also asymmetric.

6.3. Suitable Reinforcing Material

Previous sections have drawn attention to the limitations of ZFRPs regarding their inadequate transverse strength. This raises concerns about the suitability of Zylon fibers for pulsed high-field magnets. Is there any other fiber that could replace Zylon? Kevlar and other arylon fibers are associated with similar challenges related to insufficient transverse strength as Zylon; Glass fibers have a longitudinal tensile elastic modulus of less than 90 GPa, which is not sufficient to effectively constrain the wire deformation; Carbon and boron fibers do not meet the requirement of electrical insulation. As a result, a synergistic combination of various fibers emerges as the most promising approach. The carbon fiber component should serve as the "sandwich layer", which is responsible for bearing external loads, while the glass fiber component fully wraps the carbon fiber composites to provide electrical insulation.

Additionally, it is recommended to utilize the wet-winding technique (as opposed to the current semi-wet winding method) to mitigate the splashing of carbon fibers during winding under high pre-stress conditions. Wet-winding can also address the limitations of unsatisfactory impregnation between the fiber and epoxy in semi-wet winding. This approach offers greater flexibility in selecting epoxy materials, without the constraint of viscosity. It allows for the choice of epoxy with the best interfacial properties with the fiber. Furthermore, by employing wet winding in conjunction with the "poly layer assembly" technique [40,41], variable winding pitches of reinforcements, akin to the structure of laminated pressure vessels, can be achieved. The axial strength, stiffness and structural stability could be further enhanced with a carefully designed layup sequence.

7. Conclusions

The objective of this study was to investigate the mechanical behavior of ZFRPs reinforcements within high-field pulsed magnets. The study begins with mechanical testing of ZFRPs, followed by the development of its constitutive model, which incorporates the plasticity and progressive damage. The failure modes, damage evolution process, and plasticity deformation of ZFRP can be matters for concern. This model is more advanced than the ideal elastic model, which adopts the maximum equivalent stress failure criterion in the traditional analysis and design of pulsed magnets. Based on this, a comprehensive analysis of the mechanical behavior of ZFRP reinforcements was performed on a failed 95-T prototype from the perspective of composite materials for the first time. The findings revealed a significant attenuation of about 45% in both the radial and axial stiffness of ZFRPs, and the primary cause of failure was identified as local buckling occurring at the end of the inner magnet. Additionally, two successful 90-T magnet systems underwent thorough analyses, and the results exhibited good alignment with the experiments. Finally, the influence of the transverse strength of the reinforcement and axial Lorentz forces on the structural performance of magnets were discussed.

In general, this study provides valuable insight into the composite within pulsed magnets, thereby establishing a solid foundation for further advancements. Future work will focus on the optimization and construction of 100-T-magnet reinforcements with a laminated structure. The response method and multi-disciplinary optimization will be carried out to determine the layup of reinforcements under radial-axial bidirectional Lorentz loads. Then, we will construct a new 100-T magnet with laminated reinforcements using the wet winding and poly layer assembly technique.

Author Contributions: Conceptualization, methodology, software, writing—original draft preparation, S.C.; methodology, resources, T.P.; methodology, validation, X.H.; writing—review and editing, Q.C.; investigation, writing—review and editing, H.X.; supervision, project administration, writing—review and editing, L.L. All authors have read and agreed to the published version of the manuscript.

Funding: This research was funded by the National Natural Science Foundation of China, grant number 51821005.

Institutional Review Board Statement: Not applicable.

Data Availability Statement: Data are contained within the article.

Acknowledgments: The authors thank Zhipeng Lai and Xiaoxiang Li from Huazhong University of Science and Technology for their help.

Conflicts of Interest: The authors declare no conflicts of interest.

References

1. Chen, K.W.; Zheng, G.X.; Zhang, D.C.; Chan, A.; Zhu, Y.; Jenkins, K.; Yu, F.; Shi, M.; Ying, J.; Xiang, Z.; et al. Magnetic breakdown and spin-zero effect in quantum oscillations in kagome metal CsV_3Sb_5. *Commun. Mater.* **2023**, *4*, 96. [CrossRef]
2. Ye, Y.H.; Yamada, A.; Kinoshita, Y.; Wang, J.H.; Nie, P.; Xu, L.C.; Zuo, H.; Tokunaga, M.; Harrison, N.; McDonald, R.D.; et al. High-field immiscibility of electrons belonging to adjacent twinned bismuth crystals. *Npj Quantum Mater.* **2024**, *9*, 12. [CrossRef]
3. Zhang, S.Z.; Lee, S.Y.; Woods, A.J.; Peria, W.K.; Thomas, S.M.; Movshovich, R. Electronic and magnetic phase diagrams of the Kitaev quantum spin liquid candidate $Na_2Co_2TeO_6$. *Phys. Rev. B* **2023**, *108*, 064421. [CrossRef]
4. Nguyen, D.N.; Michel, J.; Mielke, C.H. Status and development of pulsed magnets at the NHMFL Pulsed Field Facility. *IEEE Trans. Appl. Supercond.* **2016**, *26*, 4300905. [CrossRef]
5. Zherlitsyn, S.; Wustmann, B.; Herrmannsdörfer, T.; Wosnitza, J. Status of the pulsed-magnet-development program at the Dresden High Magnetic Field Laboratory. *IEEE Trans. Appl. Supercond.* **2012**, *22*, 4300603. [CrossRef]
6. Peng, T.; Liu, S.B.; Pan, Y.; Lv, Y.L.; Ding, H.F.; Han, X.T.; Xiao, H.X.; Wang, S.; Jiang, S.; Li, L. A novel design of multi-coil pulsed magnet system for 100T. *IEEE Trans. Appl. Supercond.* **2022**, *32*, 4300104. [CrossRef]

7. Béard, J.; Billette, J.; Ferreira, N.; Frings, P.; Lagarrigue, J.; Lecouturier, F.; Nicolin, J.-P. Design and tests of the 100-T triple coil at LNCMI. *IEEE Trans. Appl. Supercond.* **2018**, *28*, 4300305. [CrossRef]
8. Jaime, M.; Daou, R.; Crooker, S.A.; Weickert, F.; Uchida, A.; Feiguin, A.E.; Batista, C.; Da, H.; Gaulin, B. Magnetostriction and magnetic texture to 100.75 Tesla in frustrated $SrCu_2(BO_3)_2$. *Proc. Natl. Acad. Sci. USA* **2017**, *109*, 12404–12407. [CrossRef]
9. Zherlitsyn, S.; Wustmann, B.; Herrmannsdörfer, T.; Wosnitza, J. Magnet-technology development at the Dresden High Magnetic Field Laboratory. *J. Low Temp. Phys.* **2013**, *170*, 447–451. [CrossRef]
10. Swenson, C.A.; Rickel, D.G.; Sims, J.R. 80 T magnet operational performance and design implications. *IEEE Trans. Appl. Supercond.* **2008**, *18*, 604–607. [CrossRef]
11. Michel, J.; Nguyen, D.N.; Lucero, J.D. Design, construction, and operation of new duplex magnet at pulsed field facility-NHMFL. *IEEE Trans. Appl. Supercond.* **2020**, *30*, 0500105. [CrossRef]
12. Michel, J.; Betts, S.B.; Lucero, J.D.; Bhardwaj, A.; Nguyen, L.N.; Nguyen, D.N. Design and construction of the new 85 T duplex magnet at NHMFL-Los Alamos. *IEEE Trans. Appl. Supercond.* **2024**, *34*, 4900305. [CrossRef]
13. Tardieu, S.; Béard, J.; Mesguich, D.; Lonjon, A.; Ferreira, N.; Chevallier, G.; Estournès, C.; Laurent, C.; Lecouturier-Dupouy, F. Scale-up of silver-copper composite wires by spark plasma sintering and room temperature wire-drawing for use in 100 T triple coil at LNCMI. *IEEE Trans. Appl. Supercond.* **2014**. *early access*. [CrossRef]
14. Talreja, R.; Waas, A.M. Concepts and definitions related to mechanical behavior of fiber reinforced composite materials. *Compos. Sci. Technol.* **2022**, *217*, 109081. [CrossRef]
15. Almeida, J.H.S.; St-Pierre, L.; Wang, Z.H.; Ribeiro, M.L.; Tita, V.; Amico, S.C.; Castro, S.G.P. Design, modeling, optimization, manufacturing and testing of variable-angle filament-wound cylinders. *Compos. Part B Eng.* **2021**, *225*, 109224. [CrossRef]
16. Eleiwi, M.; Pinto, F.S.; Botez, R.M.; Dao, T.M. Multidisciplinary optimization for weight saving in a variable tapered span-morphing wing using composite materials-application to the UAS-S4. *Actuators* **2022**, *11*, 121. [CrossRef]
17. Fallahi, H.; Taheri-Behrooz, F.; Asadi, A. Nonlinear mechanical response of polymer matrix composites: A review. *Polym. Rev.* **2020**, *60*, 42–85. [CrossRef]
18. Bru, T.; Olsson, R.; Gutkin, R.; Vyas, G.M. Use of the Iosipescu test for the identification of shear damage evolution laws of an orthotropic composite. *Compos. Struct.* **2017**, *174*, 319–328. [CrossRef]
19. Fallahi, H.; Taheri-Behrooz, F. Phenomenological constitutive modeling of the non-linear loading-unloading response of UD fiber-reinforced polymers. *Compos. Struct.* **2022**, *292*, 115671. [CrossRef]
20. Walsh, R.P.; Swenson, C.A. Mechanical properties of Zylon/epoxy composite laminates at 295K and 77K. *IEEE Trans. Appl. Supercond.* **2006**, *16*, 1761–1764. [CrossRef]
21. Huang, Y.K.; Frings, P.H.; Hennes, E. Exploding pressure vessel test on zylon/epoxy composite. *Compos. Part B Eng.* **2002**, *33*, 117–123. [CrossRef]
22. Jiang, F.; Sun, Q.Q.; Lai, Z.P.; Luo, B.; Pan, Y.; Li, L.; Peng, T. Electromagnetically driven expanding ring test for the strength study of the Zylon/epoxy composite. *IEEE Trans. Appl. Supercond.* **2016**, *26*, 9000106. [CrossRef]
23. Huang, Y.K.; Frings, P.H.; Hennes, E. Mechanical properties of Zylon/epoxy composite. *Compos. Part B Eng.* **2002**, *33*, 109–115. [CrossRef]
24. Chen, S.Y.; Peng, T.; Han, X.T.; Cao, Q.L.; Xiao, H.X.; Li, W.Z.; Li, L. Transverse tensile, compression and in-plane shear behaviors of Zylon fiber-reinforced polymer. *IEEE Trans. Appl. Supercond.* **2024**, *34*, 7400105. [CrossRef]
25. *ASTM D3039*; Standard Test Method for Tensile Properties of Polymer Matrix Composite Materials. ASTM International: West Conshohocken, PA, USA, 2017.
26. *ASTM D6641*; Standard Test Method for Compressive Properties of Polymer Matrix Composite Materials Using a Combined Loading Compression (CLC) Test Fixture. ASTM International: West Conshohocken, PA, USA, 2023.
27. *ASTM D3518*; Standard Test Method for In-Plane Shear Response of Polymer Matrix Composite Materials by Tensile Test of a ±45° Laminate. ASTM International: West Conshohocken, PA, USA, 2018.
28. Huang, Z.M. A bridging model prediction of the ultimate strength of composite laminates subjected to biaxial loads. *Compos. Sci. Technol.* **2004**, *64*, 395–448. [CrossRef]
29. Chen, S.Y.; Li, L. An anisotropic damage-plasticity constitutive model of continuous fiber-reinforced polymers. *Polymers* **2024**, *16*, 334. [CrossRef]
30. Chen, J.F.; Morozov, E.V.; Shankar, K. A combined elastoplastic damage model for progressive failure analysis of composite materials and structures. *Compos. Struct.* **2012**, *94*, 3478–3489. [CrossRef]
31. Puck, A.; Schürmann, H. Failure analysis of FRP laminates by means of physically based phenomenological models. *Compos. Sci. Technol.* **2002**, *62*, 1633–1662. [CrossRef]
32. Ladeveze, P.; LeDantec, E. Damage modelling of the elementary ply for laminated composites. *Compos. Sci. Technol.* **1992**, *43*, 257–267. [CrossRef]
33. Rajaneesh, A.; Ponthot, J.P.; Bruyneel, M. High velocity impact response of composite laminates using modified meso-scale damage models. *Int. J. Impact Eng.* **2021**, *147*, 103701. [CrossRef]
34. Zhu, T.Q.; Ren, Z.Y.; Xu, J.; Shen, L.L.; Xiao, C.L.; Zhang, C.; Zhou, X.; Jian, X. Damage evolution model and failure mechanism of continuous carbon fiber-reinforced thermoplastic resin matrix composite materials. *Compos. Sci. Technol.* **2023**, *244*, 110300. [CrossRef]

35. Puck, A.; Mannigel, M. Physically based non-linear stress–strain relations for the inter-fibre fracture analysis of FRP laminates. *Compos. Sci. Technol.* **2007**, *67*, 1955–1964. [CrossRef]
36. Chen, S.Y.; Lv, Y.L.; Peng, T.; Li, X.X.; Li, L. Finite-element analysis of pulsed high-field magnets with precise thermal and structural calculations. *IEEE. Trans. Magn.* **2020**, *56*, 7511904. [CrossRef]
37. Nguyen, Q.V.M.; Torres, L.; Nguyen, D.N. Electromagnetic interaction between the component coils of multiplex magnets. *IEEE Trans. Appl. Supercond.* **2018**, *28*, 4300504. [CrossRef]
38. Hua, Y.; Yang, Y.C.; Yamanaka, A.; Ni, Q.Q. Low friction coefficient property of super fiber-reinforced composites. *Adv. Compos. Mater.* **2011**, *20*, 133–147. [CrossRef]
39. Chen, S.Y.; Peng, T.; Li, Y.H.; Lv, Y.L.; Wang, S.; Li, L. Structural analysis of pulsed magnets considering interface characteristics. *Rev. Sci. Instrum.* **2020**, *91*, 085113. [CrossRef] [PubMed]
40. Marshal, W.S.; Swenson, C.A.; Gavrilin, A.; Schneider-Muntau, H.J. Development of "fast cool" pulsed magnet coil technology at NHMFL. *Phys. B* **2004**, *346*, 594–598. [CrossRef]
41. Swenson, C.A.; Marshall, W.S.; Gavrilin, A.; Han, K.; Schillig, J.; Sim, J.R.; Schneider-Muntau, H.J. Progress of the insert coil for the US-NHMFL 100 T multi-shot pulse magnet. *Phys. B* **2004**, *346*, 561–565. [CrossRef]

polymers

Article

An Investigation of the Energy-Absorption Characteristics of Thin-Walled Polymer Composite C-Channels: Experiment and Stacked Shell Simulation

Xiaomin Zhang [1,2], Haolei Mou [3], Shanshan Song [4] and Zhenyu Feng [3,*]

1 College of Safety Science and Engineering, Civil Aviation University of China, Tianjin 300300, China; xm-zhang@cauc.edu.cn
2 Engineering Techniques Training Center, Civil Aviation University of China, Tianjin 300300, China
3 Science and Technology Innovation Research Institute, Civil Aviation University of China, Tianjin 300300, China
4 Sino-European Institute of Aviation Engineering, Civil Aviation University of China, Tianjin 300300, China
* Correspondence: caucstructure@163.com

Abstract: Polymer composite materials are increasingly used in civil aircraft structures. The failure mode and energy-absorption characteristics of polymer composite structures have garnered significant attention from academia and industry. For thin-walled polymer composite C-channels with layups of $[0/90]_{3s}$, $[45/\text{-}45]_{3s}$, and $[45/90/\text{-}45/0]_3$, low-speed axial compression tests were performed to investigate the failure modes, failure mechanisms, and energy-absorbing characteristics. After parametric studies using [0] and [90] single-element models, stacked shell models of thin-walled composite C-channels were established using the Lavadèze single-layer damage constitutive model, Puck 2000, and Yamada Sun failure criteria. The results show that these thin-walled composite C-channels exhibit a stable progressive crushing process with a local buckling failure mode, encompassing local buckling, fiber break-age, matrix cracks, delamination, and corner cracking. The stacked shell model demonstrates reasonable agreement with the progressive crushing process of thin-walled composites, accurately capturing interlayer matrix failure and interface delamination cracking behavior. A comparison of the specific energy absorption (*SEA*) and mean crushing force (F_{mean}) between the simulation and test results yields a difference of less than 6%, indicating a strong correlation between the simulation results and the experimental energy-absorbing characteristics. It also shows that a deep understanding of the parameters is helpful for accurate numerical modeling.

Keywords: thin-walled polymer composite; axial collapse; failure mechanism; energy absorption; finite-element model

Citation: Zhang, X.; Mou, H.; Song, S.; Feng, Z. An Investigation of the Energy-Absorption Characteristics of Thin-Walled Polymer Composite C-Channels: Experiment and Stacked Shell Simulation. *Polymers* **2024**, *16*, 2099. https://doi.org/10.3390/polym16152099

Academic Editor: Alexey V. Lyulin

Received: 24 April 2024
Revised: 15 July 2024
Accepted: 17 July 2024
Published: 23 July 2024

1. Introduction

Due to the advantages of low density, high specific strength, high specific modulus, and great potential for the integrated design of structural materials, polymer composite structures are increasingly used in the automobile and aerospace industries [1–4]. In recent years, polymer composite structures have played a major role in the primary load-bearing structures and energy-absorbing structures of large aircraft. For composite fuselage structures, an equivalent level of crashworthiness must be provided compared to similar traditional metal fuselage structures under emergency landing conditions [5,6], which mainly ensures that the acceleration and loads experienced by occupants during impact events are consistent with those specified in FAR/CCAR 25.562 (b) and provide sufficient living space for occupants.

The most effective and economical means to meet the crashworthiness requirements is to arrange thin-walled polymer composite structures in the fuselage section to absorb large amounts of impact kinetic energy in a controlled manner [7–9]. Taking the Boeing 787 [10], a series of thin-walled polymer composite C-channels were installed under the cargo subfloor, affecting the failure modes and crashworthiness of the fuselage section. Therefore, much research has been carried out by the Boeing Company, the Federal Aviation Administration, and others to verify and improve crashworthiness, achieving great success [11,12]. In recent decades, various thin-walled polymer composite structures with different shapes have been put forward and their axial collapse behaviors studied, including circular tubes [13–15], square tubes [16,17], corrugated plates [18–20], conical shells [21], composite-reinforced aluminum tubes [22,23], and filled structures [24,25]. Factors such as the trigger [25–27], geometric features [28–30], mechanical properties of the fiber and matrix [31,32], and layup [33] have also been studied.

With the development of finite-element codes, such as PAM-CRASH [34], LS-DYNA [35], and ABAQUS [36], the combination method, i.e., many finite-element simulations with a small number of experiments, has become an effective way to study the energy-absorbing characteristics of composite structures and to design composite energy-absorbing structures [37–39]. Researchers have carried out extensive numerical simulation work, and various failure criteria are employed to predict material failure under different loading conditions. The most commonly used failure criteria include the maximum stress criterion [40,41], maximum strain criterion [42], Tsai–Hill criterion [43], Hashin criterion [44], Puck's failure criterion [45], LaRC criterion [46], and Yamada–Sun criterion [47]. In addition, the finite-element modeling of composite thin-wall tubes has been researched using single-layer shell models and multi-layer shell models. Palanivelu et al. [48] developed single-layer and two-layer shell models of glass fiber composite thin-wall cylindrical and square tubes, respectively. In the two-layer model, solid adhesive units were established between the two shell elements, and predefined cracks were set in the shell elements to simulate crack propagation. The results show that the single-layer shell model could not capture delamination failure accurately, leading to inaccurate results, whereas the multi-layer shell model could simulate the failure morphology and energy-absorbing characteristics consistent with the test results. Siromani et al. [49] developed the multi-layer shell modeling method to numerically simulate the crushing process of composite cylindrical tubes using the LS-DYNA MAT 54 material model, achieving relatively accurate simulation of the chamfered end and boundary conditions. The average compressive load and *SEA* obtained by the simulation were relatively in accordance with the test results, though the initial peak load was 8–20% higher. Johnson et al. [50] numerically simulated the quasi-static crush test of $[0/90]_8$ thin-walled structures using the stacked shell model. A rigid element was built at the top of the model, and offset nodes were situated in the upper part of the two shell elements that were located in the middle. In this way, the interlayer separation during the initial compression phase can be simulated. The simulated average compressive load was similar to the experimental results, which verified the effectiveness of the modeling method.

In addition, Feraboli et al. [51,52] explored the LS-DYNA MAT 54 material model and the Chang–Chang failure criteria [53,54], developed by Chang and Chang, which distinguish between tension and compression failures in the fiber and matrix of composite materials. The criteria include fiber tensile failure, fiber compressive failure, matrix tensile failure, matrix compressive failure, and shear failure. The key parameters of the constitutive relations, damage evolution, and failure criteria were identified, and their sensitivity was analyzed. The results exhibit that some model parameters, which are either non-physical or cannot be measured experimentally, enormously influence the simulation. It is expected to lack predictive capability before parameter calibration. However, the study was confined to LS-dyna MAT54, and similar research on other material models is still rarely seen.

In the present work, three groups of thin-walled polymer composite C-channels with different layups, i.e., $[0/90]_{3s}$, $[45/-45]_{3s}$, and $[45/90/-45/0]_3$, were analyzed for failure modes, failure mechanisms, and energy-absorbing characteristics by conducting low-speed axial compression tests. The [0] and [90] single-element Lavadèze models were developed to conduct parametric analysis, investigate the sensitivity of various input parameters, and determine the most appropriate parameters based on the stress–strain relationship.

Subsequently, the stacked shell models of thin-walled polymer composite C-channels were established using the Lavadèze single-layer damage constitutive model. Owing to the advantages in matrix failure and transverse tensile and shear stress interaction evaluation, the Puck 2000 failure criteria and Yamada–Sun failure criteria have been chosen. These models are further verified based on test results and energy-absorbing characteristics. These findings enhanced the understanding of the behavior of composite materials under axial compression and provided a strategy for calibrating finite-element models.

2. Specimens and Experimental

2.1. Specimens Design and Manufacture

In the sub-cargo of aircraft, thin-walled polymer composite C-channels were designed to support the cargo floor crossbeam. During a crash event, they can transfer and absorb the impact load. Thin-walled polymer composite C-channels were fabricated using the T700/MTM28 carbon fiber reinforced epoxy composite provided by AVIC Manufacturing Technology Institute (Beijing, China) through a hot-pressing process. The fiber volume content was approximately 60% according to the standard method in DIN EN ISO 7822 [55]. T700 carbon fibers provide a high specific strength and modulus, making them ideal for aerospace applications where weight reduction and mechanical performance are critical; furthermore, extensive research has validated the effectiveness of T700/MTM28 composites in energy absorption and crashworthiness [18,19,51], making them a well-understood and reliable choice for further study. The geometric configuration of the C-channels is as shown in Figure 1. The total height of the C-channels is 100 mm, with a width of 50 mm. The left and right flanges measure 20 mm in width. The transition arc between flat segment and flange has an inner radius of 1.5 mm, while the outer radius is equivalent to 1.5 mm plus the thickness of the C-channel. A 45° outer chamfer was milled at one end of the C-channel, as shown in Figure 1a. θ is the fiber angle, as shown in Figure 1b. The fiber direction is consistent with the axial direction of the C-channel, if θ = 0°. The thin-walled polymer composite C-channel specimens, as shown in Figure 1c, were designed with a thickness of 1.8 mm, corresponding to 12 layers, i.e., $[0/90]_{3S}$, $[45/\text{-}45]_{3S}$, and $[45/90/\text{-}45/0]_3$. Each type of C-channel was tested at least 3 times to ensure the consistency and reliability of the results.

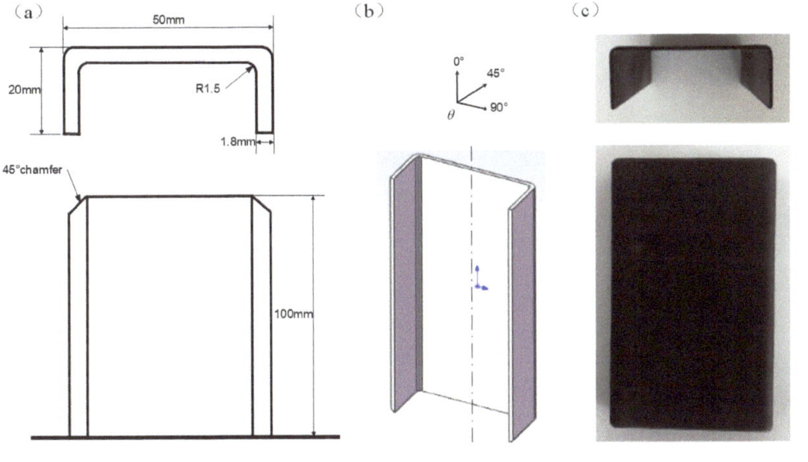

Figure 1. Schematic diagram of the polymer composite C-channels (**a**), three-dimensional stereogram (**b**), and C-channel specimen (**c**).

2.2. Experimental Procedures

Low-speed axial compression tests of thin-walled composite C-channels were conducted at room temperature using the INSTRON VHS 160/100-20 (Norwood, MA, USA).

The 45° chamfer of the thin-walled composite C-channel was set at the top and free, while the other end was fixed using a clamp mounted on the plate by two bolts. A load cell was installed between the plate and the lower fixed platen to capture the load–time history curve, as shown in Figure 2. The upper moving platen moved downward at a constant velocity of 50 mm/s. The entire process of the low-speed axial compression tests was recorded using PHOTRAN high-speed video (Tokyo, Japan) at 2000 frames per second. Following the experiments, the C-channels were scanned using the ZEISS METROTOM (Deutschland, GmbH, Düsseldorf, Germany) industrial computed tomography (CT) scanning machine.

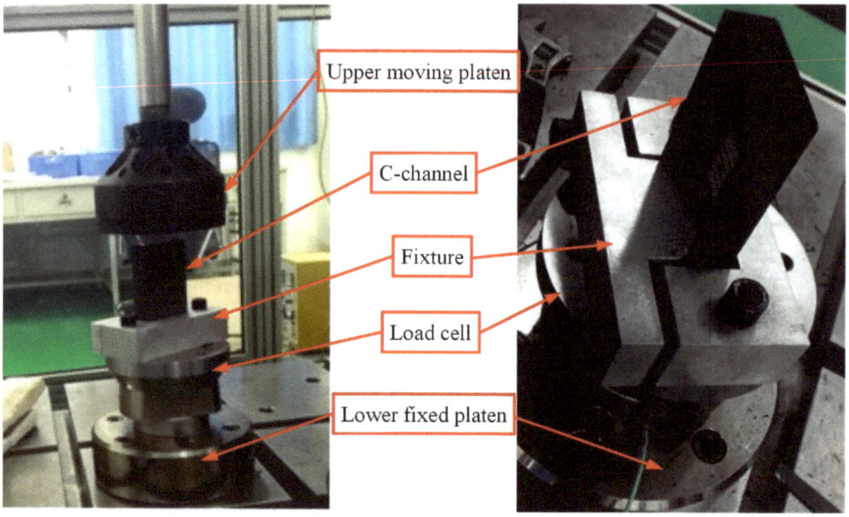

Figure 2. Test scheme, fixture and C-channel.

2.3. Energy-Absorbing Metrics

(1) Energy absorption (EA, unit: kJ): The total energy absorbed by the integration of the crushing force over the entire crushing distance.

$$EA = \int F\mathrm{d}l \tag{1}$$

where F is the crushing force.

(2) Specific energy absorption (SEA, unit: kJ/kg): The energy absorbed per unit of mass of the structure. The SEA is an important parameter to measure the energy-absorbing capabilities. The ratio of the total energy to the mass is the SEA, which can be calculated as follows:

$$SEA = \frac{EA}{\rho AL} \tag{2}$$

where ρ is the density, A is the effective cross-sectional area, and L is the total crushing distance.

(3) Peak crushing force (F_{max}, unit: kN): The threshold value of structural damage, which is used to evaluate the structural failure under an external force. This parameter is the peak crushing force of the load–displacement curve.

(4) Mean crushing force (F_{mean}, unit: kN): The mean crushing force during the entire crushing process, which can be calculated as follows:

$$F_{mean} = \frac{EA}{L} \tag{3}$$

3. Numerical Study

3.1. Constitutive Model

The Lavadèze orthotropic single-layer composite model was used, which judges the matrix damage by using the Puck inter fiber failure (IFF) criteria combined with the Yamada–Sun fiber failure criteria. The failure criteria of the combination of the fiber and matrix can take the failure of the fiber under tensile and compressive loads, the formation of matrix microcracks under a transverse tensile load, and the debonding of the fiber and the matrix under the shear load into consideration. The constitutive relation of the Lavadèze single-layer composite model is shown in Equation (4):

$$\begin{Bmatrix} \varepsilon_{11}^e \\ \varepsilon_{22}^e \\ 2\varepsilon_{12}^e \\ 2\varepsilon_{23}^e \\ 2\varepsilon_{13}^e \end{Bmatrix} = \begin{bmatrix} \frac{1}{E_1} & -\frac{\nu_{12}^0}{E_1} & 0 & 0 & 0 \\ -\frac{\nu_{12}^0}{E_1} & \frac{1}{E_2} & 0 & 0 & 0 \\ 0 & 0 & \frac{1}{G_{12}} & 0 & 0 \\ 0 & 0 & 0 & \frac{1}{G_{23}^0} & 0 \\ 0 & 0 & 0 & 0 & \frac{1}{G_{13}} \end{bmatrix} \begin{Bmatrix} \sigma_{11} \\ \sigma_{22} \\ \sigma_{12} \\ \sigma_{23} \\ \sigma_{13} \end{Bmatrix} \tag{4}$$

where Direction 1 of the natural coordinates is the fiber direction; Direction 2 is the vertical fiber direction; Direction 3 is the normal direction of the unidirectional composite layer; ε_{11} is the strain in the direction of the fiber; ε_{22} is the strain in the vertical direction of the fiber; ε_{12}, ε_{23}, and ε_{13} are the shear strains in the corresponding directions; σ_{11} is the stress in the direction of the fiber; σ_{22} is the stress in the vertical direction of the fiber; σ_{12}, σ_{23}, and σ_{13} are the shear stresses in the corresponding directions; E_1 and E_2 are the elastic moduli in the direction of the fiber and the vertical fiber direction; G_{12} is the shear modulus of the plane 1, 2; G_{23} is the shear modulus of the plane 2, 3; G_{13} is the shear modulus of the plane 1, 3; and ν_{12} is the Poisson's ratio.

3.2. Failure Criteria

The Puck's IFF criterion in natural coordinates is shown in Equation (5):

$$\lambda^M = \begin{cases} \sqrt{\left(\frac{\sigma_{12}}{R_{12}}\right)^2 + \left(1 - P_{12}^+ \frac{R_{22}^+}{R_{12}}\right)^2 \left(\frac{\sigma_{22}}{R_{22}^+}\right)} + P_{12}^+ \frac{\sigma_{22}}{R_{12}} (\sigma_{22} \geq 0) \\ \frac{1}{R_{12}} \left(\sqrt{(\sigma_{12})^2 + (p_{12}^- \sigma_{22})^2} + p_{12}^- \sigma_{22} \right) (\sigma_{22} < 0,\ 0 \leq \left|\frac{\sigma_{22}}{\sigma_{12}}\right| \leq \frac{R_{22}^A}{|\sigma_{12}^c|}) \\ \left[\left(\frac{\sigma_{12}}{2(1+p_{22}^-)R_{12}} \right)^2 + \left(\frac{\sigma_{22}}{R_{22}^-} \right) \right] \frac{R_{22}^-}{(-\sigma_{22})} (\sigma_{22} < 0,\ 0 \leq \left|\frac{\sigma_{12}}{\sigma_{22}}\right| \leq \frac{|\sigma_{12}^c|}{R_{22}^A}) \end{cases} \tag{5}$$

where λ^M is the matrix damage judgment factor; when λ^M exceeds 1, matrix failure will occur; R_{22}^+, R_{22}^-, and R_{12} represent, respectively, the transverse tensile strength, transverse compressive strength, and matrix shear strength; p_{12}^+ and p_{12}^- are matching parameters of the curve slope, which need to be determined by a multi-axis test.

The Yamada–Sun fiber failure criteria are shown in Equation (6):

$$\begin{aligned} Fail &= \sqrt{\left(\frac{\varepsilon_{fib}}{\varepsilon_{11}^+}\right)^2 + \left(\frac{\varepsilon_{12}}{\varepsilon_{12}^+}\right)^2 + \left(\frac{\varepsilon_{13}}{\varepsilon_{13}^+}\right)^2}, \varepsilon_{fib} > 0 \\ Fail &= \sqrt{\left(\frac{\varepsilon_{fib}}{\varepsilon_{11}^-}\right)^2 + \left(\frac{\varepsilon_{12}}{\varepsilon_{12}^-}\right)^2 + \left(\frac{\varepsilon_{13}}{\varepsilon_{13}^-}\right)^2}, \varepsilon_{fib} < 0 \\ \varepsilon_{fib} &= \varepsilon_{11} + \nu_{12}(1-d)\varepsilon_{22} \end{aligned} \tag{6}$$

where ε_{11}^+, ε_{12}^+, and ε_{13}^+ represent the fiber tensile failure strain, positive in-plane shear failure strain, and positive out-of-plane shear failure strain, respectively; ε_{11}^-, ε_{12}^-, and ε_{13}^- represent the fiber compression failure strain, negative in-plane shear failure strain, and negative out-of-plane shear failure strain, respectively.

3.3. Parametric Analysis

To characterize the basic elasticity, initial failure, and post-failure behavior of a single-element model, the [0] and [90] single-element models are established to obtain the stress–strain curves, and the influence of each parameter on the material performance is determined. Combined with the performance parameters of T700/MTM28, the set values of the material model parameters are determined by inverting the relevant parameters of the material model, which can provide the basis values of the input parameters for the finite-element modeling of thin-walled composite C-channels. A summary of the parametric studies performed is shown in Table 1.

Table 1. Summary of the parametric studies performed.

Parameter	Physical Significance	Unit	Parametric Variation			
E0t1	Young's modulus in Direction 1	GPa	50	100	200	300
EPSIfti	Tensile fiber initial strain	-	0.005	0.01	0.018	0.02
EPSIftu	Tensile fiber ultimate strain	-	0.018	0.02	0.03	0.035
Dftu	Tensile fiber ultimate damage	-	0.2	0.4	0.8	1
Xt11	Tensile fiber strength/strain for failure	-	0.05	0.1	0.3	0.4
E0c1	Compressive fiber Young's modulus in Direction 1	GPa	50	100	200	300
GAMMA	Compressive factor of modulus correction	-	0	0.2	0.6	0.8
Dfcu	Compressive fiber ultimate damage	-	0.2	0.4	0.8	1
Xc11	Compressive tensile fiber strength/strain for failure	-	0.05	0.08	0.2	0.3
E0t2	Young's modulus in Direction 2	GPa	6	7	9	10
Ycp	Critical transverse damage limit	$GPa^{1/2}$	0.02	0.025	0.035	0.04
Y0p	Initial transverse damage limit	$GPa^{1/2}$	0.005	0.0075	0.0125	0.015
R0	Initial yield stress	GPa	0.005	0.01	0.05	0.075
BETA	Hardening law multiplier	GPa	1.2	1.5	3	3.5
m	Hardening law exponent		0.6	0.7	0.8	0.9
DAMpost	Post damage of matrix	-		1	0.9	0.85

3.3.1. [0] Single-Element Model

A square single-element model of a Belytschko–Tsay shell is developed, with a side length of 1 mm and a thickness of 0.15 mm. The tension/compression load and boundary conditions are shown in Figure 3, and the stress–strain curve can be extracted.

For the [0] single-element model, the typical stress–strain curve, shown in Figure 4, can be divided into the following four stages: the linear elastic stage (0–1), degradation stage (1–2), platform stage (2–3), and element deleting stage (3–4).

(a) (b)

Figure 3. Single-element shell models with tension load (**a**) and compression load (**b**).

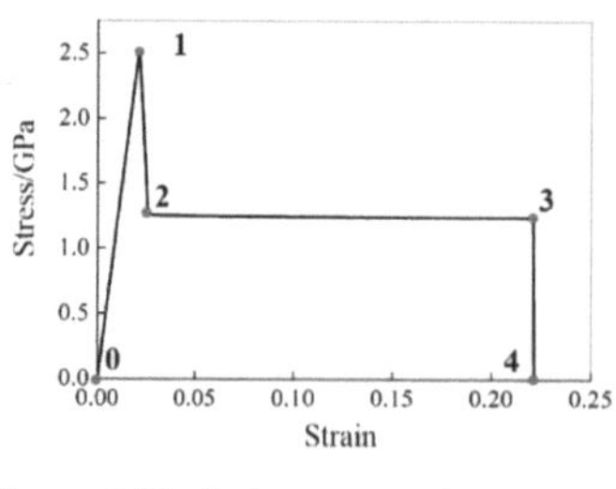

Figure 4. Typical stress–strain curve of the [0] single-element model, point 0 represents the initial state of the material before any load is applied, point 1 indicates the yield point of the material, point 2

represents the stress value after material degradation, point 3 represents the end of the platform stage, and point 4 represents the element deletion.

The effects of E0t1 and E0c1 on the stress–strain curve are similar, which can determine the slope of the linear elastic stage (0–1) in the stress–strain curve during the tension and compression processes, respectively, as shown in Figure 5a,b. The EPSIfti can determine the maximum strain value of the linear elastic stage (0–1) in the stress–strain curves, as shown in Figure 5c. The EPSIftu can determine the strain value (Point 2 in Figure 4 at the junction between the degradation stage (1–2) and the platform stage (2–3). The greater the EPSIftu, the larger the strain value and the greater the strength after failure, as shown in Figure 5d.

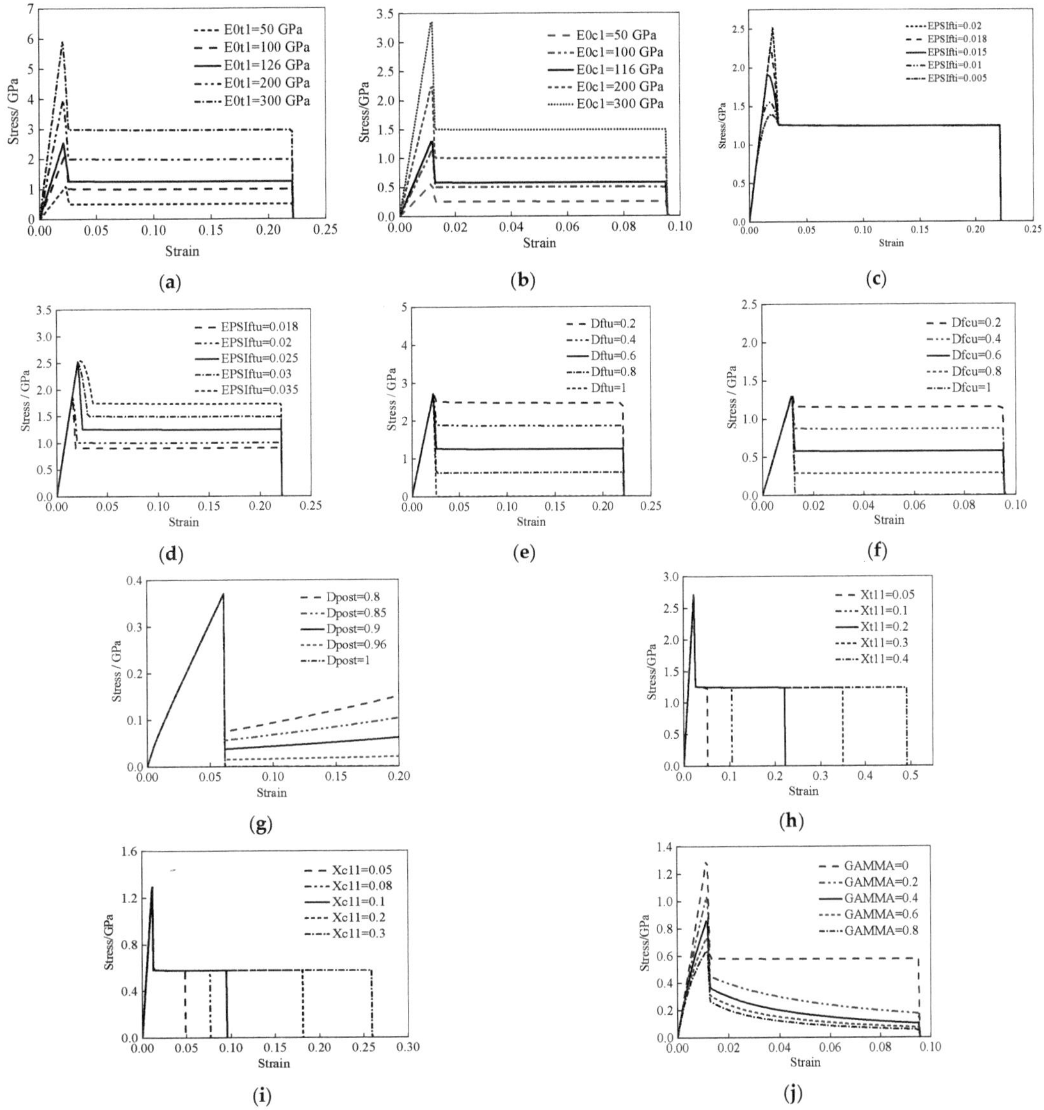

Figure 5. Stress–strain curves of the [0] single-element model, E0t1 (**a**), E0c1 (**b**), EPSIfti (**c**), EPSIftu (**d**), Dftu (**e**), Dfcu (**f**), Dpost (**g**), Xt11 (**h**), Xc11 (**i**), and GAMMA (**j**).

The effects of Dftu and Dfcu on the stress–strain curve are similar, which can determine the strength value (Point 3 in Figure 4) of the platform stage (2–3) in the stress–strain curve during the tension and compression processes, respectively, as shown in Figure 5e,f. The greater the Dftu and Dfcu, the smaller the strength after degradation.

The Dpost can mainly affect the strength of the matrix after degradation. The smaller the Dpost, the greater the strength after the degradation of the matrix, as shown in Figure 5g.

The Xt11 and Xc11 can determine the strain value (Point 4 in Figure 4) of the element deleting stage (3–4) in the stress–strain curve, as shown in Figure 5h,i. The greater the Xt11 and Xc11, the later the element is deleted.

The GAMMA can affect the compressive elastic modulus and further affect the strength limit and the post-degradation compressive strength, as shown in Figure 5j. The value range of GAMMA is [0, 1], which can adjust the linearity of the compressive modulus before failure and the corresponding strength level.

3.3.2. [90] Single-Element Model

The tension/compression load and boundary conditions are shown in Figure 6, and the stress–strain curve can be extracted. For the [90] single-element model, the typical stress–strain curve, shown in Figure 7, can be divided into the following two stages: the elastic stage (0–2) and the degradation and element deleting stage (2–3), which can demonstrate the stress–strain properties of the matrix's elastic to plasticity.

Figure 6. Tension load (**a**) and compression load (**b**).

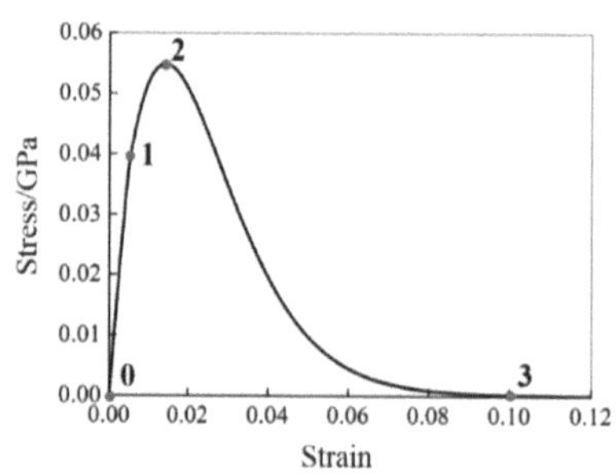

Figure 7. Typical stress–strain curve of the [90] single-element model, point 0 represents the initial state of the material before any load is applied, point 1 is in the early stages of the loading process where the material exhibits linear elastic behavior, point 2 represents the peak stress, also known as the ultimate strength of the material, and point 3 indicates the failure of the material.

The E0t2 can determine the slope of the linear elastic stage in the stress–strain curves, as shown in Figure 8a. The larger the E0t2, the larger the slope of the linear elastic stage. At the same time, the E0t2 can affect the variation in the tensile stress–strain curve of the matrix from the elastic stage to the failure stage to a certain extent.

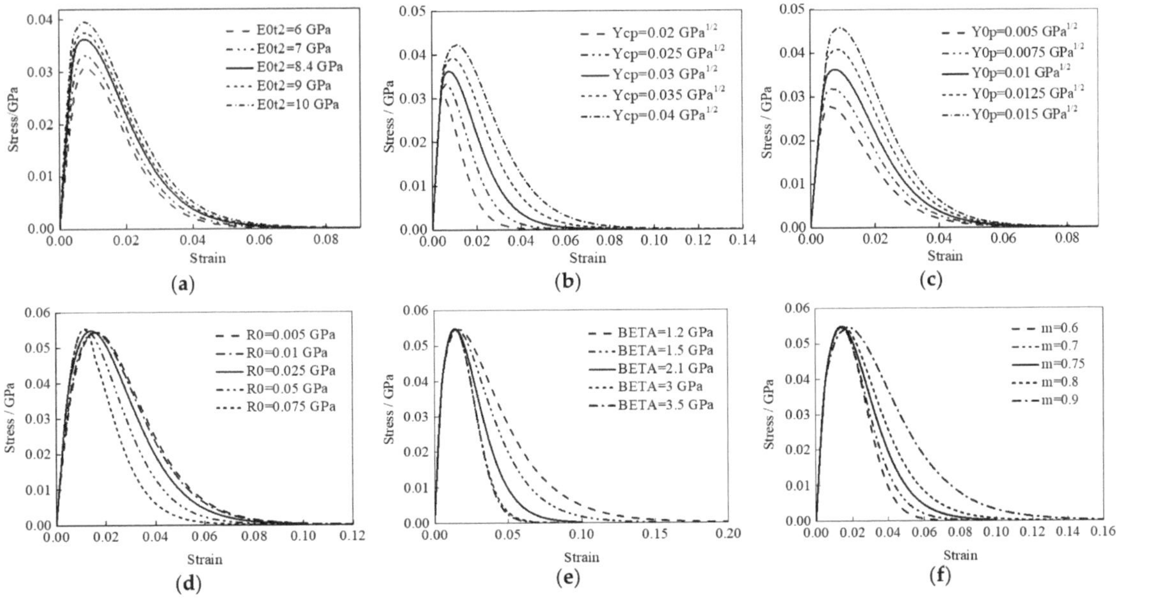

Figure 8. Stress–strain curves of the [90] single-element model, E0t2 (**a**), Ycp (**b**), Y0p (**c**), R0 (**d**), BETA (**e**), and m (**f**).

The Ycp and Y0p mainly affect the degradation stage in the stress–strain curve, as shown in Figure 8b,c, and both Ycp and Y0p can affect the tensile strength of the matrix. The larger the Ycp and Y0p, the greater the tensile strength of the matrix.

The effects of R0, BETA, and m on the stress–strain curves are shown in Figure 8d–f. By changing the values of the parameters, the downward trend of the degradation stage in the stress–strain curve can be found, and the failure strain, which causes the element deletion, is changed to varying degrees.

3.3.3. Material Model Parameters

The parameters' physical meanings and the numerical meanings of the Lavadèze material constitutive model, the Puck IFF failure criteria, and the Yamada–Sun fiber failure criteria were identified through the parametric analysis of a single-element model. Finally, the parameters are presented in Table 2.

Table 2. Material model parameters.

Parameters	Definition	Unit	Value	Measurement
RHO	Mass density of ply material	g/cm^3	1.52	ASTM D792 [56]
E0t1	Young's modulus in Direction 1 of fiber	GPa	126	ASTM D3039 [57]
E0t2	Young's modulus in Direction 2 of fiber	GPa	8.40	ASTM D3039
E0c1	Compressive Young's modulus of fiber	GPa	116	ASTM D3410 [58]
G012	Shear modulus of 1,2-plane	GPa	3.69	ASTM D3518 [59]
NU12	Poisson's ratio in 1,2-plane	-	0.3	ASTM D3039
R22+	Matrix tensile transverse strength	GPa	0.0554	ASTM D638 [60]
R22-	Matrix compressive transverse strength	GPa	0.225	ASTM D695 [61]
R12	Matrix shear strength	GPa	0.2	ASTM D5379 [62]
EPSIfti	Tensile fiber initial strain	-	0.0217	

Table 2. *Cont.*

Parameters	Definition	Unit	Value	Measurement
Ycp	Critical transverse damage limit	$GPa^{1/2}$	0.0596	Calibrated by material coupon level test results.
Y0p	Initial transverse damage limit	$GPa^{1/2}$	0.01	
Dmax	Maximum allowed damage value for shear damage and transverse damage	-	1	
GAMMA	Compressive factor of modulus correction	-	0	
R0	Initial yield stress	GPa	0.025	
BETA	Hardening law multiple	GPa	2.1	
m	Hardening law exponent	-	0.75	
Dpost	Post damage of matrix	-	0.96	
Dftu	Tensile fiber ultimate damage		0.7~0.85	
Dfcu	Compressive fiber ultimate damage		0.7~0.85	
EPSIftu	Tensile fiber ultimate strain	-	0.025	
Xt11	Tensile fiber strain for failure	-	0.52	
Xc11	Compressive tensile fiber strain for failure	-	0.38	

3.4. Stacked Shell Model

Based on the actual dimensions of the composite C-channels, a stacked shell FE model was developed in PAM-CRASH (2G version VPS 2020), i.e., each ply was represented by individual shell elements and stacked together using the cohesive elements to form the entire composite laminate, as shown in Figure 9. Due to stress concentration at the interface between the web and the flanges, the corner region of the C-channel was modeled by introducing a column of corner elements. The 45° outer chamfer section was created by employing a height-by-layer decrement approach. The mesh size of the web and flanges in the model was approximately 1.2 mm × 1.4 mm, while the mesh size of the corner elements was approximately 1.4 mm × 1.4 mm. This mesh density can ensure the accuracy of simulation results without significantly increasing the calculation time. The Belytschko–Tsay shell element was used, and each layer contained 4608 shell elements. The rigid wall had no deformation during the crushing process and provided a compulsive uniform velocity. Therefore, the element size, which was set to 10 mm × 10 mm, was relatively large and comprised a total of 219 elements.

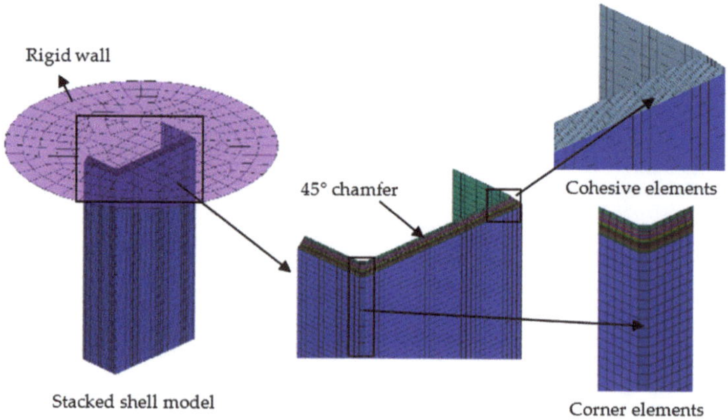

Figure 9. Stacked shell model of C-channels.

To accurately simulate interlayer failure and ensure the reliability of the finite-element model, the cohesive elements were used to connect each layer, which can effectively simulate the load transmission of interlamination and delamination failure. The parameters of the cohesive element are shown in Table 3.

Table 3. Cohesive element parameters.

Parameters	Definition	Unit	Values
hcont	Distance for kinematic computation	mm	0.3
E0	Normal modulus	GPa	4
G0	Shear modulus	GPa	2.5
SIGprpg	Normal stress to continue delamination	GPa	0.098
GAMAprpg	Shear stress to continue delamination	GPa	0.094
G_u^I	Mode I fracture energy	J/mm^2	0.00047
G_u^{II}	Mode II fracture energy	J/mm^2	0.002
SIGstrt	Normal stress to initiate delamination	GPa	0.1
GAMAstrt	Shear stress to initiate delamination	GPa	0.1
Ncycle	Stress reduction cycle number		100

G_u^I and G_u^{II} represent Type I and Type II fracture energy, which can be measured by a double cantilever beam test. These two parameters determine the interlaminar strength. In the model, the Pickett criteria [49] was used to couple the Type I and Type II cracks in a linear superposition, and the criteria calculation is shown in Equation (7).

$$\frac{G_i^I}{G_u^I} + \frac{G_i^{II}}{G_u^{II}} = 1 \tag{7}$$

G_i^I and G_i^{II} represent Type I and Type II crack propagation energy. In addition, the initial delamination stress, the continuous delamination stress and the stress reduction cycle number were utilized to regulate the relationship between the interlaminar crack length and the interlaminar stress. Taking the Type I crack as an example, the stress–crack length curve is shown in Figure 10.

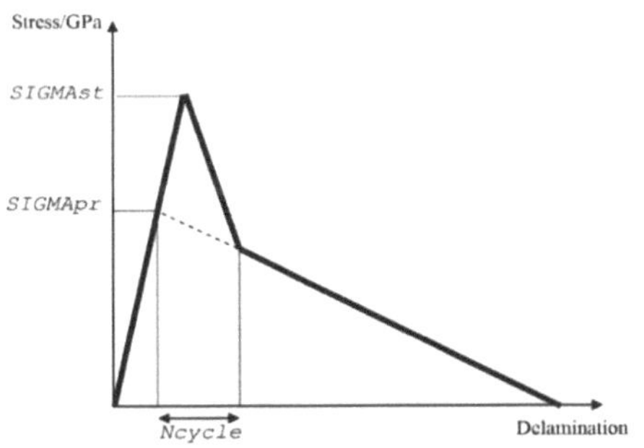

Figure 10. Stress-delamination propagation of cohesive element.

To prevent the interlayer penetration, which can lead to inaccuracies in the simulation results, the self-contact method was employed for each layer. Accordingly, the node-to-surface contact method was used to avoid penetration between the rigid wall and the composite C-channel. For the finite-element model, the C-channel was immobilized by constraining all degrees of freedom of the bottom row of nodes situated opposite the 45° outer chamfer, while the remaining nodes were entirely free.

The strain rate effect was not considered in the finite-element models. Therefore, to enhance computational efficiency while maintaining simulation accuracy, a loading speed of 6000 mm/s was selected. Accordingly, the crushing distance was set to 44 mm, which was consistent with the test conditions. It took approximately 2–3 h to complete the simulation using a dual-core X5675 CPU and a workstation operating at a speed of 3.06 GHz.

4. Results and Discussion

4.1. Failure Morphology and Mechanism Analysis

The C-channels experienced a progressive crushing process under low-speed axial compression. The failure morphology of the $[0/90]_{3s}$ C-channel, depicted in Figure 11a, shows repeated folding and collapse, leading to large intervals in the load–displacement curve (Figure 12). Horizontal cracks appeared in the flat segment and flange, with some 0° ply detaching within the flat segment. Delamination occurred at the top side of the C-channel and both flange sides. Stress concentration in the corner caused complete fiber and matrix breakage, as 0° fibers failed under axial compression, leaving the 90° fibers unable to provide support.

For the $[45/\text{-}45]_{3s}$ C-channel, shown in Figure 11b, no folding occurred due to ±45° fiber reinforcement, which provided shear force resistance. This C-channel displayed significant elastic deformation without buckling failure or fiber fracture. However, local buckling in the flat segment caused matrix slippage along the fiber direction and numerous 45° cracks. The flange exhibited 45° matrix cracks and inward bending in the folding zone. The $[45/\text{-}45]_{3s}$ C-channel's fibers withstood shearing forces, preventing cracking in the corner area, but 45° direction cracks led to fiber and matrix yielding, causing the flat segment to bend outward and causing slight twisting of the C-channel.

The $[45/90/\text{-}45/0]_3$ C-channel, illustrated in Figure 11c, exhibited more compact folding. Fiber and matrix fractures, along with delamination, were observed on the top and flange sides. There was 0° layer debonding on the outside of the flat segment, and cracking occurred in the corner area. This C-channel also showed significant elastic deformation due to matrix cracks and fiber in the flat segment region.

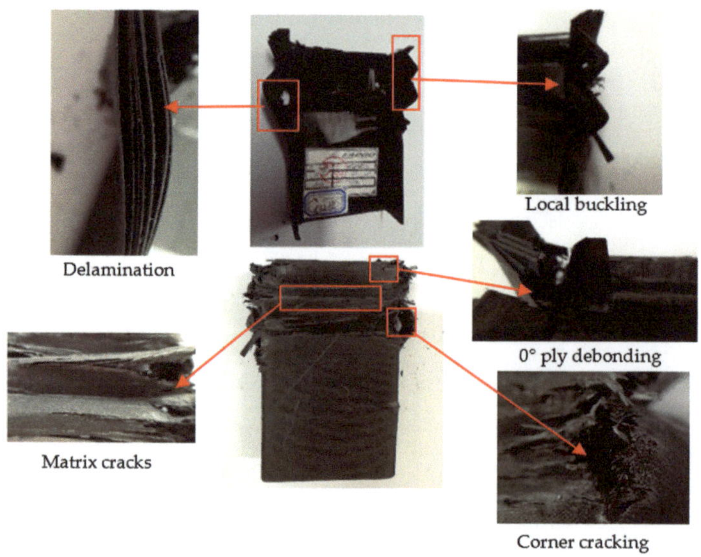

(**a**)

Figure 11. *Cont.*

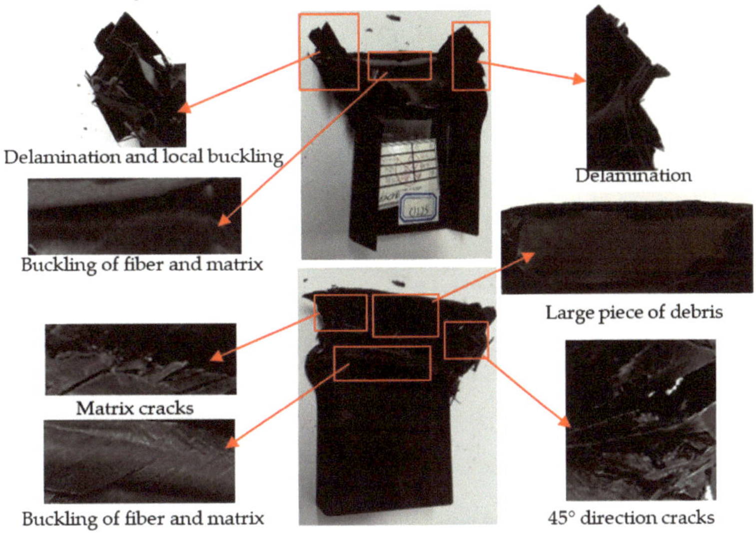

(**b**)

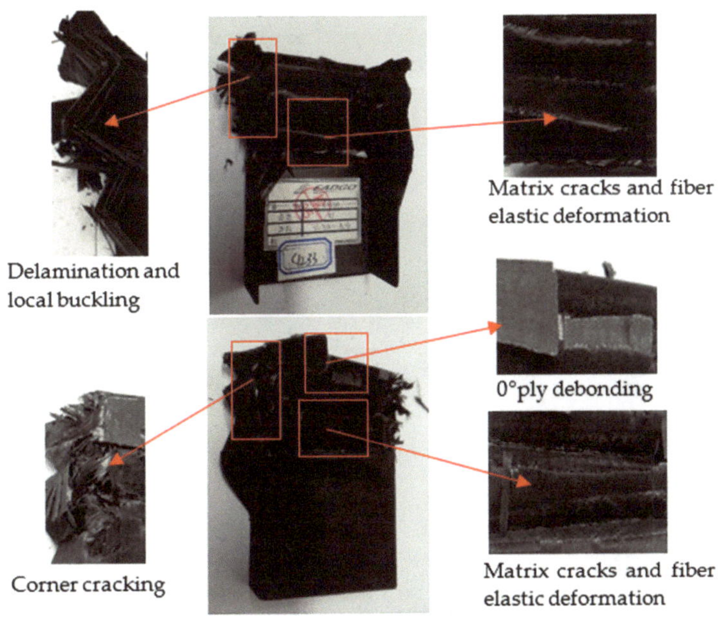

(**c**)

Figure 11. Failure morphology of the $[0/90]_{3s}$ C-channel (**a**), $[45/\text{-}45]_{3s}$ C-channel (**b**), and $[45/90/\text{-}45/0]_3$ C-channel (**c**).

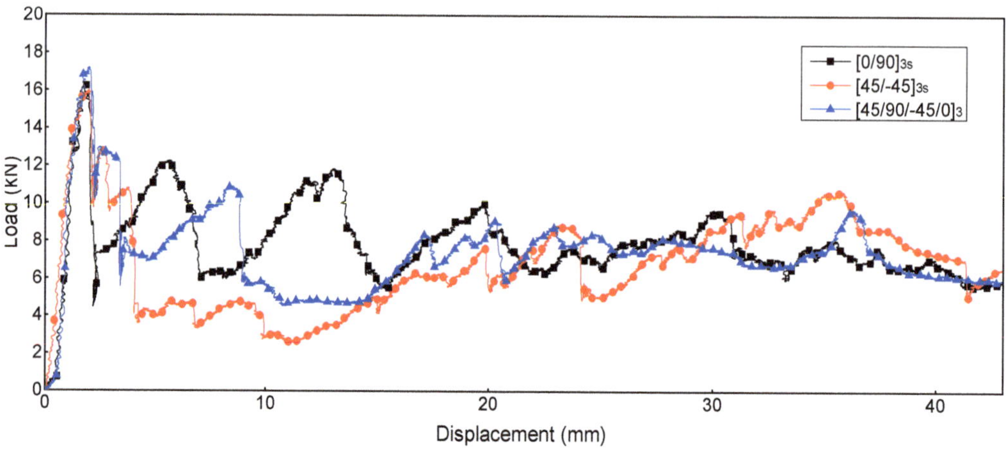

Figure 12. Load-displacement curves: $[0/90]_{3s}$, $[45/\text{-}45]_{3s}$, and $[45/90/\text{-}45/0]_3$ C-channels.

The thin-walled composite C-channels were scanned by using CT tomography measuring machine. The three-dimensional image of $[0/90]_{3s}$ C-channel is shown in Figure 13a. There are multiple buckling zones in the region of the flat segment and flange, many fibers and matrix are buckling without breaking, and a small amount of fiber and matrix is fracturing. The failure mode is a local buckling failure mode, resulting in superposition, as shown in Figure 13b. In the initial section, delamination occurs in the chamfer, and it is gradually crushed. Under the combination of debris and the indenter, local buckling failure occur, with fibers breaking or deforming elastically. The interlaminar shear stress are increased because of the buckling failure of fibers, and interlaminar cracks are formed, which causes the delamination and local brittle fracture of the matrix in the buckling zone. The failure mechanisms are intralaminar and interlaminar delamination, the elastic deformation of the fibers, the buckling of the fibers and matrix, the debonding and fracture of the fibers, the deformation cracking of the matrix, and corner cracking.

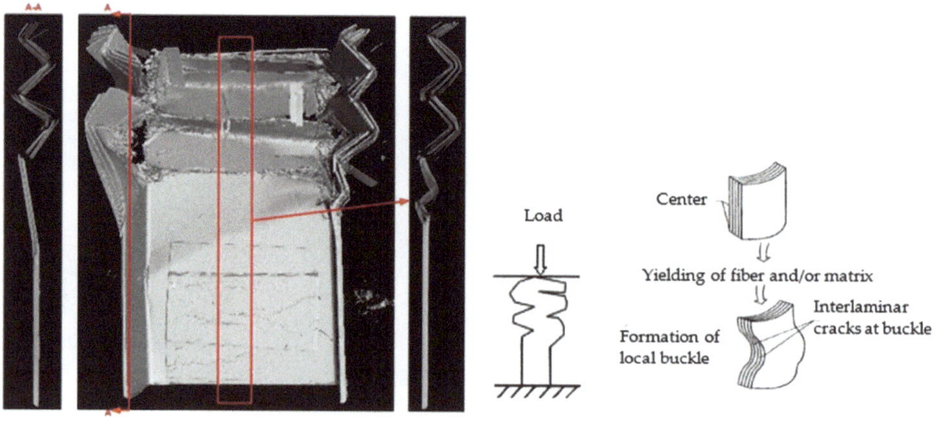

(a) (b)

Figure 13. Three-dimensional image of the $[0/90]_{3s}$ C-channel (**a**) and the local buckling failure mode (**b**).

4.2. Energy-Absorbing Characteristics Analysis

Figure 12 shows the load–displacement curves of thin-walled composite C-channels under low-speed axial compression tests. For three samples with the same layer numbers and layups, the load–displacement curves are very similar, so only one of the three curves is selected to analyze the energy-absorbing characteristics. The energy-absorbing metrics are listed in Table 4.

Table 4. Energy-absorbing metrics of C-channels.

Layups	F_{max}/kN	F_{mean}/kN	*EA*/kJ	*SEA*/(kJ/kg)
$[0/90]_{3s}$	16.294	7.801	345.058	41.793
$[45/\text{-}45]_{3s}$	16.006	6.776	300.236	36.364
$[45/90/\text{-}45/0]_3$	17.225	7.329	324.625	39.318

For the thin-walled composite C-channels with different layups, the crushing loads increase approximately linearly from zero to F_{max}: 16.294 kN for the $[0/90]_{3s}$ C-channel, 16.006 kN for the $[45/\text{-}45]_{3s}$ C-channel, and 17.225 kN for the $[45/90/\text{-}45/0]_3$ C-channel. The chamfers of composite C-channels have been fully crushed, and the crushing loads suddenly drop, reaching a lower sustained crushing load for the remainder of crushing displacement. The $[45/\text{-}45]_{3s}$ C-channel experiences the largest load drop, and the F_{mean} is the minimum. The $[0/90]_{3s}$ C-channel and $[45/90/\text{-}45/0]_3$ C-channel experience load drops less severe than exhibited by the $[45/\text{-}45]_{3s}$ C-channel. The F_{mean} of the $[0/90]_{3s}$ C-channel is the maximum, so the SEA value is also the maximum, which are 14.93% and 6.30% higher than the $[45/\text{-}45]_{3s}$ C-channel and $[45/90/\text{-}45/0]_3$ C-channel, respectively.

4.3. Simulation and Model Verification

The simulated failure morphologies of the 0° layer, 90° layer, and 45° layer are shown in Figure 14. During the process of the axial compression test, the 0° layer was mainly subjected to an axial compressive load, which can provide the axial stiffness for thin-walled composite C-channels. The circumferential strength and stiffness were weak, which resulted in axial bending and circumferential cracking. During the process of the axial compression simulation, brittle fracture failure mainly occurs for the 0° layer. Most of the elements were deleted after reaching the failure strain. A small part of the elements, which were not deleted, were separated from the whole structure, and a few of the elements had larger deformations, as shown in Figure 14a.

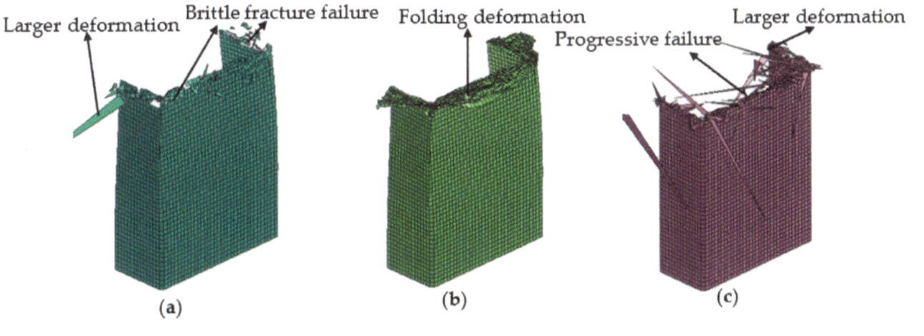

Figure 14. Simulation failure morphologies, 0° layer (**a**), 90° layer (**b**), and 45° layer (**c**).

During the axial compression test, the 90° layer was primarily supported by the matrix, leading to numerous cracks forming along the fiber direction, ultimately resulting in failure. During the process of the axial compression test, the 90° layer was primarily supported by the matrix, resulting in numerous cracks developing along the direction of the fibers, ultimately leading to failure. During the process of the axial compression simulation, the thin walls folded outward or inward, subsequently resulting in damage in the form of progressive collapse, as shown in Figure 14b.

During the process of the axial compression test, the 45° layer was mainly subjected to shear load, transverse shear load, and adjacent interlaminar forces caused by the axial compressive loads. During the process of the axial compression simulation, a large number of elements were deleted after reaching the failure strain, demonstrating the progressive failure behavior, and some elements exhibited larger deformations, as shown in Figure 14c.

The crushing process of the $[0/90]_{3s}$ C-channel between the test and simulation is shown in Figure 15. With the axial movement of the steel indenter, the $[0/90]_{3s}$ C-channel was continuously folded and exhibited progressive compression failure, resulting in a plurality of transverse cracks, and the fiber and the matrix at the corner area completely broke. The simulated progressive compression failure process is in good agreement with the test results.

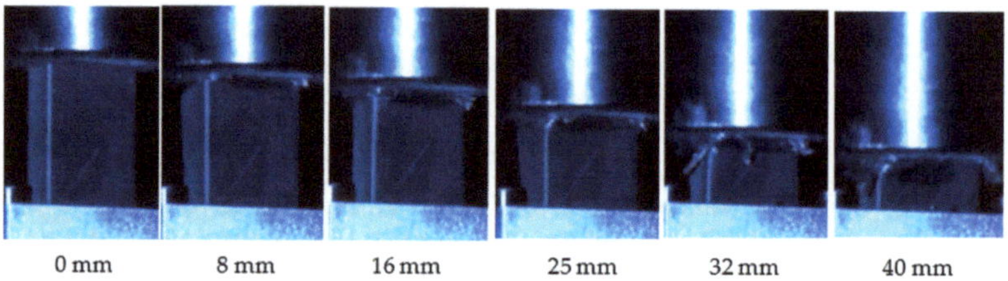

(**a**)

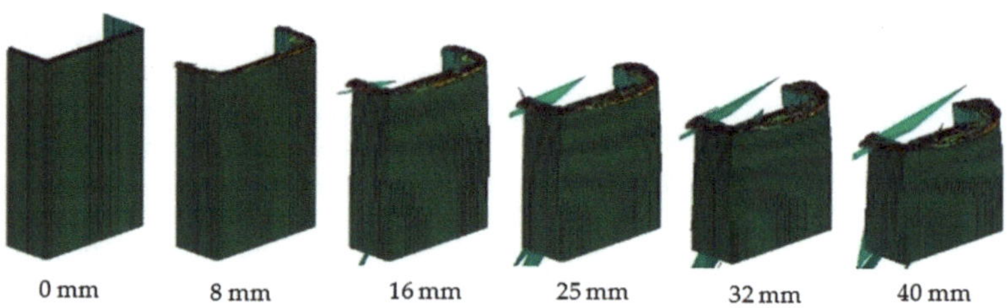

(**b**)

Figure 15. Test crushing process (**a**) and simulation crushing process (**b**).

For the stacked shell models of composite C-channels, the outer chamfer is firstly crushed under the action of the rigid wall, and delamination can be observed, as shown in Figure 16a. With the progressive crushing process, a small number of elements fail to be deleted, and many elements are deformed and stacked, as shown in Figure 16b.

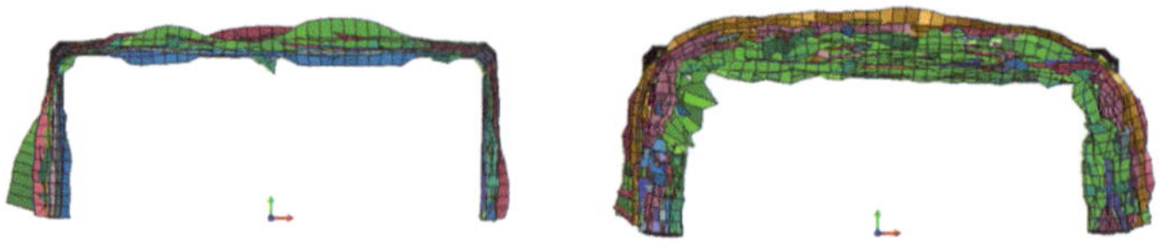

(**a**) (**b**)

Figure 16. Delamination (**a**) and folding deformation (**b**).

A comparison of the load–displacement curves of C-channels with three different layers subject to low-speed axial compression was obtained. A low-pass digital filter (SAE 1000 Hz) was applied to filter the simulation results during post-processing. The simulated load–displacement curves aligned well with the test load–displacement curves, as shown in Figure 17. In the early stage of the axial compression process, the load increases linearly until the initial peak is reached. Subsequently, the bearing capacity of the C-channel decreases, and the load value of the load–displacement curve correspondingly decreases. Then, the compression load fluctuates up and down within a certain range.

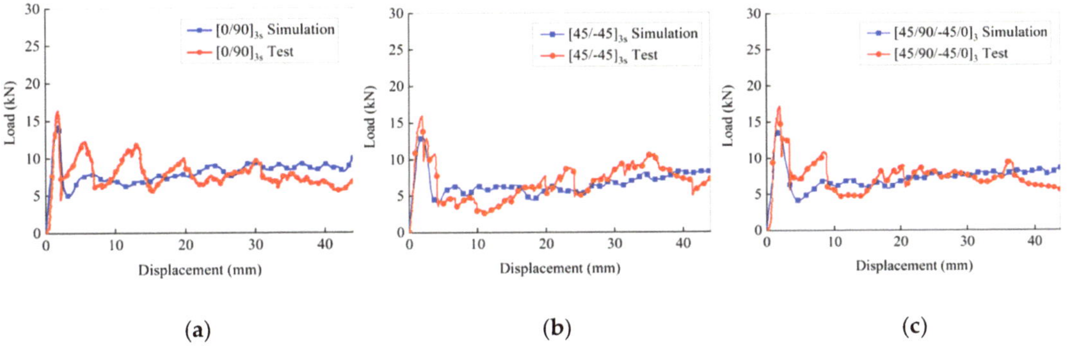

Figure 17. Comparison of load–displacement curves between test and simulation: $[0/90]_{3S}$ (**a**), $[45/\text{-}45]_{3S}$ (**b**), and $[45/90/\text{-}45/0]_3$ (**c**).

A comparison of the energy-absorbing characteristics of the C-channels from the test and simulation is shown in Figure 18. The simulation Fmax values are lower by approximately 5% to 30% than the test results, as shown in Figure 18a. This discrepancy is likely due to the exclusion of the strain rate effect in the material constitutive model. Additionally, material processing defects and the manufacturing process contribute to significant dispersion in the Fmax values. The failure behavior of the corner area is complex, and the finite-element model, while simplified, further complicates the accurate simulation of Fmax. However, the differences in F_{mean} and *SEA* between the simulation and test results are within 6%, as shown in Figure 18b,c. The model can effectively calculate the energy-absorbing characteristics of specimens and reflect their failure modes.

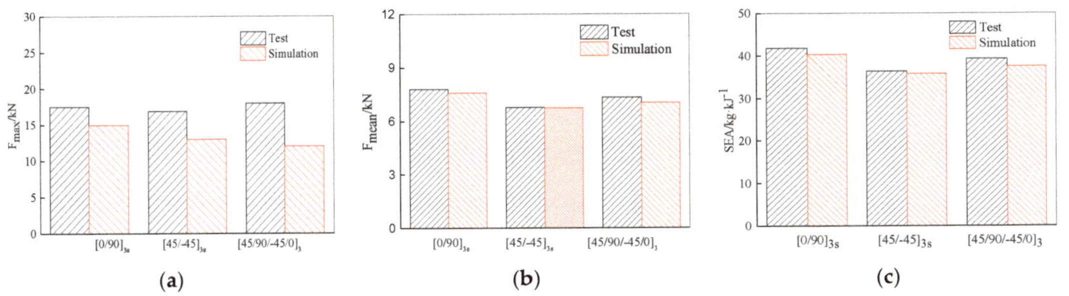

Figure 18. Comparison of energy-absorbing characteristics, F_{max} (**a**), F_{mean} (**b**), and *SEA* (**c**).

5. Conclusions

In this paper, low-speed axial compression tests were conducted on carbon fiber T700/MTM28 composite C-channels with different layups. Load–displacement curves and failure damage modes were obtained. Based on the parametric analysis for the [0] and [90] single-element models, some key parameters were determined and stacked shell models of thin-walled composite C-channels were established. The failure mechanism and energy-absorbing characteristics were analyzed, and the reliability of the stacked shell model was verified against the test results. Some conclusions were drawn as follows:

(1) The thin-walled composite C-channels with $[0/90]_{3s}$, $[45/\text{-}45]_{3s}$, and $[45/90/\text{-}45/0]_3$ exhibit a stable progressive crushing progress characterized by local buckling. This stability is crucial for energy-absorption applications in crashworthiness.

(2) The primary failure mechanisms identified include intralaminar and interlaminar delamination, elastic deformation of fibers, buckling of fiber and matrix, debonding and fracture of fibers, deformation cracking of matrix, and corner cracking. These failure modes interact to dissipate energy during compression, enhancing the energy-absorption capacity of the composite structures.

(3) The physical and numerical meanings of the Lavadèze single-layer model, the Puck IFF failure criteria, and the Yamada–Sun fiber failure criteria are clarified by studying the stress–strain relationship of the single-element model. The proposed values of the material model are given by parametric analysis.

(4) The specific energy absorption (SEA) values for the different layups are 18.5kJ/kg ([0/90]3s), 15.2kJ/kg([45/-45]3s), and 17.1kJ/kg([45/90/-45/0]3). The mean crushing force (Fmean) values are 14.5kN([0/90]3s), 12.8kN([45/-45]3s), and 13.9kN([45/90/-45/0]3). The stacked shell model can accurately capture the interlayer matrix failure and interface delamination cracking behavior observed in the experiments. The simulations exhibit a discrepancy of less than 6% when compared to the experimental results, indicating a strong correlation between the two sets of data. This modeling approach provides a reliable tool for predicting the energy-absorption characteristics of composite structures.

(5) The methods used in this study, including parametric analysis and finite-element modeling, offer robust tools to assess the service performance of materials produced on a large scale for high-performance applications.

The results obtained in this study demonstrate that the molecular-level interactions and matrix–fiber interactions are fundamental in determining the mechanical properties and failure modes of composite materials. A thorough understanding of these interactions enables more accurate predictions of composite behavior under load and facilitates the design of composites with optimized performance characteristics. Our models, which incorporate these molecular-level considerations, provide a robust framework for analyzing and predicting the energy absorption and failure mechanisms of composite structures.

Author Contributions: Conceptualization, X.Z. and Z.F.; methodology, X.Z. and S.S.; investigation, S.S.; resources, H.M.; data curation, X.Z. and S.S.; writing—original draft preparation, X.Z. and H.M.; supervision, H.M. and Z.F.; project administration, H.M.; funding acquisition, Z.F. All authors have read and agreed to the published version of the manuscript.

Funding: This research was funded by National Key R&D Program of China, grant number 2022YFB43010 Tianjin Applied Basic Research Multi-Input Fund Project, grant number 21JCYBJC00690, 23JCYBJC00070; and the Fundamental Research Funds for the Central Universities, grant number 3122017020.

Institutional Review Board Statement: Not applicable.

Data Availability Statement: The original contributions presented in the study are included in the article, further inquiries can be directed to the corresponding author.

Conflicts of Interest: The authors declare no conflicts of interest.

References

1. Sharifah, R.; Ali, J.S.; Shaik, I.; Adib, R. Influence of Composite Skin on the Energy Absorption Characteristics of an Aircraft Fuselage. *Int. J. Eng. Mater. Manuf.* **2023**, *8*, 13–20. [CrossRef]
2. Guida, M.; Marulo, F.; Abrate, S. Advances in crash dynamics for aircraft safety. *Prog. Aerosp. Sci.* **2018**, *98*, 106–123. [CrossRef]
3. Jacob, G.C.; Fellers, J.F.; Simunovic, S.; Starbuck, J.M. Energy Absorption in Polymer Composites for Automotive Crashworthiness. *J. Compos. Mater.* **2016**, *36*, 813–850. [CrossRef]

4. Di Palma, L.; Di Caprio, F.; Chiariello, A.; Ignarra, M.; Russo, S.; Riccio, A.; De Luca, A.; Caputo, F. Vertical Drop Test of Composite Fuselage Section of a Regional Aircraft. *AIAA J.* **2020**, *58*, 474–487. [CrossRef]
5. Di Caprio, F.; Ignarra, M.; Marulo, F.; Guida, M.; Lamboglia, A.; Gambino, B. Design of Composite Stanchions for the Cargo Subfloor Structure of a Civil Aircraft. *Procedia Eng.* **2016**, *167*, 88–96. [CrossRef]
6. Mou, H.; Xie, J.; Feng, Z. Research status and future development of crashworthiness of civil aircraft fuselage structures: An overview. *Prog. Aerosp. Sci.* **2020**, *119*, 100644. [CrossRef]
7. David, M.; Johnson, A.F.; Voggenreiter, H. Analysis of Crushing Response of Composite Crashworthy Structures. *Appl. Compos. Mater.* **2012**, *20*, 773–787. [CrossRef]
8. Elahi, S.A.; Rouzegar, J.; Niknejad, A.; Assaee, H. Theoretical study of absorbed energy by empty and foam-filled composite tubes under lateral compression. *Thin-Walled Struct.* **2017**, *114*, 1–10. [CrossRef]
9. Reuter, C.; Tröster, T. Crashworthiness and numerical simulation of hybrid aluminium-CFRP tubes under axial impact. *Thin-Walled Struct.* **2017**, *117*, 1–9. [CrossRef]
10. Bolukbasi, A.O.; Baxter, T.R.; Nguyen, T.A.; Rassaian, M.; Davis, K.R.; Koch, W.; Firth, L.C. Energy Absorbing Structure for Aircraft. USA. GB2444645A, 11 June 2008.
11. FAA. Special Conditions: Boeing Model 787-8 Airplane. Crashworthiness. In *5-362-SC*; Federal Aviation Administration: Washington, DC, USA, 2007.
12. FAA. Special Conditions: Airbus A350-900 Airplane. Crashworthiness, Emergency Landing Conditions. In *25-537-SC*; Federal Aviation Administration: Washington, DC, USA, 2014.
13. Mahdi, E.; Sebaey, T.A. An experimental investigation into crushing behavior of radially stiffened GFRP composite tubes. *Thin-Walled Struct.* **2014**, *76*, 8–13. [CrossRef]
14. Böhm, H.; Weck, D.; Hornig, A.; Langkamp, A.; Adam, F.; Gude, M. Experimental and Numerical Study on the Axial Crushing Behavior of Textile-Reinforced Thermoplastic Composite Tubes. *Adv. Eng. Mater.* **2016**, *18*, 437–443. [CrossRef]
15. Wang, Z.; Jin, X.; Li, Q.; Sun, G. On crashworthiness design of hybrid metal-composite structures. *Int. J. Mech. Sci.* **2020**, *171*, 105380. [CrossRef]
16. Rozylo, P.; Debski, H. Stability and load carrying capacity of thin-walled composite columns with square cross-section under axial compression. *Compos. Struct.* **2024**, *329*, 117795. [CrossRef]
17. Bussadori, B.P.; Schuffenhauer, K.; Scattina, A. Modelling of CFRP crushing structures in explicit crash analysis. *Compos. Part B: Eng.* **2014**, *60*, 725–735. [CrossRef]
18. Jiang, H.; Ren, Y.; Gao, B. Research on the progressive damage model and trigger geometry of composite waved beam to improve crashworthiness. *Thin-Walled Struct.* **2017**, *119*, 531–543. [CrossRef]
19. Ren, Y.; Jiang, H.; Ji, W.; Zhang, H.; Xiang, J.; Yuan, F.-G. Improvement of Progressive Damage Model to Predicting Crashworthy Composite Corrugated Plate. *Appl. Compos. Mater.* **2017**, *25*, 45–66. [CrossRef]
20. Mou, H.L.; Su, X.; Xie, J.; Feng, Z.Y. Parametric analysis of composite sinusoidal specimens under quasi-static crushing. *Aeronaut. J.* **2018**, *122*, 1244–1262. [CrossRef]
21. Lin, J.S.; Wang, X.; Fang, C.Q.; Huang, X. Collapse loading and energy absorption of fiber-reinforced conical shells. *Compos. Part B Eng.* **2015**, *74*, 178–189. [CrossRef]
22. Ying, L.W.; Yang, F.P.; Wang, X. Analytical method for the axial crushing force of fiber-reinforced tapered square metal tubes. *Compos. Struct.* **2016**, *153*, 222–233. [CrossRef]
23. El-Hage, H.; Mallick, P.K.; Zamani, N. A numerical study on the quasi-static axial crush characteristics of square aluminum–composite hybrid tubes. *Compos. Struct.* **2006**, *73*, 505–514. [CrossRef]
24. Paz, J.; Díaz, J.; Romera, L.; Costas, M. Crushing analysis and multi-objective crashworthiness optimization of GFRP honeycomb-filled energy absorption devices. *Finite Elem. Anal. Des.* **2014**, *91*, 30–39. [CrossRef]
25. Bi, J.; Fang, H.; Wang, Q.; Ren, X. Modeling and optimization of foam-filled thin-walled columns for crashworthiness designs. *Finite Elem. Anal. Des.* **2010**, *46*, 698–709. [CrossRef]
26. Siromani, D.; Henderson, G.; Mikita, D.; Mirarchi, K.; Park, R.; Smolko, J.; Awerbuch, J.; Tan, T.-M. An experimental study on the effect of failure trigger mechanisms on the energy absorption capability of CFRP tubes under axial compression. *Compos. Part A Appl. Sci. Manuf.* **2014**, *64*, 25–35. [CrossRef]
27. Feraboli, P. Development of a Modified Flat-plate Test Specimen and Fixture for Composite Materials Crush Energy Absorption. *J. Compos. Mater.* **2009**, *43*, 1967–1990. [CrossRef]
28. Palanivelu, S.; Paepegem, W.V.; Degrieck, J.; Kakogiannis, D.; Ackeren, J.V.; Hemelrijck, D.v.; Wastiels, J.; Vantomme, J. Comparative study of the quasi-static energy absorption of small-scale composite tubes with different geometrical shapes for use in sacrificial cladding structures. *Polym. Test.* **2010**, *29*, 381–396. [CrossRef]
29. Tran, T.; Hou, S.; Han, X.; Chau, M. Crushing analysis and numerical optimization of angle element structures under axial impact loading. *Compos. Struct.* **2015**, *119*, 422–435. [CrossRef]
30. Różyło, P. Experimental-numerical test of open section composite columns stability subjected to axial compression. *Arch. Mater. Sci. Eng.* **2017**, *84*, 58–64. [CrossRef]
31. Wysmulski, P. Non-linear analysis of the postbuckling behaviour of eccentrically compressed composite channel-section columns. *Compos. Struct.* **2023**, *305*, 116446. [CrossRef]

32. Wysmulski, P.; Debski, H.; Falkowicz, K. Sensitivity of Compressed Composite Channel Columns to Eccentric Loading. *Materials* **2022**, *15*, 6938. [CrossRef]
33. Friedrich, L.; Loosen, S.; Liang, K.; Ruess, M.; Bisagni, C.; Schröder, K.-U. Stacking sequence influence on imperfection sensitivity of cylindrical composite shells under axial compression. *Compos. Struct.* **2015**, *134*, 750–761. [CrossRef]
34. *PAM-CRASH User Manual Version 2014*; ESI Group Paris: Paris, France, 2014.
35. LSDKU Manual. *Version 971, Livermore Software Technology Corporation (LSTC)*; LSDKU Manual: Livermore, CA, USA, 2009.
36. *Abaqus Theory Guide, Version 6.14*; Dassault Systèmes Simulia. Inc.: Providence, RI, USA, 2014.
37. Rozylo, P.; Falkowicz, K.; Wysmulski, P.; Debski, H.; Pasnik, J.; Kral, J. Experimental-Numerical Failure Analysis of Thin-Walled Composite Columns Using Advanced Damage Models. *Materials* **2021**, *14*, 1506. [CrossRef]
38. Debski, H.; Rozylo, P.; Teter, A. Buckling and limit states of thin-walled composite columns under eccentric load. *Thin-Walled Struct.* **2020**, *149*, 106627. [CrossRef]
39. Wysmulski, P.; Debski, H. Post-buckling and limit states of composite channel-section profiles under eccentric compression. *Compos. Struct.* **2020**, *245*, 112356. [CrossRef]
40. Vasiliev, V.; Lurie, S.; Salov, V. Estimation of the strength of plates with cracks based on the maximum stress criterion in a scale-dependent generalized theory of elasticity. *Phys. Mesomech.* **2019**, *22*, 456–462. [CrossRef]
41. Zhao, L.; Qin, T.; Zhang, J.; Shenoi, R.A. Modified maximum stress failure criterion for composite π joints. *J. Compos. Mater.* **2013**, *47*, 2995–3008. [CrossRef]
42. Mirsayar, M. Maximum principal strain criterion for fracture in orthotropic composites under combined tensile/shear loading. *Theor. Appl. Fract. Mech.* **2022**, *118*, 103291. [CrossRef]
43. Rao, P.M.; Subba Rao, V. Degradation model based on Tsai-Hill factors to model the progressive failure of fiber metal laminates. *J. Compos. Mater.* **2011**, *45*, 1783–1792. [CrossRef]
44. Hashin, Z. Finite thermoelastic fracture criterion with application to laminate cracking analysis. *J. Mech. Phys. Solids* **1996**, *44*, 1129–1145. [CrossRef]
45. Lee, C.-S.; Kim, J.-H.; Kim, S.-k.; Ryu, D.-M.; Lee, J.-M. Initial and progressive failure analyses for composite laminates using Puck failure criterion and damage-coupled finite element method. *Compos. Struct.* **2015**, *121*, 406–419. [CrossRef]
46. Zheng, J.; Maharaj, C.; Liu, J.; Chai, H.; Liu, H.; Dear, J.P. A comparative study on the failure criteria for predicting the damage initiation in fiber-reinforced composites. *Mech. Compos. Mater.* **2022**, *58*, 125–140. [CrossRef]
47. Mohamadzadeh, M.; Haftkhani, A.R.; Ebrahimi, G.; Yoshihara, H. Numerical and experimental failure analysis of screwed single shear joints in wood plastic composite. *Mater. Des.* **2012**, *35*, 404–413. [CrossRef]
48. Palanivelu, S.; Van Paepegem, W.; Degrieck, J.; Kakogiannis, D.; Van Ackeren, J.; Van Hemelrijck, D.; Wastiels, J.; Vantomme, J. Parametric study of crushing parameters and failure patterns of pultruded composite tubes using cohesive elements and seam, Part I: Central delamination and triggering modelling. *Polym. Test.* **2010**, *29*, 729–741. [CrossRef]
49. Siromani, D.; Awerbuch, J.; Tan, T.-M. Finite element modeling of the crushing behavior of thin-walled CFRP tubes under axial compression. *Compos. Part B Eng.* **2014**, *64*, 50–58. [CrossRef]
50. Johnson, A.F.; David, M. Failure mechanisms in energy-absorbing composite structures. *Philos. Mag.* **2010**, *90*, 4245–4261. [CrossRef]
51. Feraboli, P.; Wade, B.; Deleo, F.; Rassaian, M.; Higgins, M.; Byar, A. LS-DYNA MAT54 modeling of the axial crushing of a composite tape sinusoidal specimen. *Compos. Part A Appl. Sci. Manuf.* **2011**, *42*, 1809–1825. [CrossRef]
52. Feraboli, P.; Wade, B. Simulating Laminated Composite Materials Using LS-DYNA Material Model MAT54: Single-Element Investigation. Federal Aviation Administration Technical Report. 2015. Available online: https://www.researchgate.net/publication/322152587_Simulating_Laminated_Composite_Materials_Using_LS-DYNA_Material_Model_MAT54_Single-Element_Investigation (accessed on 23 April 2024).
53. Chang, F.-K.; Chang, K.-Y. Post-failure analysis of bolted composite joints in tension or shear-out mode failure. *J. Compos. Mater.* **1987**, *21*, 809–833. [CrossRef]
54. Chang, F.-K.; Lessard, L.B. Damage tolerance of laminated composites containing an open hole and subjected to compressive loadings: Part I—Analysis. *J. Compos. Mater.* **1991**, *25*, 2–43. [CrossRef]
55. *DINENISO7822-2000*; Textile Glass Reinforced Plastics—Determination of Void Content—Loss on Ignition, Mechanical Disintegration and statistical Counting Methods. German version 2000. Deutsches Institut für Normung e.V.: Berlin, Germany, 1990; pp. 1–8.
56. *ASTM D792-13*; Standard Test Methods for Density and Specific Gravity (Relative Density) of Plastics by Displacement, American Society for Testing and Materials. ASTM: West Conshohocken, PA, USA, 2013.
57. *ASTM D3039/D3039M-17*; Standard Test Method for Tensile Properties of Polymer Matrix Composite Materials. ASTM: West Conshohocken, PA, USA, 2015.
58. *ASTM D3410*; Standard Test Method for Compressive Properties of Unidirectional or Crossply Fiber-Resin Composites. ASTM: West Conshohocken, PA, USA, 1987.
59. *ASTM D3518/D3518M-18*; Standard Test Method for In-Plane Shear Response of Polymer Matrix Composite Materials by Tensile Test of a ±45° Laminate EUA. ASTM: West Conshohocken, PA, USA, 2010.
60. *ASTM D638*; Standard Test Method for Tensile Properties of Plastics. ASTM: West Conshohocken, PA, USA, 2008.

61. *ASTM D695*; Standard Test Method for Compressive Properties of Rigid Plastics. ASTM: West Conshohocken, PA, USA, 2010.
62. *ASTM D5379*; Standard Test Method for Shear Properties of Composite Materials by the V-Notched Beam Method. ASTM: West Conshohocken, PA, USA, 2015.

polymers

MDPI

Article

Performance and Damage Study of Composite Rotor Blades under Impact

Guorui Yu, Xiaobin Li and Wenjun Huang *

AVIC China Helicopter Design and Research Institute, Jingdezhen 333001, China; yugr@avic.com (G.Y.); lxbin1990@163.com (X.L.)
* Correspondence: chrdihwj@163.com

Abstract: A military helicopter is easily attacked by bullets in a battlefield environment. The composite blade is the main lifting surface and control surface of the helicopter. Its ballistic performance directly determines the vulnerability and survivability of the helicopter in the battlefield environment. To study the ballistic performance of the composite helicopter blade, the damage characteristics of the impacted composite rotor blade are obtained by experiments. A numerical simulation model is established by applying Abaqus software to predict the blade ballistic damage. The three-dimensional progressive damage failure model is used to analyze the ballistic damage under the experimental conditions. The effectiveness and accuracy of the numerical simulation model are verified through a comparison with the experimental results. The ballistic damage of composite blades under three experimental conditions was investigated. The results show that the ballistic damage type of composite blade mainly includes delamination, fiber breakage, and foam collapse. The damage to the composite material at the position of bullet incidence is mainly local shear fracture, while the damage to the composite material at the exit position is mainly fiber tensile fracture. The ballistic damage size of the composite blade is closely related to the ballistic position, incident angle, and structure characteristics along the ballistic path. The larger the incident angle, the smaller the ballistic damage size of the blade. The greater the structural stiffness of the structure near the exit, the greater the damage size of the exit. The numerical simulation model presented in this paper can provide a reference for research on the ballistic performance of composite helicopter blades.

Keywords: composite blade; failure criteria; finite element analysis; ballistic experiment

Citation: Yu, G.; Li, X.; Huang, W. Performance and Damage Study of Composite Rotor Blades under Impact. *Polymers* **2024**, *16*, 623. https://doi.org/10.3390/polym16050623

Academic Editors: Miguel Ángel López Manchado and Fahmi Zaïri

Received: 21 December 2023
Revised: 3 February 2024
Accepted: 21 February 2024
Published: 25 February 2024

1. Introduction

Military helicopters are faced with different threats of ground weapons and air weapons in the battlefield environment. To avoid destructive destruction after the ballistic attack and improve the survivability of the aircraft, the military helicopter needs to have certain anti-ballistic capabilities. As an important part of a helicopter, the rotor blade is the main lifting surface and control surface of the helicopter, so the anti-ballistic capability of the blade should be paid more attention to.

Fiber-reinforced composite materials show excellent mechanical properties and better design ability compared with traditional metal. Composite rotor blades are widely used in modern helicopter design. Fiber breakage, delamination, and foam collapse of the composite rotor blades usually emerge under ballistic impact. And the bearing capacity of the composite material will be reduced, which poses a threat to helicopter safety. Therefore, the damage to the composite rotor blade under high-speed bullet impact will directly affect the helicopter. The composite rotor blades of common military helicopters, such as Black Hawk, V22, Apache, Comanche, NH90, and CH47, can resist one bullet of 12.7 mm or 23 mm bullets without affecting the safety of returning to the aircraft base [1–7].

Edward T. Keast [8] investigated the improvement method for the damage tolerance of the composite rotor blade. A variety of helicopter profile structures were designed, and the residual stiffness and performance of the blade after impact were analyzed. The damage and failure mechanisms of the impacted composite blade are studied by J. Aubry and P. Navarro [9]. A steel ball was used as an alternative to the bullet. Normal and oblique impacts were considered. Zach [10] established a macro-mechanical model of composite rotor blades to simulate bullet damage. Normal and lateral bullet damage were simulated. However, only unidirectional fiber composite was considered; the woven fiber composite and foam were neglected.

The composite rotor blade is essentially similar to a composite sandwich panel. In recent years, the high-speed impact characteristics of sandwich panels have been extensively studied. Four failure modes of composite foam sandwich structure, including delamination and buckling of composite panels, matrix cracking and fiber breaking, Degumming of panel/core interface, and Collapse of foam cores, mainly occur under impact load [11]. Wang H [12] proposed that the flexible matrix composites always had higher perforation resistance but larger deformation than the rigid matrix counterparts. Yashiro [13] and Bernard [14] studied the influence of layer sequence and angle on impact damage, and concluded that the delamination damage to the panels first occurs between layers with different mat angles. The size of the panels tends to increase gradually from the impacted side to the interface. Both the delamination damage between layers and the damage between fibers and substrates were related to the direction of the material. Lang [15] concluded that the interface adhesive layer of the panel/core was a weak link of the sandwich structure, which was easy to fail under impact load. This reduced the overall strength and stiffness of the sandwich structure. The effect of sandwich density changes on the high-speed impact properties of composite foam sandwich structures was investigated by Nasirzadeh [16]. The impact resistance of sandwich panels did not increase with increasing foam density. Mehran [17] reached a similar conclusion. In their study of high-speed impact energy absorption of polyurethane foam sandwich panels, it was found that the high-strength and high-modulus panels were the main energy-absorbing components, and increasing the density of foam sandwich did not significantly improve the energy-absorbing performance of the whole structure. Acanfora [18] proposed that the polypropylene honeycomb core and CFRP composite external skins sandwich configurations are characterized by better overall crashworthiness performances. Albayrak [19] showed that using rubber intermediate, the energy absorption ability of curved sandwich composite can be increase by a significant improvement.

The high-speed impact characteristics of curved composite sandwich panels were studied by Usta [20]. It was concluded that the curvature of the panels changed the energy absorption rates of sandwich panels, and the doubly curved sandwich panels had increased high-velocity impact resistance. Alonse [21–23] further studied the ballistic characteristics of composite foam sandwich panels. A continuum damage model was proposed. The energy absorption mechanism and the influence of different composite laminates on ballistic limits were analyzed. The damaged area of the front face skin was smaller than the damage produced in the rear face skin, and bending effects were notable in the latter, considering the largest fraction of the energy was absorbed by the out-of-plane.

Although the current research has been done to experiment with the ballistic performance of composite blades, only their anti-ballistic indicators have been verified, and no research has been conducted on the ballistic damage mechanism of composite blades. Comprehensive numerical research on the performance of the composite rotor blade has yet to be conducted. The high-speed impact analysis of the composite sandwich panels can reveal a certain mechanism of the impact resistance of the sandwich panels. However, it cannot reflect the effect of the impact position and impact angle of the bullet while investigating the full-size composite blade with a curved surface shape and stiffening ribs.

In order to study the ballistic performance of a full-scale composite blade, a type of composite rotor blade configuration with strong impact resistance is designed. The performance of composite blades under different ballistic impact conditions is studied by ballistic experiments. Based on this, a numerical simulation model considering the three-dimensional progressive damage failure model of the composite blade airfoil section is established. The simulation analysis of the damage of the composite blade under the condition of high-speed impact is carried out. The simulation results are compared with the experimental results to verify the accuracy of the presented numerical simulation model. And through the above analysis model, the damage mechanism of the composite rotor blade under bullet impact was studied. At the same time, the effects of ballistic position, incident angle, and ballistic path on ballistic damage were studied. The research results of this paper can provide a theoretical basis and reference for the anti-ballistic design and ballistic performance analysis of composite blades.

2. Specimen and Experiments

2.1. Design and Manufacture of the Experimental Specimen

The size and cross-section of the composite rotor blade airfoil section are shown in Figure 1. The spanwise length of the experiment specimen is 1300 mm. The chordwise length is 500 mm. The cross-section of the experiment specimen is a multi-closed cavity structure, mainly consisting of skin, spar, reinforcing rib, and foam filling. The skin and spar are the main load-bearing structures, which bear the centrifugal force and bending moment. The skin is mainly composed of carbon fiber fabric CF3502/epoxy and glass fiber fabric EW250F/epoxy (Beijing Institute of Aerial Materials, Beijing, China). The layer configuration and material properties of the skin are provided in Figure 2 and Table 1, respectively. The spar and the trailing edge strip consist of S_4C10-800 glass fiber/epoxy prepreg tape (Beijing Institute of Aerial Materials, Beijing, China). Two closed box cavities in the D shape of the spar at the leading edge are formed. The material properties of S_4C10-800 are shown in Table 1. The reinforcing rib consists of one layer of CF3502 and one layer of EW250F, which are laid in the rear edge of the mid-foam filling in a U shape. The foam filling selects polymethacrylimide (PMI) foam.

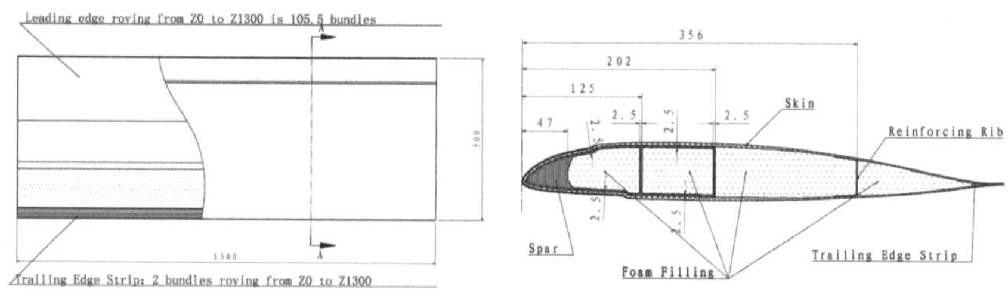

(a) (b)

Figure 1. Composite blade airfoil section experiment specimen. (**a**) Schematic diagram of top view (unit: mm). (**b**) Schematic diagram of cross-section (unit: mm).

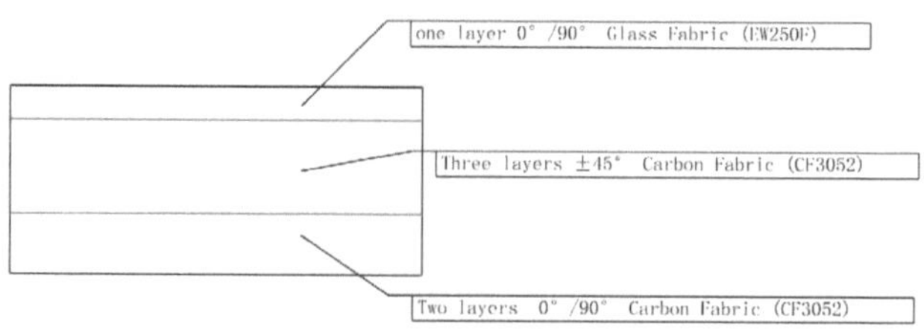

Figure 2. Schematic diagram of the skin layer.

Table 1. Material properties of composite materials.

Property	EW250F/Epoxy	CF3052/Epoxy	S4C10-800/Epoxy	Property	EW250F/Epoxy	CF3052/Epoxy	S4C10-800/Epoxy
E_1 (GPa)	25	50.4	57	X_T (Mpa)	380	520	1350
E_2 (GPa)	25	50.4	13	X_C (Mpa)	380	380	700
E_3 (GPa)	3	6.8	13	Y_T (Mpa)	380	520	50
ν_{12}	0.14	0.08	0.3	Y_C (Mpa)	380	380	200
ν_{13}	0.2	0.35	0.3	Z_T (Mpa)	33.6	33.6	50
ν_{23}	0.2	0.35	0.33	Z_C (Mpa)	93.6	93.6	200
G_{12} (GPa)	4.5	3.13	5.5	S_{12} (Mpa)	80	80	52
G_{13} (GPa)	2.4	2.78	5.5	S_{13} (Mpa)	49	62	52
G_{23} (GPa)	2.4	2.78	5.5	S_{23} (Mpa)	49	62	50

2.2. Experimental Process and Equipment

As shown in Figure 3, the blade airfoil specimen is fixed on a steel structure mounting bracket arranged on the ground. Two angle adjustment fixtures are arranged on the fingernails, which are used for adjusting and fixing the blade airfoil specimen. Both fixtures help to change the ballistic impact angle relative to the blade airfoil specimen. As shown in Figure 4, the bullet used in the experiment is a 12.7 mm API. The 12.7 mm linear bore ballistic gun is applied to fire the bullet. A pair of light-screen targets are arranged on the ballistic line in order to measure the velocity of the bullet during the experiment process.

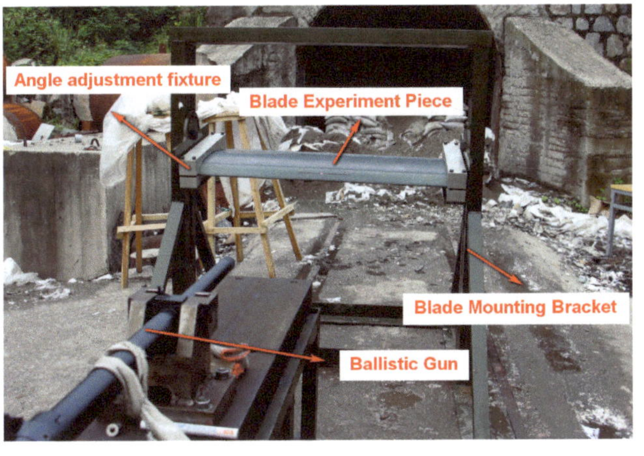

Figure 3. Test scheme and fixture.

Figure 4. Ballistic gun and bullet.

In order to verify the ballistic performance of the composite blade airfoil specimen, three experimental cases were developed. The specific experiment conditions are shown in Figure 5. The impact angle of the bullet relative to the blade was set by the angle-adjusting fixture. The deviation of the incidence angle was within 1 degree. The accuracy of the impact point was ensured by the sighting instrument of the ballistic gun. The muzzle is 1.5 m away from the impacted point of the blade. The speed of the bullet is controlled to be 650 +/− 20 m/s.

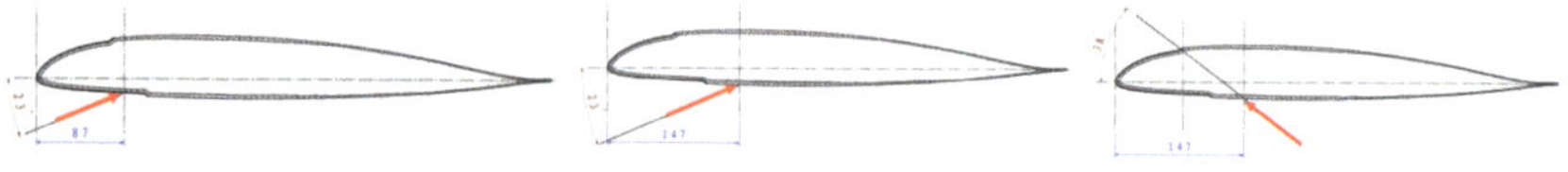

(**a**) Case 1 (**b**) Case 2 (**c**) Case 3

Figure 5. Cases of the projectile experiment (The red arrow represents the direction of the impact) (unit: mm).

3. Numerical Simulation of Composite Blade under Ballistic Impact

3.1. Finite Element Model and Boundary Conditions

In order to study the failure mechanism of the composite blade under ballistic impact, a numerical simulation model including skin, spar, foam filling, and reinforcing ribs was established using Abaqus software (Version 6.14); the analysis process is shown in Figure 6. The finite element model of the bullet is established according to the actual size of the bullet and is regarded as a rigid body (Figure 7). The 8-node hexahedral linear reduced integration element C3D8R was selected to analyze the damage characteristics of the composite blade to save computational time. The bonding units COH3D8 were introduced between the skin and foam interface and each layer of skin, considering the simulation of the delamination of composite blades under high-speed impact. The mechanical properties of bonding units are provided in Table 2. PMI adopts the crushable foam model, and the material properties are shown in Table 3.

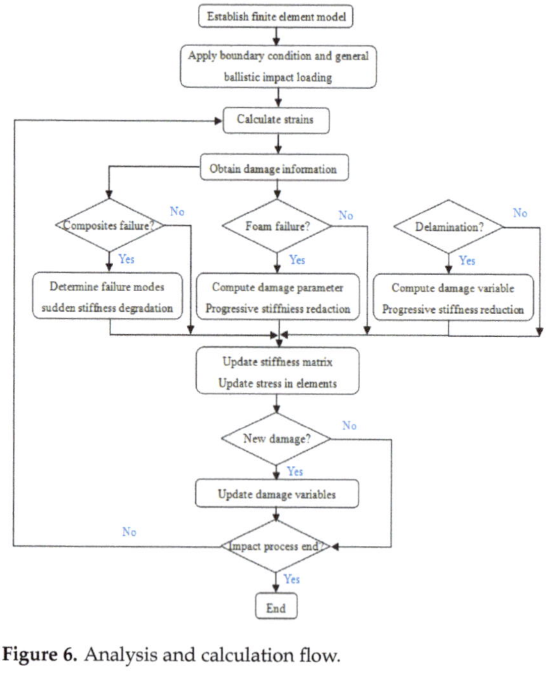

Figure 6. Analysis and calculation flow.

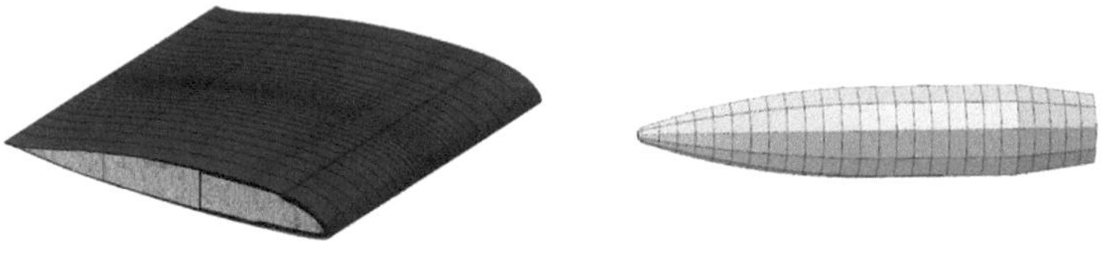

(**a**) Blade airfoil segment (**b**) 12.7 mm bullet

Figure 7. Schematic of the finite element.

Table 2. Mechanical properties of bonding units.

Property	Value	Property	Value
E (GPa)	5	t_s^0 (MPa)	30
G (GPa)	5	G_n^c (N/mm)	0.6
t_n^0 (MPa)	30	G_s^c (N/mm)	2.1

Table 3. Material properties of PMI.

Property	Value	Property	Value
E (GPa)	92	δ(%)	30
G (GPa)	29	ρ (kg/m^3)	75
σ_t (Mpa)	2.8	ν	0.3
σ_c (Mpa)	1.5	k_s	0
S (MPa)	1.3	θ_s	1.5

During the process of high-speed impact by the bullet, the deformation of the blade mainly occurs at the contact zone between the bullet and the blade. The grid density decreases from the impact zone to the edge of the blade. This kind of grid not only ensures the accuracy of the numerical results but also improves the computational efficiency of the model. The element mesh size dependency analysis was conducted on the impact contact area as shown in Figure 8, and based on the analysis results, a element size of 1mm was selected in the impact contact area.

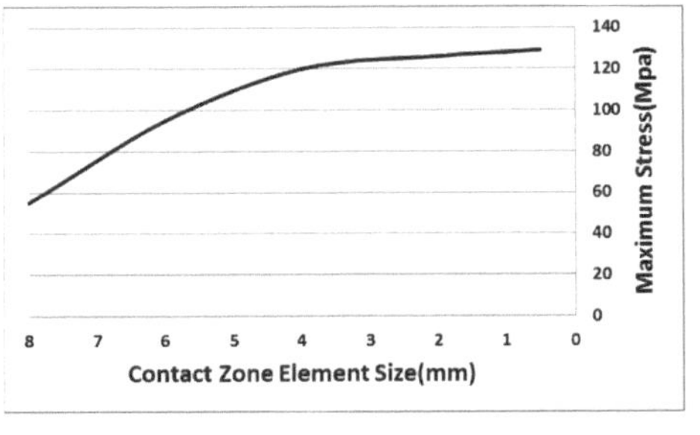

Figure 8. Element mesh size dependency analysis.

The fixed boundary conditions (shown in Figure 9) are applied to the two ends of the composite blade, and all the degrees of freedom are constrained to simulate the clamping fixation of the impact experiment. The initial velocity of the bullet is set to 650 m/s, which is the same as the actual experiment condition.

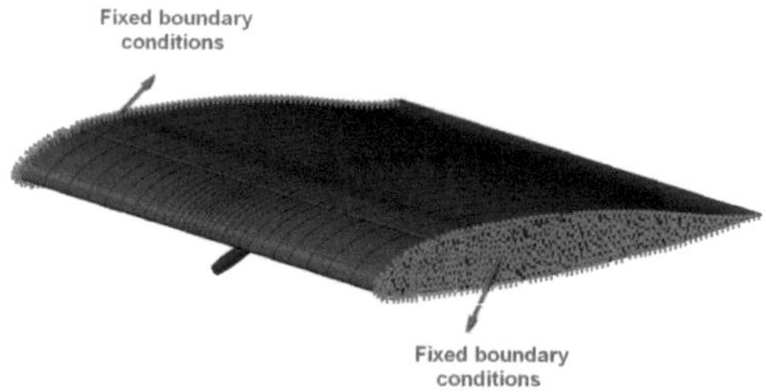

Figure 9. The fixed boundary conditions.

The contact response between the bullet and the blade is simulated by surface contact. The contact force is calculated by a finite sliding penalty contact algorithm.

3.2. Material Damage Model

Capturing the damage modes seen in the experiment is a key requirement of the model. Selecting proper failure criteria is essential to accurately detecting the onset of material damage.

3.2.1. Damage Model of Composite Materials

The failure strength of composites is mainly related to the properties of materials, stress state, strain state, and strain energy under loading. The failure modes of composite materials are mainly as follows: fiber tensile failure, fiber compression failure, matrix tensile failure, and matrix compression failure. The Hashin failure criterion is one of the classical failure criteria for composite materials. It is based on the stress failure criterion. According to the different failure mechanisms produced by tensile and compressive failure, the failure of the fiber and matrix of composite materials is classified into tensile and compressive failure [24,25].

There are two kinds of composite materials used in the blade design of this work. One is the warp-knitting fabric used for the spar, which belongs to the unidirectional fiber composite material. The other is the composite plain fabric used for skin, which belongs to a bidirectional fiber-reinforced composite material. The failure criteria used for the two composites are different.

The failure criteria used for spar are as follows:

Fiber tensile failure ($\sigma_1 \geq 0$)

$$\left(\frac{\sigma_1}{X_T}\right)^2 + \left(\frac{\sigma_{12}}{S_{12}}\right)^2 \geq 1 \tag{1}$$

Fiber compression failure ($\sigma_1 < 0$)

$$\left(\frac{\sigma_1}{X_C}\right)^2 \geq 1 \tag{2}$$

Matrix tensile failure ($\sigma_2 + \sigma_3 \geq 0$)

$$\left(\frac{\sigma_2 + \sigma_3}{Y_T}\right)^2 + \left(\frac{\sigma_{12}}{S_{12}}\right)^2 + \left(\frac{\sigma_{13}}{S_{13}}\right)^2 \geq 1 \tag{3}$$

Matrix compression failure ($\sigma_2 + \sigma_3 \geq 0$)

$$\left(\frac{\sigma_2 + \sigma_3}{Y_C}\right)^2 \geq 1 \tag{4}$$

The failure criteria used for skin are as follows:

Warp tensile failure ($\sigma_1 \geq 0$)

$$(\frac{\sigma_1}{X_T})^2 + \alpha(\frac{\sigma_{12}}{S_{12}})^2 + \alpha(\frac{\tau_{13}}{S_{13}})^2 \geq 1 \quad (5)$$

Warp compression failure ($\sigma_1 < 0$)

$$(\frac{\sigma_1}{X_C})^2 \geq 1 \quad (6)$$

Weft tensile failure ($\sigma_2 \geq 0$)

$$(\frac{\sigma_2}{Y_T})^2 + \alpha(\frac{\sigma_{12}}{S_{12}})^2 + \alpha(\frac{\tau_{23}}{S_{23}})^2 \geq 1 \quad (7)$$

Weft compression failure ($\sigma_2 < 0$)

$$(\frac{\sigma_2}{Y_C})^2 \geq 1 \quad (8)$$

Matrix tensile failure ($\sigma_3 \geq 0$)

$$(\frac{\sigma_3}{Z_T})^2 + (\frac{\sigma_{13}}{S_{13}})^2 + (\frac{\sigma_{23}}{S_{23}})^2 \geq 1 \quad (9)$$

Matrix compression failure ($\sigma_3 \geq 0$)

$$(\frac{\sigma_3}{Z_C})^2 \geq 1 \quad (10)$$

where X_T, X_C, Y_T, and Y_C represent the longitudinal tensile strength, the transverse tensile strength, and the compressive strength of the fiber on the surface of the composite material, respectively. Z_T and Z_C are the tensile strength and the compressive strength of the matrix. S_{12}, S_{13}, and S_{23} are in-plane shear strength and inter-laminar shear strength, respectively, and α is the shear correction factor.

The failure of composite material occurs through the whole high-speed penetration process. When the composite material is subjected to a high-speed dynamic impact load, the bearing capacity of the damage direction will be reduced. But not all the failures directly. Therefore, a reasonable stiffness reduction scheme should be adopted to properly reduce the stiffness. When both the composite fiber and the matrix lose their load-bearing capacity, the element will be deleted from the finite element model. The stiffness reduction scheme [26] of this paper is shown in Table 4.

Table 4. Composite stiffness reduction scheme.

Material	Damage Model	Stiffness Reduction Scheme
Spar	Fiber failure	$E_1 = 0.1E_1$
	Matrix failure	$E_2 = 0.2E_2$, $E_3 = 0.2E_3$, $G_{12} = 0.2G_{12}$, $G_{23} = 0.2G_{23}$
Skin	Warp failure	$E_1 = 0.1E_1$, $G_{12} = 0.1G_{12}$, $G_{13} = 0.1G_{13}$, $\upsilon_{12} = 0.1\upsilon_{12}$, $\upsilon_{13} = 0.1\upsilon_{13}$
	Weft failure	$E_2 = 0.1E_2$, $G_{12} = 0.1G_{12}$, $G_{23} = 0.1G_{23}$, $\upsilon_{12} = 0.1\upsilon_{12}$, $\upsilon_{23} = 0.1\upsilon_{23}$
	Matrix failure	$E_3 = 0.2E_3$, $G_{12} = 0.2G_{12}$, $G_{23} = 0.2G_{23}$

3.2.2. Damage Model of a Cohesive Unit

Delamination failure is also an important failure mode for composites under high-speed impact loads. Since cohesive elements are widely used to model adhesives between two components, interfacial debonding, gaskets, and small adhesive patches. In this paper, the cohesive elements are inserted into the interfaces between the adjacent layers inside the composite skin and between the skin and the filled foam, which aims at effectively simulating the delamination mode under ballistic impact.

The coordinate system of cohesive elements is displayed in Figure 10a. It is assumed that only a normal stress t_1 and two shear stresses t_2,t_3 exist, since the cohesive interface is extremely thin. The interface behavior is governed by a bilinear traction-separation law, which adopts an initially linear elastic response followed by the linear evolution of damage (Figure 10b).

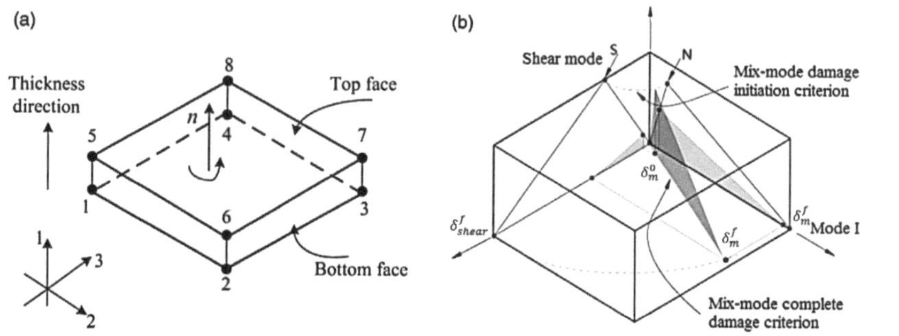

Figure 10. Cohesive zone model of interface: (**a**) coordinate system of cohesive element and (**b**) mix-mode bilinear constitutive model of cohesive element.

The initial elastic response of the cohesive elements can be expressed as

$$\begin{Bmatrix} t_1 \\ t_2 \\ t_3 \end{Bmatrix} = \begin{bmatrix} k_1 & 0 & 0 \\ 0 & k_2 & 0 \\ 0 & 0 & k_3 \end{bmatrix} \begin{Bmatrix} \delta_1 \\ \delta_2 \\ \delta_3 \end{Bmatrix} \tag{11}$$

where k_i is the initial stiffness and δ_i is the separation displacement.

The quadratic nominal stress criterion is used as the damage initiation of the cohesive elements [26]:

$$\left(\frac{\langle t_1 \rangle}{\mathrm{N}}\right)^2 + \left(\frac{t_2}{\mathrm{S}}\right)^2 + \left(\frac{t_3}{\mathrm{T}}\right)^2 = 1 \tag{12}$$

where N, S, and T are normal and two tangential interface strengths, respectively.

Normal compressive stress does not contribute to the damage initiation of cohesive elements. Hence, the Macaulay operator in Equation (12) is defined by

$$\langle a \rangle = \begin{cases} a & a > 0 \\ 0 & a \leq 0 \end{cases} \tag{13}$$

The material stiffness needs to be degraded once the initiation criterion is satisfied. After damage initiation, a scalar damage variable d is used to represent the linear damage evolution of the interface:

$$d = \frac{\delta_m^f(\delta_m - \delta_m^0)}{\delta_m(\delta_m^f - \delta_m^0)} \; (d \in [0,1]) \tag{14}$$

Here, δ_m is the equivalent displacement determined by Equation (15), including normal and tangential displacement. δ_m^0 represents the equivalent displacement at the beginning of interface damage. δ_m^f is the equivalent displacement at the complete failure of the interface.

$$\delta_m = \sqrt{\langle \delta_1 \rangle^2 + \delta_2{}^2 + \delta_3{}^2} \tag{15}$$

The damage variable d is introduced into the constitutive relationship of the cohesive unit:

$$\begin{Bmatrix} \hat{t_1} \\ \hat{t_2} \\ \hat{t_3} \end{Bmatrix} = \begin{bmatrix} 1-d\frac{\langle t_1 \rangle}{t_1} & 0 & 0 \\ 0 & 1-d & 0 \\ 0 & 0 & 1-d \end{bmatrix} \begin{bmatrix} k_1 & 0 & 0 \\ 0 & k_2 & 0 \\ 0 & 0 & k_3 \end{bmatrix} \begin{Bmatrix} \delta_1 \\ \delta_2 \\ \delta_3 \end{Bmatrix} \quad (16)$$

Here, the symbol (^) denotes actual stress.

3.2.3. Foam Damage Model

The crushable foam model [27] is used to solve the problems of material strength and damage by making the filled foam equivalent to macroscopic homogeneous continuous material. In this paper, as shown in Figure 11, an isotropic hardening compressible foam model is used to describe the mechanical behavior of the filled foam under a high-speed impact load. The yield stress expression is as follows: it is an ellipse centered at the stress plane p–q origin and evolves in a self-similar manner.

$$F = \sqrt{q^2 + \beta^2 (p - p_0)^2} - B = 0 \quad (17)$$

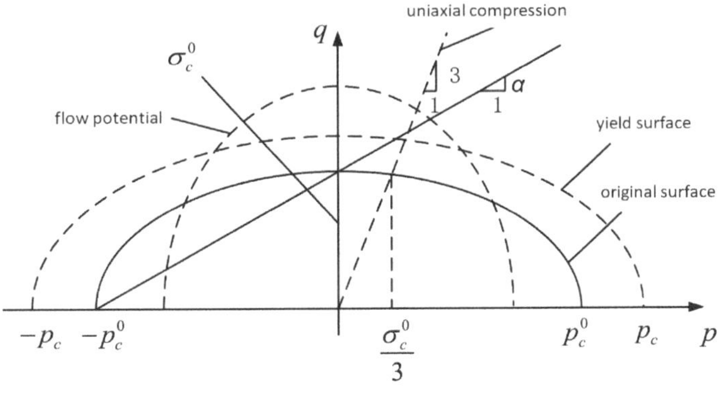

Figure 11. The compression process of the model foam of each isotropic.

Here, p is hydrostatic pressure; q is Mises stress; β is the ellipse shape factor; B is ellipse q axis length; and p_0 is an ellipse major axis center.

Shear plugging is the main damage mode of foam under high-speed impact load [28], so the shear damage criterion is adopted as the damage criterion. This criterion is a high strain rate deformation design, using equivalent plastic strain as the failure measure, and can be used in conjunction with Mises or Johnson–Cook plastic models. The shear stress ratio in this guideline is defined as follows:

$$\theta_S = \frac{q + k_S p}{\tau_{max}} \quad (18)$$

Here, q is Mises equivalent stress; k_s is a material parameter; the value of closed-cell foam is 0 [27]; p is hydrostatic pressure; and τ_{max} is the maximum shear stress.

4. Results and Discussion

4.1. Experiment Result Analysis

As can be seen from the ballistic results of Figure 12, the bullet passes through the blade airfoil section quickly, leaving holes in the upper and lower airfoil surfaces. The failure of composite material blades is mainly divided into three parts. Part one is that the composite skin is torn apart, and the main failure mode is fiber breakage. Part two is the delamination caused by bonding failure, including the delamination between skins and between skins and foams. Part three is the foam collapse, which is in the shape of powders and loses its bearing capacity.

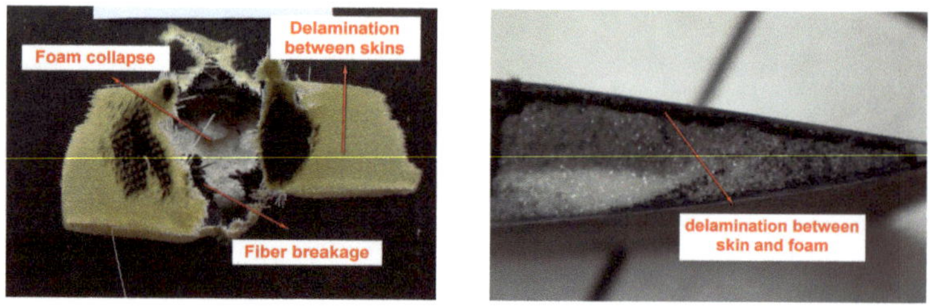

Figure 12. Ballistic failure model of the composite blade.

4.2. Model Verification

4.2.1. Comparison of Damage Morphology

Figures 13 and 14 show the comparison between the damage experimental results and numerical results at the exit of the blade after being impacted by a bullet under three experimental conditions. The simulation results can predict the damage mode of the blade. According to the experimental results, the damage to the blade is mainly in the form of skin delamination, fiber breakage, and foam collapse. The damage characteristics at the exit of the skin obtained by simulation are in good agreement with the results obtained by the ballistic experiment.

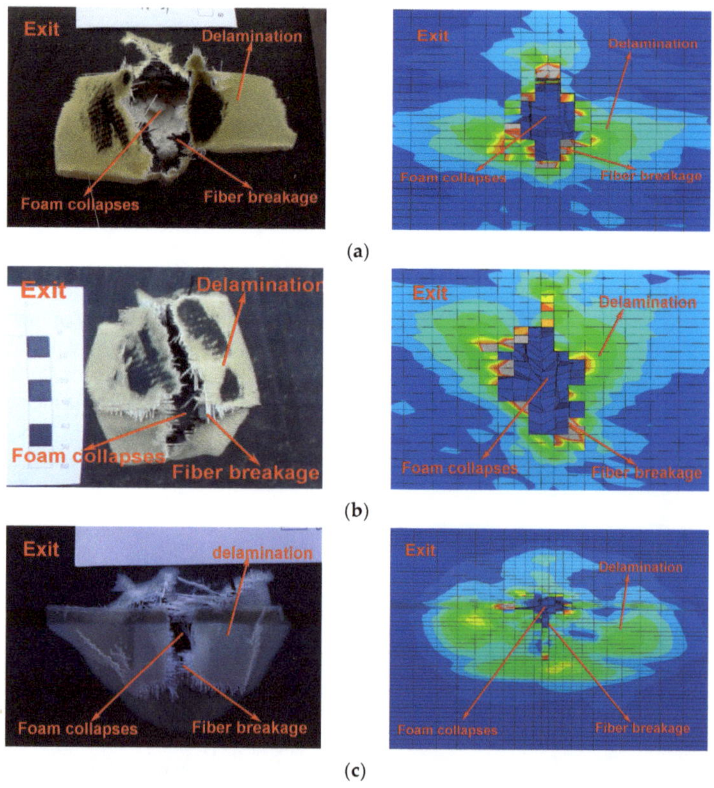

Figure 13. Comparison of damage morphology: (**a**) case 1, (**b**) case 2, and (**c**) case 3.

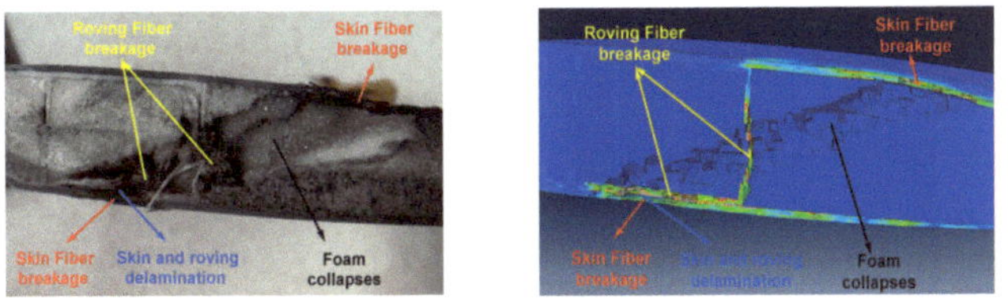

Figure 14. Cross-sectional damage comparison.

4.2.2. Comparative Analysis of the Damage Range

As shown in Figure 15, the size of the damage notch in the simulation analysis and the ballistic experiment are compared. The damage notches on the surface of the skin show that the simulation results are in good agreement with the experiment results. The numerical simulation model can accurately predict the damage range of composite blades subjected to ballistic impact.

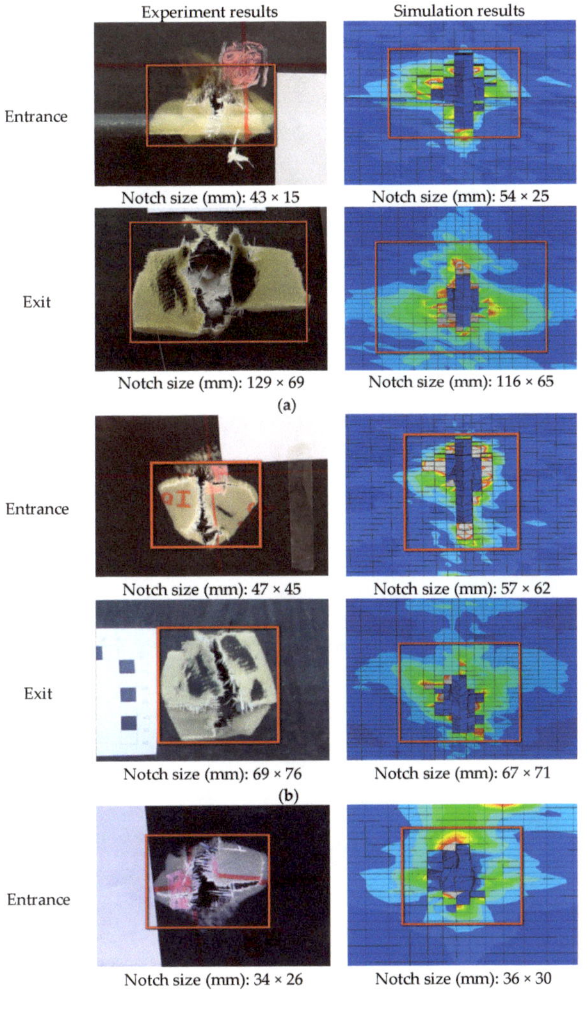

Figure 15. *Cont.*

Exit

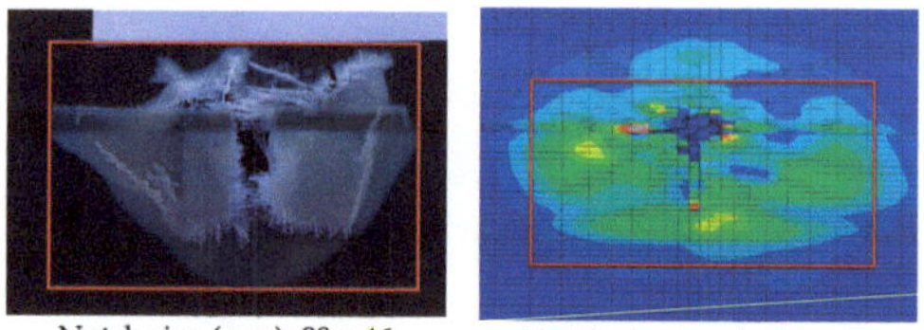

Notch size (mm): 89 × 46 Notch size (mm): 110 × 56

(c)

Figure 15. Comparison of damage range between experiment and simulation results: (**a**) case 1, (**b**) case 2, and (**c**) case 3.

5. Analysis of the Bullet Penetration Process

The ballistic process of composite blades is a transient dynamic process. The damage to the blade is determined by the local properties of the structure near the impact zone. There are different failure modes of the composite blade during ballistic penetration, including fiber tensile failure, delamination failure, and foam collapse. Through the finite element model, the deformation, damage expansion, and delamination of the blade under the bullet penetration process can be analyzed in detail. Figures 16–18 show the penetration process of bullets under three experimental conditions. With a speed of 650 m/s, the duration of the bullet penetration process is less than 400 μs. During the projectiles, different damage modes occur, promote, and couple in succession.

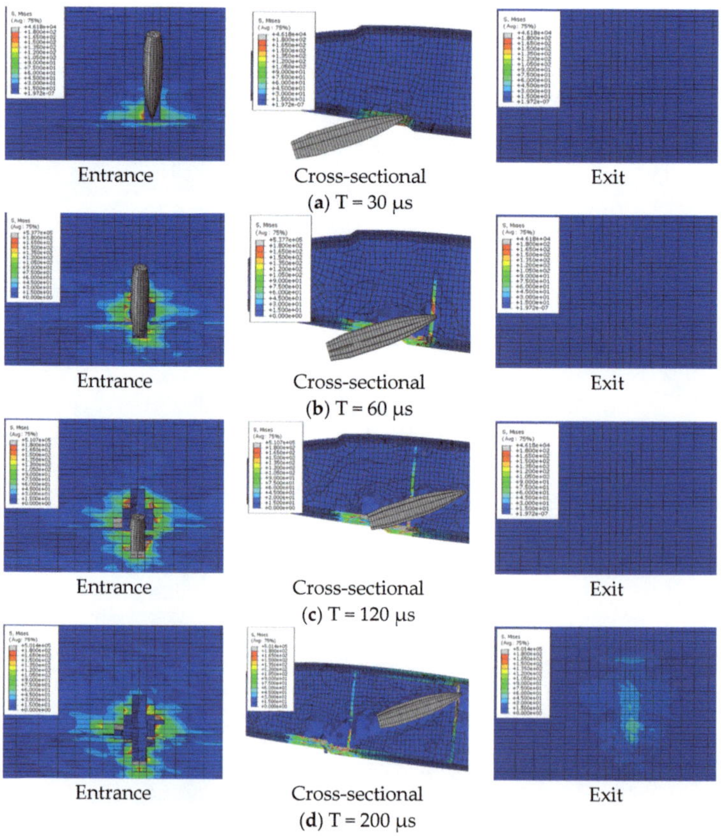

Figure 16. *Cont.*

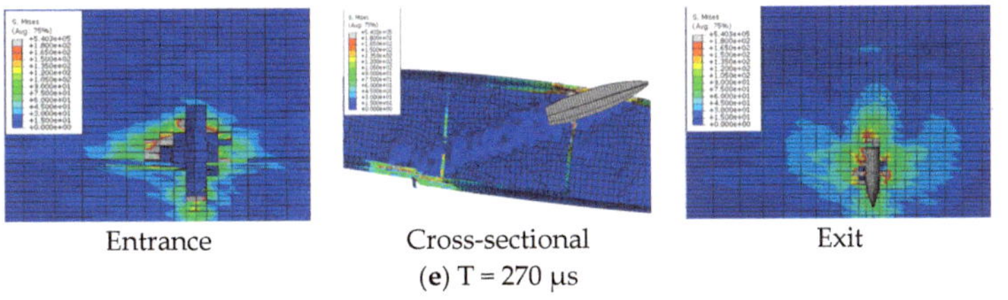

Figure 16. Ballistic penetration process of case 1.

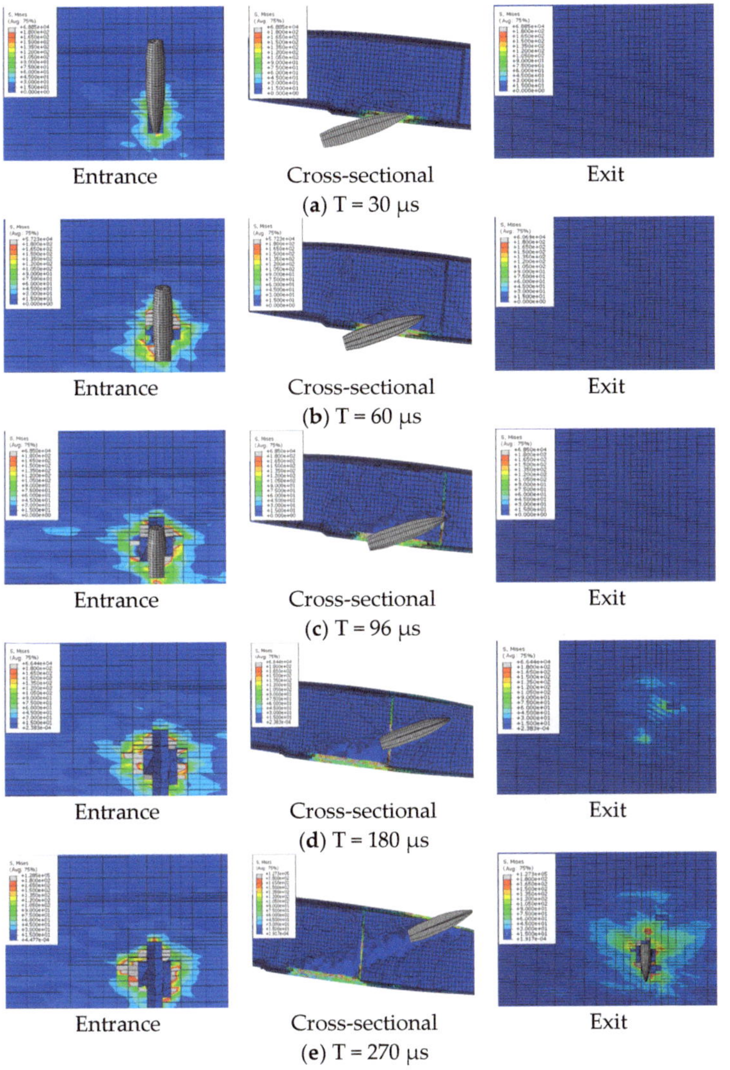

Figure 17. Ballistic penetration process of case 2.

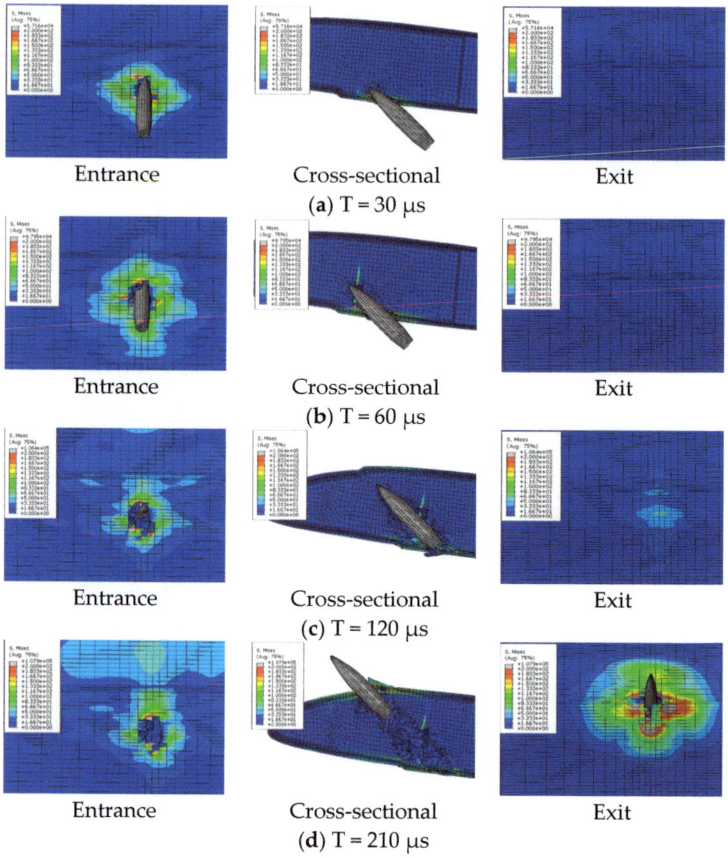

Figure 18. Ballistic penetration process of case 3.

5.1. Experiment Result Analysis

The bullet penetration process under three conditions can be divided into four stages: incident skin failure stage, foam failure stage, reinforced rib failure stage, and exit skin failure stage.

(1) The incident skin failure stage is mainly the failure damage and damage expansion of the lower wing surface skin. The bullet with high kinetic energy exerts strong pressure on the skin, but due to the dispersion of the load by the foam core, the skin at the entrance experiences local shearing-type failure only in the contact area. Meanwhile, the foam can be observed to produce a small range of collapse, which is due to the fiber block of the fractured skin at the entrance due to the extrusion under the push of the projectile.

(2) The foam sandwich core failure stage is mainly caused by foam press damage. The material stress in a very small strain range increases rapidly. The foam generates collapse and fails to damage. The ballistic hole of the foam core part is almost as wide as the bullet body at this time. It indicates that the failure mode is mainly local shear fracture.

(3) The failure stage of the reinforcing rib is mainly the destruction and damage expansion of the reinforcing rib and foam. The bullet passes through the foam to impact the reinforcing rib, and the reinforcing rib bears strong tensile stress. The failure mode is fiber breakage.

(4) The failure stage of the skin at exit is mainly the failure and damage expansion of the skin. The skin at the exit deformed greatly in the process of destruction before the bullet contacted the skin. Due to the accumulation of foam on the front side of the ballistic

and extruding the outer skin and the transmission of impact stress waves, the skin at exit already has protrusion deformation, which leads the back fiber to bear strong compressive stress. The stress state leads to the origin of the outer skin damage, and the failure mode is different from the incident skin. The maximum stress appears on the outer surface. The skin begins to break from the outermost side, and the fiber tensile fracture is the main damage.

Because the skin at the exit has certain stress and deformation before the bullet arrives, and stress is further increased and diffused when the bullet penetrates the skin, the damage size of the skin at the exit is much larger than that of the skin at the entrance. As can be seen from Figure 13, in the three projectile conditions, the ballistic damage size of the exit port can reach six times the ballistic damage size of the entrance port.

As can be seen from the impact penetration paths of the three experimental conditions, the bullet always passes obliquely through the leading and trailing edges of the blade profile. Due to the large bullet size (diameter 12.7 mm, length 65 mm) and blade size (chord length 500 mm, airfoil thickness 60 mm), one impact will only cause local damage to the upper and lower airfoil surfaces. The distribution design of the upper and lower airfoil surfaces should be considered in the design of the main bearing structure. The spar and the skin-reinforcing cloth should not be concentrated on the front edge of the paddle. To avoid large damage to the bearing structure caused by one bullet, the spar and the skin reinforcing cloth should be distributed to the upper wing surface and the lower wing surface as much as possible. More reinforcing ribs bearing shear load should be designed in the section to form multiple closed cavities to ensure that all shear load-bearing structures will not be damaged by one shot.

5.2. The Transmission of Stress Waves during Bullet Penetration

During the bullet penetration process, the blade structure experiences both the direct impact effect on the bullet penetrating path and the transmission of stress waves caused by the high-speed impact response within the blade. As shown in Figure 19, these stress waves propagate through spars, reinforcing ribs, and other components, resulting in significant stress within the non-shock penetration region. While the stress values may not immediately cause structural damage or invalidation, it is important to consider that the blade is in a non-loaded state during simulation analysis. Therefore, under actual operational conditions, the loads may cause damage outside the trajectory.

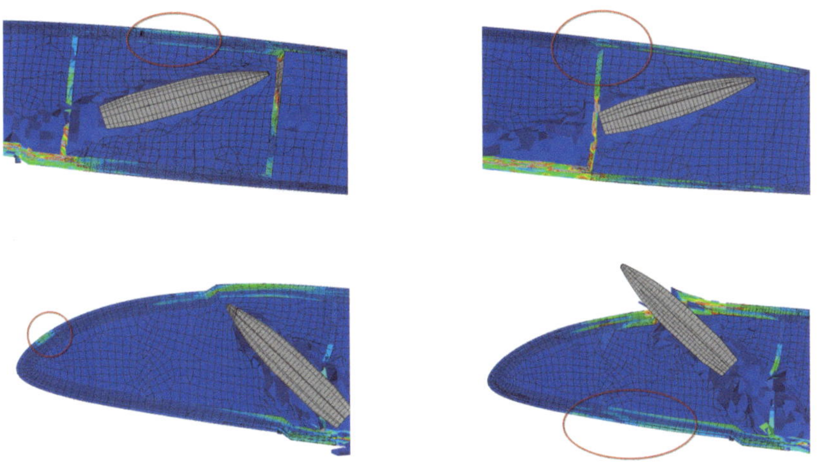

Figure 19. Stress on the non-impact path.

5.3. Analysis of Factors Influencing the Range of Ballistic Damage

5.3.1. Analysis of Incident Angles

Cases 2 and 3 are in the same incident position but have different incident angles. Due to the different impact angles, the damage to the incident port is different (shown in Figure 20). From the angle of the incident port damage size, the larger the incident angle formed with the skin plane, the greater the damage caused. Relatively speaking, the vertical entry results in minimal damage.

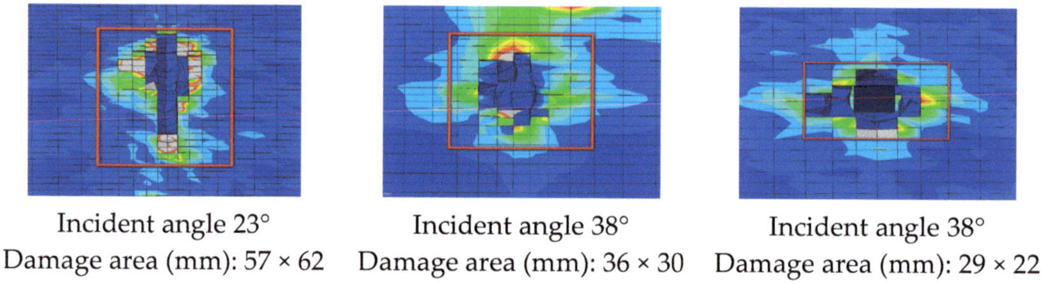

Figure 20. Comparison of the damage at the exit of different incident angles.

Due to the curved surface shape and complex internal structure of the blade, it is not similar to a foam sandwich board. The interior is filled with foam but features structures such as spars and reinforcing ribs. Therefore, different incident angles directly lead to different ballistic paths. As a result, the damage caused by different load-bearing structures along the path varies (shown in Figure 21). In case 2, the incident angle is obliquely pointed towards the rear edge, so the ballistic path of the bullet is only the reinforcing rib in the middle of the blade section. This results in a small damage range during the bullet impact process. On the other hand, in case 3, the incident angle points to the position of the front edge spar, which is the main load-bearing structure, causing more damage and a wider trajectory.

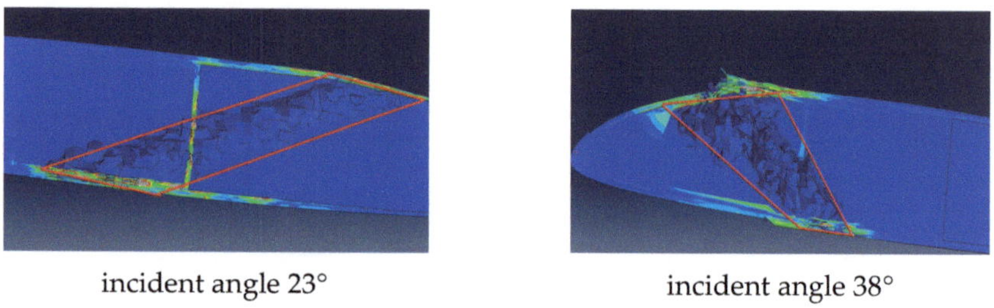

Figure 21. Comparison of trajectory damage at different impact angles.

5.3.2. Analysis of Structural Characteristics

As can be seen from Figure 22, the incident angle in case 1 and case 2 is the same, but the skin damage range of the exit opening is quite different. The damage size of the exit opening in case 1 is 116 mm × 65 mm, while it is 67 mm × 71 mm in case 2.

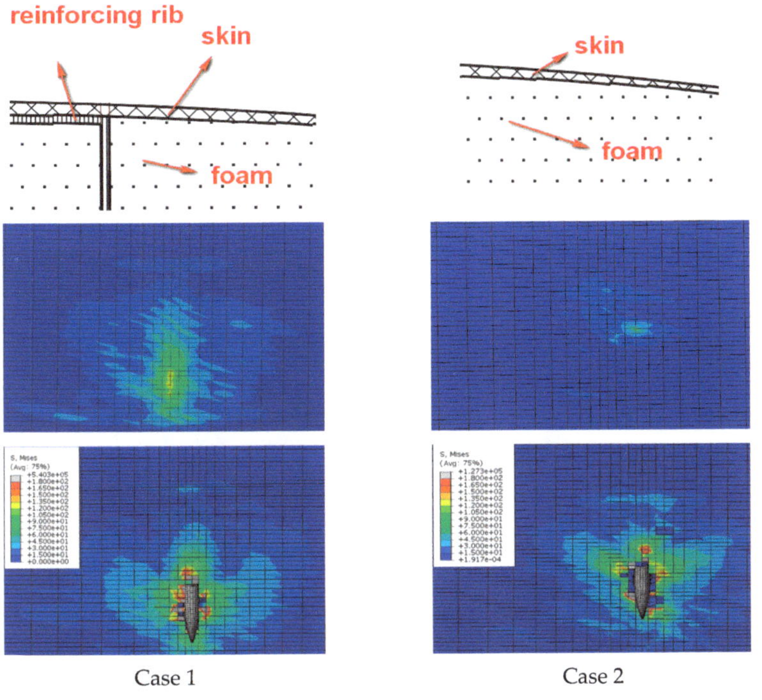

Figure 22. Comparison of ballistic damage at different structural characteristics.

The comparative analysis of the structure at the exit in case 1 and case 2 shows that the structure at the exit of case 1 consists of skin and spar. The structure at the exit of case 2 consists of only skin. The structure at the exit of case 1 has greater stiffness than that of case 2. Therefore, in the structure at the exit of case 1, before touching the bullet, the stress caused by foam extrusion and stress wave transmission is greater, the range is wider, and the bullet can absorb more energy during bullet penetration, so the bullet damage is greater. That is to say, the more bearing structure and stiffness near the exit, the larger the range of ballistic damage.

6. Conclusions

In this paper, the damage characteristics of the composite rotor blade under ballistic impact are studied by experiments. A numerical simulation model considering a three-dimensional progressive damage failure model that can simulate and analyze the impacted blade ballistic damage is established. The simulation results are compared with the experimental results to verify the accuracy of the presented numerical simulation model. Furthermore, the ballistic performance of composite blades impacted by different incident angles and different locations is systematically studied. The conclusions are as follows:

(1) The main damage forms of the impacted composite blade by the bullet are fiber breakage, delamination, and foam collapse. The damage size and characteristics of the numerical results, which are calculated by the established numerical simulation model, are consistent with the experimental results under different ballistic conditions.

(2) Through the bullet penetration process, the form and mechanism of blade damage vary. The damage of skin at the entrance is mainly local shear failure caused by short-term strong pressure. Foam failure is mainly caused by the rapid increase of stress in a very small strain range. The damage of the reinforced rib and the skin at the exit is mainly due

to the tensile fracture of the back fiber caused by strong compressive stress. This leads to the damage size of the exit being generally larger than that of the entrance.

(3) The damage size of the impacted blade at the entrance is mainly related to the incident angle of the bullet. The larger the incident angle, the smaller the impact damage size at the entrance. The impact damage size of the exit is mainly related to the structural characteristics of the structure. The more load-bearing structures and the greater the stiffness at the exit, the larger the impact damage size at the exit.

(4) Impact by one bullet will only cause damage to the local positions of the upper and lower airfoil surfaces. Therefore, in the design of the blade structure, the distribution design of the main load-bearing structures along the airfoil surfaces should be considered, and more reinforcing ribs for bearing shear load should be designed to form a plurality of closed cavities to reduce the damage to the main bearing structure of the blade impacted by one bullet.

(5) Due to the extrusion and the transmission of the stress wave of the foam, the skin at the exit has great stress and deformation, resulting in greater damage at the exit. The material with lower consistency, such as honeycomb material, should be selected as far as possible as the filling material of the rotor blade. This makes the bullet pass through the material quickly without causing material accumulation and extrusion, thereby reducing the damage to the skin at the exit.

We have shown that the proposed numerical model is effective in predicting the ballistic performance and analyzing the failure mechanism of the composite blade under various ballistic impact conditions. It provides a theoretical basis and reference for the anti-ballistic design and ballistic performance analysis of composite blades.

Author Contributions: Conceptualization, G.Y. and W.H.; methodology, G.Y., X.L. and W.H.; software, G.Y. and X.L.; validation, G.Y. and X.L.; formal analysis, G.Y. and X.L.; investigation, G.Y., X.L. and W.H.; resources, W.H.; data curation, G.Y. and X.L.; writing—original draft preparation, G.Y., X.L. and W.H.; writing—review and editing, G.Y., X.L. and W.H.; visualization, G.Y. and W.H.; supervision, W.H.; project administration, G.Y. and W.H.; funding acquisition, W.H. All authors have read and agreed to the published version of the manuscript.

Funding: This research received no external funding.

Institutional Review Board Statement: Not applicable.

Data Availability Statement: Data are contained within the article.

Acknowledgments: We would like to thank Zhou for providing English writing assistance.

Conflicts of Interest: The authors declare no conflicts of interest.

Dual-use Research Statement: We have made the following statements to our paper titled "Performance and Damage Study of Composite Rotor Blades under Impact".

- Ø Explanation of Potential Risks: Our paper examines the performance of composite rotor blades under impact. The research is limited to providing some theoretical and experimental support for the composite rotor blades under impact only and does not pose a threat to public health or national security.
- Ø Evaluation of Benefits to the General Public: Our research is limited to the academic field, which is beneficial to the development of composite rotor blades under impact. There is no risk to the general public.
- Ø Compliance with Laws: As an ethical responsibility, we strictly adhere to relevant national and international laws about dual-use research. And we have considered and adhered to these regulations in our paper.

References

1. Jenney, D.S. *Rotor Technology for New Generation Helicopters*; SAE Technical Paper; SAE International: Warrendale, PA, USA, 1975.
2. Foulk, J.B. Survivability of the Army/Sikorsky YUH-60A Helicopter. *Preprint* **1976**, *101*, 1–21.
3. Brown, W.P. AH-64A Apache-Battle ready. In Proceedings of the Ninth European Rotor Craft Forum, Stresa, Italy, 13–15 September 1983.

4. Liardon, D.; Johnson, J.; Holtrop, J. V-22 ballistic Vulnerability Hardening Program. In Proceedings of the AHS, Annual Forum, 45th, Boston, MA, USA, 22–24 May 1989.
5. McCann, M. *Test Results of the RAH-66 Main Rotor Ballistically Damaged Flexbeam*; Defense Technical Information Center: Fort Belvoir, VA, USA, 1997.
6. Scarpati, T.; Feenan, R.; Stratton, W. The Results of Fabrication and Testing of the Prototype Composite Rotor Blades for HLH and UTTAS. In Proceedings of the Aircraft Systems and Technology Meeting, Los Angeles, CA, USA, 4–7 August 1975; Volume 1010.
7. Griffin, C. *Increased Rotor Blade Survivability*; ADA016929; Lockheed-California Co.: Burbank, CA, USA, 1975; pp. 1–88.
8. Edward, K. *Ballistically Tolerant Rotor Blade Investigation*; Eustic Directorate; US Army Air Mobility Research and Development Laboratory: Adelphi, MD, USA, 1975.
9. Aubry, J.; Navarro, P.; Tawk, I.; Marguet, S.; Ferrero, J.F.; Lemaire, S.; Rauch, P. Experimental and Numerical Study of Normal and Oblique Impacts on Helicopter Blades. In *Dynamic Failure of Composite and Sandwich Structures*; Springer: Dordrecht, The Netherlands, 2013; pp. 545–575.
10. Stahlecker, Z.; Centolanza, L. Macro-mechanical Modeling Approach for Simulating ballistic Damage to Composite Rotor blades. In Proceedings of the AHS, Annual Forum, 67th, Virginia Beach, VA, USA, 3–5 May 2011.
11. Shipsha, A.; Hallström, S.; Zenkert, D. Failure mechanisms and modelling of impact damage in sandwich beams–A 2D approach: Part I-experimental investigation. *J. Sandw. Struct. Mater.* **2003**, *5*, 7–31. [CrossRef]
12. Wang, H.; Weerasinghe, D.; Hazell, P.J.; Mohotti, D.; Morozov, E.V.; Escobedo-Diaz, J.P. Ballistic impact response of flexible and rigid UHMWPE textile composites: Experiments and simulations. *Def. Technol.* **2023**, *22*, 37–53. [CrossRef]
13. Yashiro, S.; Ogi, K. High-velocity impact damage in CFRP laminates. In *Dynamic Deformation, Damage and Fracture in Composite Materials and Structures*; Woodhead Publishing: Duxford, UK, 2023; pp. 141–164.
14. Bernard, M.; Lagace, P. Impact resistance of composite sandwich plates. *J. Reinf. Plast. Compos.* **1989**, *8*, 432–445. [CrossRef]
15. Lang, U.; John, M.; Schlimper, R.; Schäuble, R. Damage-tolerant CFRP-Foam Core Sandwich Structures. *Lightweight Des.* **2016**, *9*, 18–25. [CrossRef]
16. Nasirzadeh, R.; Sabet, A. Study of foam density variations in composite sandwich panels under high velocity impact loading. *Int. J. Impact Eng.* **2014**, *63*, 129–139. [CrossRef]
17. Ghalami-Choobar, M.; Sadighi, M. Investigation of high velocity impact of cylindrical projectile on sandwich panels with fiber–metal laminates skins and polyurethane core. *Aerosp. Sci. Technol.* **2014**, *32*, 142–152. [CrossRef]
18. Acanfora, V.; Zarrelli, M.; Riccio, A. Experimental and numerical assessment of the impact behaviour of a composite sandwich panel with a polymeric honeycomb core. *Int. J. Impact Eng.* **2023**, *171*, 104392. [CrossRef]
19. Albayrak, M.; Kaman, M.O.; Bozkurt, I. Experimental and numerical investigation of the geometrical effect on low velocity impact behavior for curved composites with a rubber interlayer. *Appl. Compos. Mater.* **2023**, *30*, 507–538. [CrossRef]
20. Usta, F.; Turkmen, H.; Scarpa, F. High-velocity impact resistance of doubly curved sandwich panels with re-entrant honeycomb and foam core. *Int. J. Impact Eng.* **2022**, *165*, 104230. [CrossRef]
21. Alonso, L.; Solis, A. High-velocity impact on composite sandwich structures: A theoretical model. *Int. J. Mech. Sci.* **2021**, *201*, 106459. [CrossRef]
22. Alonso, L.; Garcia-Gonzalez, D.; Martinez-Hergueta, F.; Navarro, C.; Teixeira-Dias, F.; García-Castillo, S.K. Modeling high velocity impact on thin woven composite plates: A non-dimensional theoretical approach. *Mech. Adv. Mater. Struct.* **2022**, *29*, 2780–2794. [CrossRef]
23. Alonso, L.; Solis, A.; García-Castillo, S. A numerical-analytical study to determine a suitable distribution of plies in sandwich structures subjected to high-velocity impact. *Compos. Struct.* **2023**, *307*, 116645. [CrossRef]
24. Hashin, Z. Failure criteria for unidirectional fiber composites. *J. Appl. Mech.* **1980**, *47*, 329–334. [CrossRef]
25. Falkowicz, K. Experimental and numerical failure analysis of thin-walled composite plates using progressive failure analysis. *Compos. Struct.* **2023**, *305*, 116474. [CrossRef]
26. Gu, X.; Xu, X.W. Numerical simulation of damage in fiber reinforced composite laminates under high velocity impact. *Acta Mater. Compos. Sin.* **2012**, *29*, 150–161.
27. Camanho, P.; Davila, C. *Mixed-Mode Decohesion Finite Elements for the Simulation of Delamination in Composite Materials*; NASA/TM-2002-211737; NASA: Hampton, VA, USA, 2002.
28. Zhou, R. *Damage Detection Stitched Foam Core Sandwich Composites under Low Velocity Impact*; Nanchang University: Nanchang, China, 2016.

polymers

MDPI

Article

Evaluation of Fatigue Damage Monitoring of Single-Lap Composite Adhesive Joint Using Conductivity

Chow-Shing Shin * and Shun-Hsuan Huang

Department of Mechanical Engineering, National Taiwan University, No. 1, Sec. 4, Roosevelt Road, Taipei 10617, Taiwan
* Correspondence: csshin@ntu.edu.tw; Tel.: +886-2-33662724

Abstract: The widely used adhesive joining technique suffers from the drawback of being unable to be dismantled to examine for degradation. To counteract this weakness, several structural health monitoring (SHM) methods have been proposed to reveal the joint integrity status. Among these, doping the adhesive with carbon nanotubes to make the joint conductive and monitoring its electrical resistance change is a promising candidate as it is of relatively low cost and easy to implement. In this work, resistance change to monitor fatigue debonding of composite single-lap adhesive joints has been attempted. The debonded area, recorded with a liquid penetrant technique, related linearly to the fatigue life expended. However, it correlates with the resistance change in two different trends. Scanning electron microscopy on the fracture surface reveals that the two trends are associated with distinct failure micromechanisms. Implications of these observations on the practical use of the resistance change for SHM are discussed.

Keywords: adhesive joint; carbon nanotubes; fatigue debonding; electrical resistance change; structural health monitoring

Citation: Shin, C.-S.; Huang, S.-H. Evaluation of Fatigue Damage Monitoring of Single-Lap Composite Adhesive Joint Using Conductivity. *Polymers* **2024**, *16*, 2374. https://doi.org/10.3390/polym16162374

Academic Editors: Xiaoquan Cheng, Wenjun Huang and Qian Zhang

Received: 2 July 2024
Revised: 14 August 2024
Accepted: 16 August 2024
Published: 22 August 2024

1. Introduction

Adhesive bonding possesses a number of advantages over conventional joining methods. It transfers load over a large area, significantly reducing stress and leading to better stiffness. There is no localized heating that may degrade the material, as in the case of weld joining. It avoids reducing the load-bearing section of the structure, introducing stress concentration, and adding extra weight, as in the case of bolt or rivet joints. For fiber-reinforced composite materials, circumventing the hole drilling helps to avoid fiber discontinuity and drilling-induced delamination. However, unlike a bolted joint, an adhesive joint cannot be dismantled to inspect for service-induced defects and degradation. Service loading conditions such as impact, occasional overload, and long-term fluctuating loading may induce joint degradation/damages that may develop into eventual structural failure.

To counteract this weakness, non-destructive evaluation (NDE) techniques have been proposed to reveal defects in adhesive joints. These include various techniques that involve acoustic waves, such as the ultrasonic pulse-echo methods [1–3], guided wave [4,5], phase array [6,7], acoustic microscopy [8,9], and electromagnetic acoustic transducer [10]. Non-acoustic techniques such as electromechanical impedance spectroscopy using external [11] or embedded piezoelectric sensors [12–14], thermography [3,15–17], and shearography [18,19] have also been proposed. These NDE techniques are helpful in detecting defects at the fabrication stage or late stage of failure but are relatively ineffective for bonds weakened by degradation during the service stage [20,21]. Moreover, periodic inspection using these techniques for large-scale adhesively joined structures will be very time-consuming and prohibitively expensive.

On the other hand, structural health monitoring (SHM) methods have been proposed to keep track of properties that may reflect the load-carrying capability or defect development. A number of principles have been employed for this purpose. The strain/stiffness at the joint has been monitored with back face strain gages [22,23], digital image correlation [24,25], and optical fibers with distributed [26,27] or discrete [28,29] sensors. Local strain perturbation by the initiation and development of internal defects may be monitored by the change in the full spectral shape of fiber Bragg gratings [30–33]. Acoustic emission offers another possibility for detecting the occurrence and growth of bond defects [15,34]. Perturbation in electrical conductivity/impedance has also been proposed as a means to monitor the integrity of adhesive joints. Among these techniques, conventional strain gages are prone to induce delamination if embedded inside a joint. When applied to the exterior, they disrupt an otherwise smooth surface and are susceptible to environmental degradation. Optical fibers have excellent fatigue endurance [35] and can be embedded inside a joint without causing adverse effects on the structural integrity. However, the sensing region is limited to a close vicinity of the fiber sensor. Moreover, the equipment involved in fiber sensing and acoustic emission is relatively expensive. Conductivity measurements offer a low-cost alternative that can surveil a relatively large region.

The use of electrical properties to monitor the integrity of a composite has been around for some time [36–42]. Early works made use of conductive carbon fiber [36] or CNT-coated glass fiber [37,38]. Conductive nanoparticles have also been dispersed in resin for this purpose [39,40]. The most popular conductive particle employed is carbon nanotubes, while graphene [41] and carbon black [42] have also been used. The latter approach has also been employed for adhesive joint monitoring. After doping with conductive particles, the adhesive was either made into a thin adhesive film [24,43–46] or applied directly [15,47–52] to fabricate the joint. Conductive sensors such as bucky paper [53], aligned carbon nanotube web [54], inkjet-printed silver nanoparticle interdigital sensors [55], and metal-pinned hybrid composite–titanium joints have also been investigated. The percentage change in D.C. resistance is normally measured in these works, but a few studies used A.C. to measure impedance [48,52]. The above works showed clearly that electrical resistance or impedance changes are associated with straining as well as the initiation and growth of a crack or debonding. The latter fact offers good potential for SHM using conductivity/resistance.

However, the considerable amount of work in this area still needs to be improved when the practical application of the technique is intended. Firstly, current results are qualitative and of a proof-of-concept nature. The practical application of this technique requires more concrete quantitative relations. Secondly, most of the current works demonstrated the phenomenon on a very limited number of specimens, and the reproducibility of results has not been thoroughly investigated. Thirdly, most of the works measured resistance under an increasing straining deformation. Resistance changes in a CNT-doped adhesive joint may result from one of the following mechanisms: (1) change in the amount of direct contact of the overlapping nanotubes [56,57]; (2) change in the tunneling resistance resulting from changes in inter-tube distance between non-touching neighboring CNTs [57–59]; (3) the intrinsic piezoresistivity of individual CNTs due to strain-dependent energy band gap opening [60,61]; (4) reduction in conductive cross-sectional area due to cracking and debonding growth. The first three mechanisms constitute the piezoresistive effects of the adhesive and will operate under straining without the existence of damage. Internal damage such as microcracking that leads to geometrical changes in the CNT network and residual stress redistribution can also operate the first three mechanisms. Debonding between the adhesive and adherend operates the fourth. When resistance measurement is made under load, both the contributions from straining and possible damages occur. For practical SHM applications, the isolation of the contribution from damages to reflect only structural degradation is needed.

The current work looks into the possibility of establishing a quantitative relationship between resistance change and debonding crack growth under cyclic loading. To understand the reproducibility of results, fatigue tests have been repeated on 30 specimens. Resistance changes are measured under load-free or a very small load condition to alleviate any piezoresistive effect. It was found that resistance change does not form a unique relationship with the debonded area. Fractographic evidence offered a reasonable explanation for this observation.

2. Materials and Methods

This investigation involved the preparation of conductive adhesive single-lap-joint specimens. The specimens were fatigue-tested while their change in conductivity was monitored. The amount of debonding at some point in these tests was also recorded using a fluorescent penetrant. Finally, the fractured surfaces are examined with a scanning electron microscope (SEM) to reveal the micromechanisms of failure and to provide more understanding of the relationship between conductivity change, debonded area, and fatigue life consumption. Experimental details about the key steps are given in the following sections.

2.1. Lap-Joint Specimen Preparation

The 210 mm × 210 mm $[0°]_{10}$ uni-directional graphite-fiber/epoxy laminates were fabricated using an autoclave vacuum bag technique at 145 °C and 10 kg/cm^2. The graphite-fiber/epoxy prepreg (Formosa Taffeta Co., Ltd., ECU431, Douliu, Taiwan) employed has a fiber volume fraction of 63%. Copper foils were pre-embedded in the mid-layer at both ends of the laminate to facilitate later conductivity measurement. From the laminates, 25.4 mm wide strips were cut to produce single-lap-joint specimens with dimensions shown in Figure 1. A batch of seven single-lap-joint specimens can be made from each composite laminate. Adhesive joint preparation followed the procedures outlined in the ASTM D5868-01 [62].

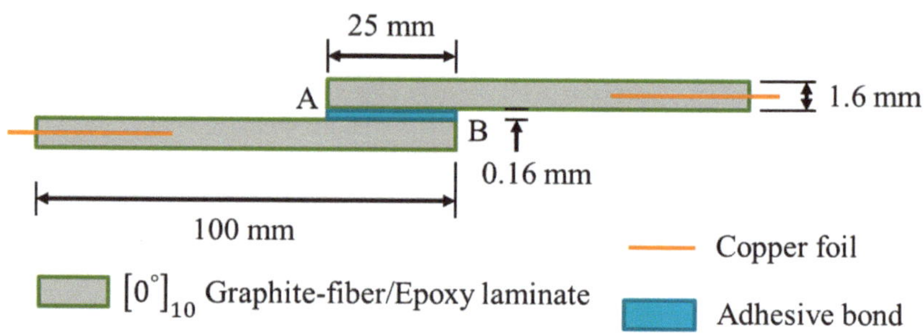

Figure 1. Dimensions and layout of the single-lap-joint specimen, A and B are the two longitudinal edges of the joint.

The area to be joined was sandblasted with carborundum (CES012, Phasic Corp., Yuanlin, Taiwan), which had a mean diameter of 106–125 μm. Three 125 μm diameter optical fibers running along the specimen loading axis were embedded as spacers to control the bond line thickness to ~160 μm. The fibers were evenly spaced across the specimen width. Masking tape was applied in the immediate vicinity beyond the boundary of the joint area to prevent excess glue from forming additional but unpredictable adhesion between the two adherends.

Room-temperature-cured epoxy resin (Swancor 2261-A/BS, Swancor, Industrial Corp., Nantou, Taiwan) was made conductive by blending 0.3 wt% of amino-functionalized multi-walled carbon nanotubes (CNTs) into the resin. The manufacturer-provided specifications of the raw resin are listed in Table S1 in the Supplementary Materials. The CNTs (MWCNT-NH3-STD, Euflex Technology Corp., New Taipei City, Taiwan) have a nominal diameter of 9.5 nm and a nominal length of 1500 nm. To ensure proper dispersion, the mixture underwent sonication using an ultrasonic homogenizer (UP200S, Hielscher Ultrasonics, Teltow, Germany) at 100 W for 5 min. It was further homogenized with a high-speed homogenizer (MiniBatch D-9, MICCRA, Heitersheim, Germany) at 21,000 rpm for 10 min. The mixture was then cooled in iced water while the hardener was added and mixed. The resulting mixture, still buffered in iced water, underwent degassing in a vacuum for one hour before application. The adhesive joints were left to cure at room temperature for 24 h. Afterwards, the specimens were post-cured for 2 h at 100 °C in an oven (DOS45, Deng-Yng Corp., New Taipei City, Taiwan).

2.2. Conductivity Measurement

Lead wires were soldered to the exposed copper foils, which were partly embedded in the composite laminate at both ends of the single-lap-joint specimens. A constant current from a source meter (Keithley 2450, Tektronix, Inc., Beaverton, OR, USA) was applied across the specimens through the lead wires. A current of 5 mA resulted in an initial voltage drop of ~3 V, suggesting that the initial resistance of the specimen from one end to the other was ~600 Ω. A 175 mm long composite strip of the same width has a resistance of ~3 Ω. Thus, the adhesive is the primary resistance contributor to the lap-joint specimen. Changes in the corresponding voltage drop were recorded, which reflected changes in the adhesive joint resistance. Preliminary tests showed that the voltage drop, or resistance, across the joint increases with increasing load. To reflect the contribution of damage and exclude the effect of load, the specimens were brought to 500 N and 0 N every time when the voltage drop was to be recorded.

2.3. Mechanical Testing

The specimens underwent either tensile or cyclic fatigue loading using a servo-hydraulic testing machine (810 Materials Testing System, MTS Systems, Eden Prairie, MN, USA).

From each batch of seven lap-joint specimens, two specimens were tested under tensile monotonic loading to failure to obtain the average batch tensile strength. Different specimen batches had average batch strengths from 7.254 to 7.986 kN. However, preliminary tests showed that within the same batch, the worst-case standard deviation of tensile strength was within 4.75%.

Sinusoidal cyclic loading between 5.5% and 55% of the average batch tensile strength at 5 Hz was employed for fatigue testing. The voltage drop across the specimens was monitored throughout testing. A total of 30 specimens were tested, with an average fatigue life of 26,651 cycles and a standard deviation of 11,310 cycles. Inherently fatigue failure is a stochastic event with considerable scatter. Treatment with a liquid penetrant at different stages of life may also introduce some differences.

2.4. Liquid Fluorescent Penetrant Treatment

For some of the specimens, the fatigue test was interrupted, and the specimens were infiltrated with a fluorescent penetrant (Metl-Chek FP-923, McGean, Cleveland, OH, USA). The purpose of this treatment is to mark the instantaneous state of joint debonding. The treatment included immersing the specimens into the penetrant for 10 min under 0 N, followed by 20 min under 50 N. Afterward, the specimen surface was wiped dry and clean using a penetrant remover (Met-l-Chek E-59, McGean, USA). Subsequently, the specimens were baked for 48 h at 100 °C in the same oven used for post-curing to dry any entrapped penetrant thoroughly.

Fatigue testing was then resumed until specimen fracture. The fracture surface was photographed under ultraviolet (UV) illumination. The resulting images were analyzed using ImageJ 1.54J [63] to compute the fluorescent area.

2.5. Fractographic Observation

Representative joint fracture surfaces were gold-coated and examined using a scanning electron microscope (TM3000 tabletop SEM, Technologies Corp., Hitachi, Tokyo, Japan) to evaluate the microscopic failure mechanisms.

3. Results

3.1. Voltage Drop across the Adhesive Joint during Fatigue Testing

The percentage voltage change is defined as the following:

$$\% \text{ voltage change} = \frac{V - V_0}{V_0} \quad (1)$$

where V_0 is the initial voltage drop at zero load, and V is the voltage drop recorded at different cycles during fatigue testing. As voltage was measured under a constant supply current, this change is equivalent to percentage resistance change.

Figure 2 shows the development of the percentage voltage change in a typical fatigue test without the penetrant treatment. The voltage drop continuously increases with the loading cycle. Figure 2a shows the voltage changes logged throughout the fatigue test. It is monotonically increasing with the loading cycle. Initially, the increase is very gradual. Towards the end of fatigue life, it increased steeply. Note that this voltage change, or the resistance change it reflects, contains the contributions from the piezoresistive effect due to the applied load as well as fatigue damages. The periodic dips in Figure 2a are temporary test interruptions to allow voltages to be measured at 0 N and 500 N. The latter results are presented in Figure 2b. When measured at 0 N, this increase can only be attributed to the occurrence of fatigue damage, such as microcracking and debonding in the adhesive joint. Measuring the voltage drop at 500 N may include some conductivity change due to the load, but the data before fatigue cycling in Figure 2b show that this was minimal initially (0.08%). The 500 N data are consistently higher than the 0 N data as the 500 N load tends to open up any defect resulting from the fatigue damage and increase the resistance. This effect is expected to be more marked towards the end of fatigue life, where damage is more extensive and is in fact borne from the increasingly prominent difference between the two curves as the specimen was approaching final failure.

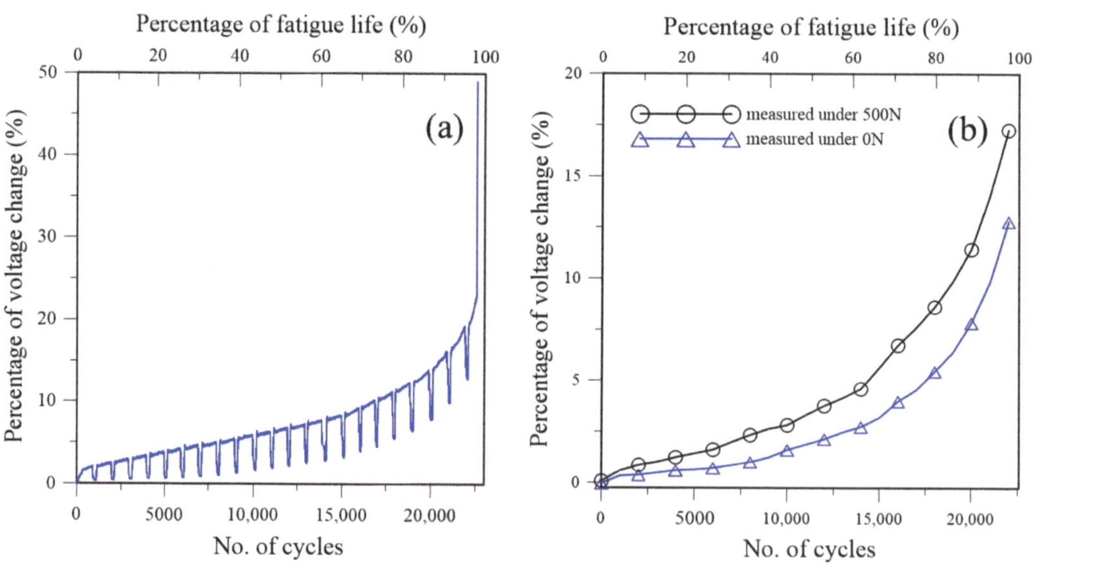

Figure 2. Development of the percentage voltage change in a typical fatigue test: (**a**) logged during the whole fatigue loading process; (**b**) measured at 0 N and 500 N.

Figure 2 suggests that through suitable calibration, the fatigue life expended, or fatigue damage incurred, may be quantified through the percentage voltage change. However, a prerequisite for this is that the relationship between voltage change and fatigue damage accumulation must be reproducible across different specimens.

3.2. Relationship between Fatigue Life, Debonded Area, and Percentage Voltage Change

A total of 22 specimens from five batches were successfully penetrant-treated to record the instantaneous debonded area. For these specimens, tests were interrupted when the 0 N percentage voltage change reached or just passed the targeted amounts of either 2%, 10%, or ~18%. The respective number of cycles was noted. It should be pointed out that only the debonded area that has an opening to the exterior can be recorded. The stress analysis of the single-lap joint [64] indicated that for each of the stress components, stress concentration occurs right at or very close to the longitudinal edges of the joint (positions corresponding to points A and B in Figure 1). Previous work [65] also showed that fiber Bragg grating sensors along the loading direction straddling the longitudinal joint edge are more sensitive than other configurations in detecting joint damage. These exterior edges will therefore be the most likely locations for the initiation of debonding.

The computed debonded area and the corresponding fractional fatigue life expended for each penetrant-treated specimen are presented in Figure 3; some typical photographs of the fractured joints taken under UV illumination are also shown as insets in Figure 3. Within experimental scatters, the data points lie remarkedly close to the fitted straight line, indicating that the debonded area is roughly directly proportional to the fatigue life expended. A debonded area of 200 mm^2 roughly corresponds to 70% expended life.

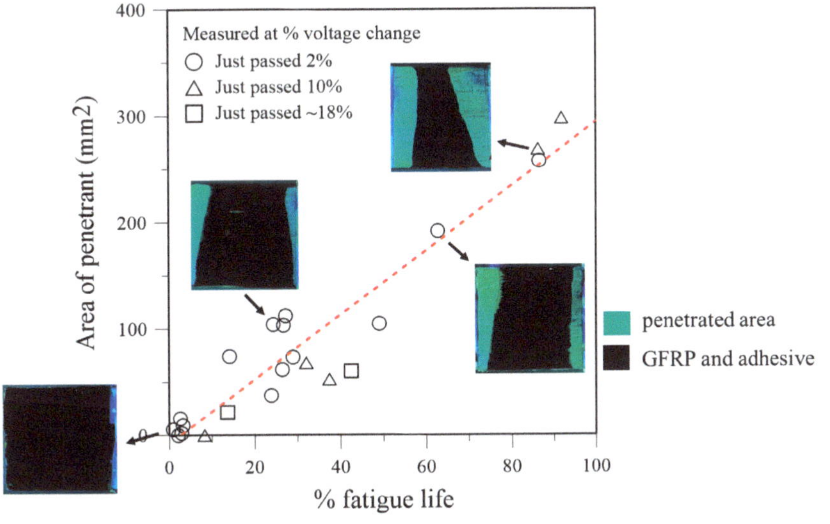

Figure 3. Debonded area versus percentage fatigue life expended for 22 specimens (the red dashed line represents the linear fit to the data).

Although the interruption for penetrant treatment aimed at the target percentage voltage changes of 2%, 10%, or ~18%, periodic measurement per the 1000-cycle increment often led to missing and going beyond the exact target. In fact, for the 18% aim, fatigue lives were close to the end, and several specimens simply fractured before interruption for penetrant treatment could be applied in time. This is the reason why the number of specimens for this category is limited, and one of them was treated when the percentage voltage drop reached 17.6%.

Figure 4 plots the actually measured percentage voltage change with the penetrant-marked debond area for the specimens. Two distinct trends can be seen in Figure 4. The first involves data points falling close to the dotted line, exhibiting a linear correlation between the percentage voltage change and debonded area. The second involves data points encircled in the red ellipse that display considerable voltage changes with small debonded areas. These two distinct sets of data suggest that different failure mechanisms might prevail. A data point labeled *X* in Figure 4 falls markedly outside these two sets of data. The non-unique relation between percentage voltage change and debonded area precluded the use of the former as a quantifying parameter to describe the latter. This further implies that percentage voltage change cannot be used to successfully quantify fatigue life expended, which has a one-to-one correspondence with the debonded area. The possible causes of the non-unique relation between the percentage voltage change and the debonded area will be further elucidated through detailed fractographic examination using an SEM.

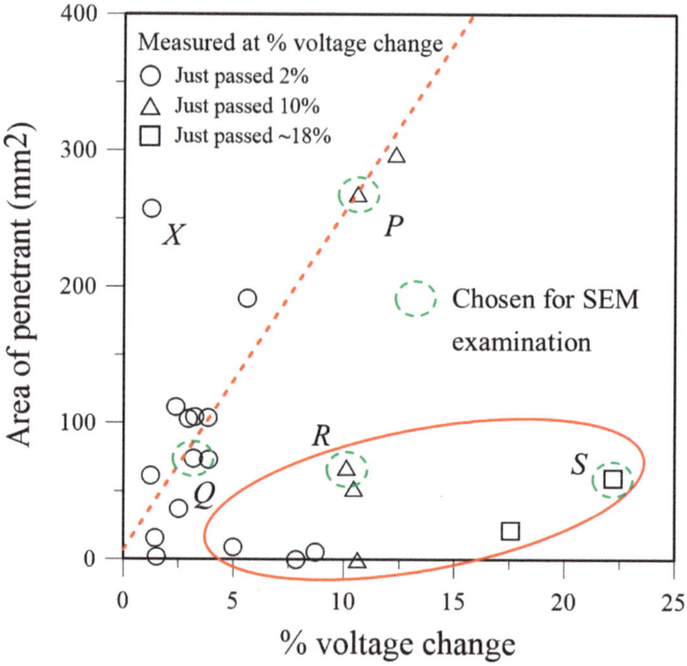

Figure 4. Measured percentage voltage change versus penetrant-marked debonded area for 22 specimens (the red dashed line and ellipse represent two different trends in the data).

3.3. Fractographic Observation of Damages in the Specimens

Specimens selected from the above two data sets, together with the specimen labeled *X*, were examined using an SEM. The chosen specimens are indicated with broken circles in Figure 4. For each specimen, the fluorescent area was thoroughly scanned using magnifications from 100× to 1000× to identify characteristic features that generally prevail on the whole area. Microscopic views of these characteristic features were recorded at a number of points for each specimen. Typical views are shown in the following discussion, and more detailed micrographs are presented in the Supplementary Materials.

Figure 5 shows magnified views of the adhesive on the joint fracture surface of specimen *X*, which has considerable debonding, but the percentage voltage change remained low. Extensive circular holes (examples of some of these are indicated by red arrows) of different sizes are evident in the adhesive. Judging from their roundness, these holes are probably residual gas pores. The presence of these gas pores reduced the cross-sectional area that is available for electric conduction. When debonding intercepts these pores, the specimen resistance will not further increase. This may explain the small change in the voltage drop even though considerable debonding has occurred.

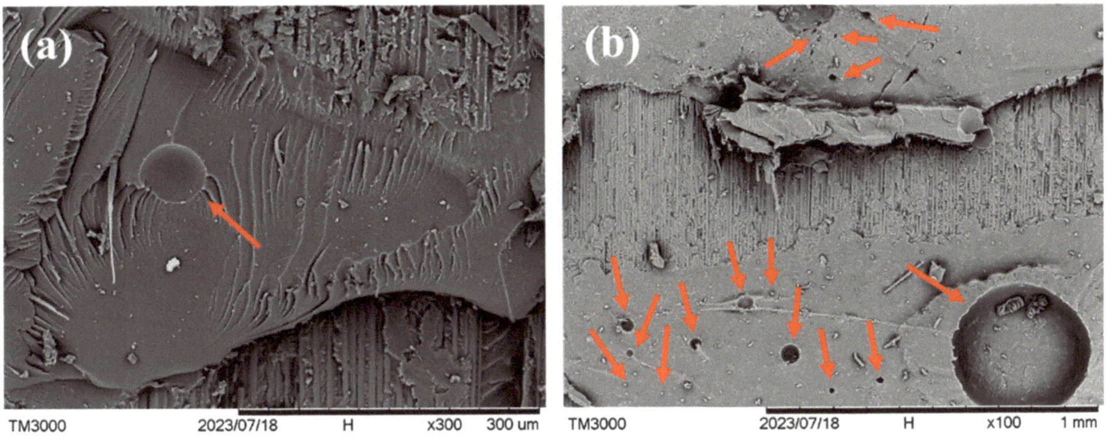

Figure 5. (**a**) and (**b**) Two positions on the joint adhesive of the fractured specimen *X* showing extensive gas pores (pointed to with red arrows).

Specimens *P* and *Q*, both lying on the dotted line in Figure 4, exhibited the same characteristic feature, which prevails over the whole fluorescent area. Figure 6 shows one typical point in specimen *P*. The location of this point is indicated in the macroscopic visible light and UV light views in the leftmost section of Figure 6a. The visible light view reveals the final fracture state of the entire joint. The green regions in the UV light view represent the fluorescent penetrant-soaked area. White arrows in the 100× micrograph in Figure 6a show the abundance of the characteristic feature, which consists of a relatively smooth area surrounded by rough areas with channel marks. One example of this feature, enclosed in the dotted rectangle, is magnified and displayed in Figure 6b. The channel marks have a width of ~7 μm. This dimension agrees with the graphite fiber diameter and was left by peeled-off fibers from the graphite/epoxy composite adherend. The smooth area shows no clear evidence of fiber imprints, suggesting that it is the adhesive for bonding the joint. Thus, the dominant fracture mechanisms involve partly debonding at the adhesive/adherend interface and partly the peel-off of fiber from the outermost lamina of the composite adherend. Typical fracture characteristics of specimen *Q* are shown in Figure 7. It shows the same interfacial debonding and outermost lamina fiber peel-off mechanisms. This feature also occurs in great abundance, as is evident by the arrows in the 100× micrograph in Figure 7a. More fractographs at different points on specimens *P* and *Q* are shown in Figures S1 and S2, respectively, in the Supplementary Materials.

A different fracture appearance is exhibited by specimens *R* and *S*, which lie in the ellipse in Figure 4. Figure 8 shows the typical fracture appearance from one point in specimen *R*. The location of this point is indicated in the macroscopic visible light and UV light views in the leftmost section of Figure 8a. In the 100× micrograph in Figure 8a, interfacial debonding between the adhesive and adherend is rarely seen across the whole fluorescent area. The fracture surface mainly consists of fiber peel-off. Moreover, from the magnified view in Figure 8b, fibers can be seen beneath the channels of this peel-off layer (see the circled regions for examples). This indicates that debonding occurred between two graphite/epoxy laminae or inside a lamina in the composite adherends. Specimen *S* shows occasional adhesive/adherend interfacial debonding. However, its occurrence (indicated by arrows in the 100× micrograph of Figure 9a) is much less extensive than that in specimens *P* and *Q*. Again, the circled region in the magnified view in Figure 9b suggests intra- or inter-laminar debonding (broken circle) instead of interfacial debonding between the adhesive and the adherend. More fractographs at different points on specimens *R* and *S* are shown in Figures S3 and S4, respectively, in the Supplementary Materials.

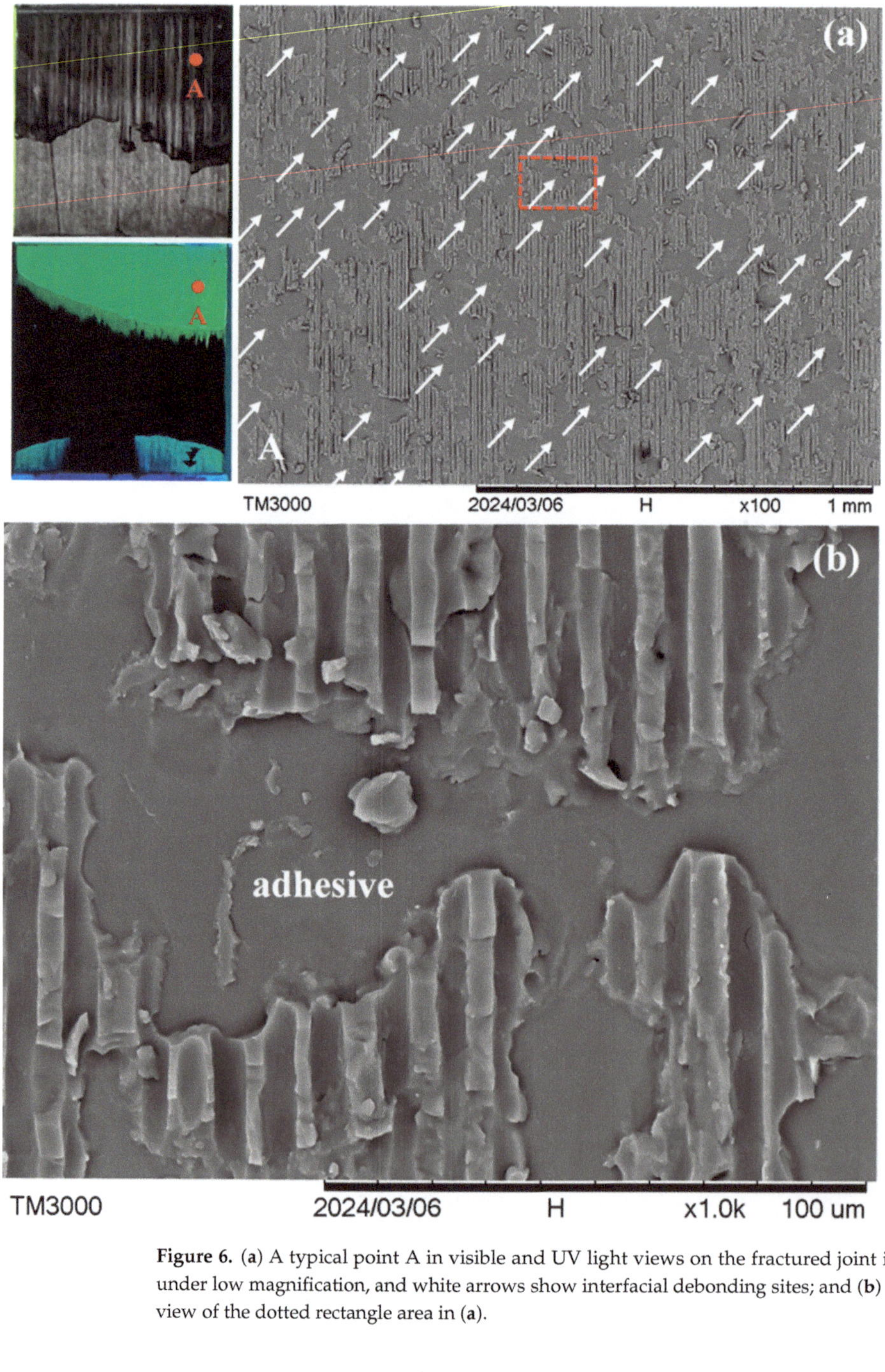

Figure 6. (**a**) A typical point A in visible and UV light views on the fractured joint in specimen *P* under low magnification, and white arrows show interfacial debonding sites; and (**b**) the magnified view of the dotted rectangle area in (**a**).

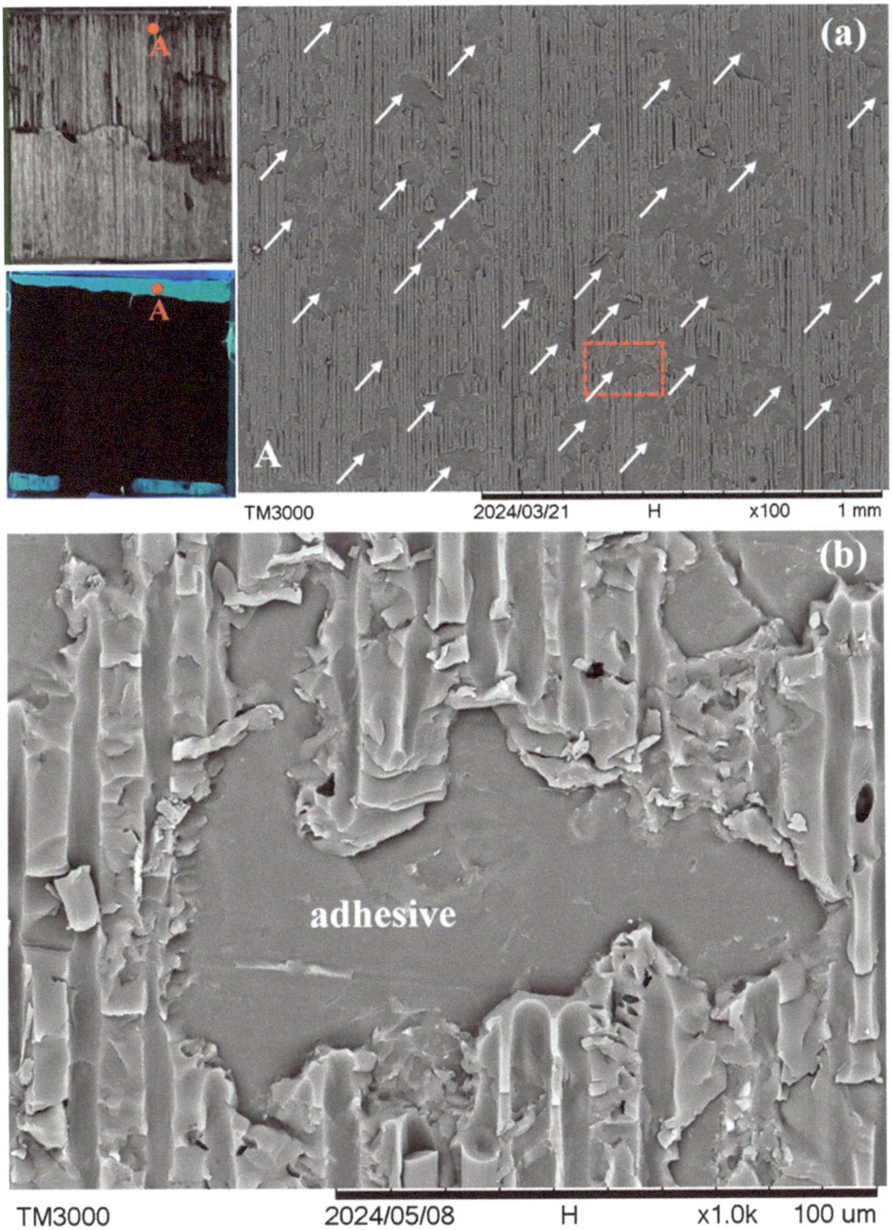

Figure 7. (**a**) A typical point A in visible and UV light views on the fractured joint in specimen *Q* under low magnification, and white arrows show interfacial debonding sites; and (**b**) a magnified view of the dotted rectangle area in (**a**).

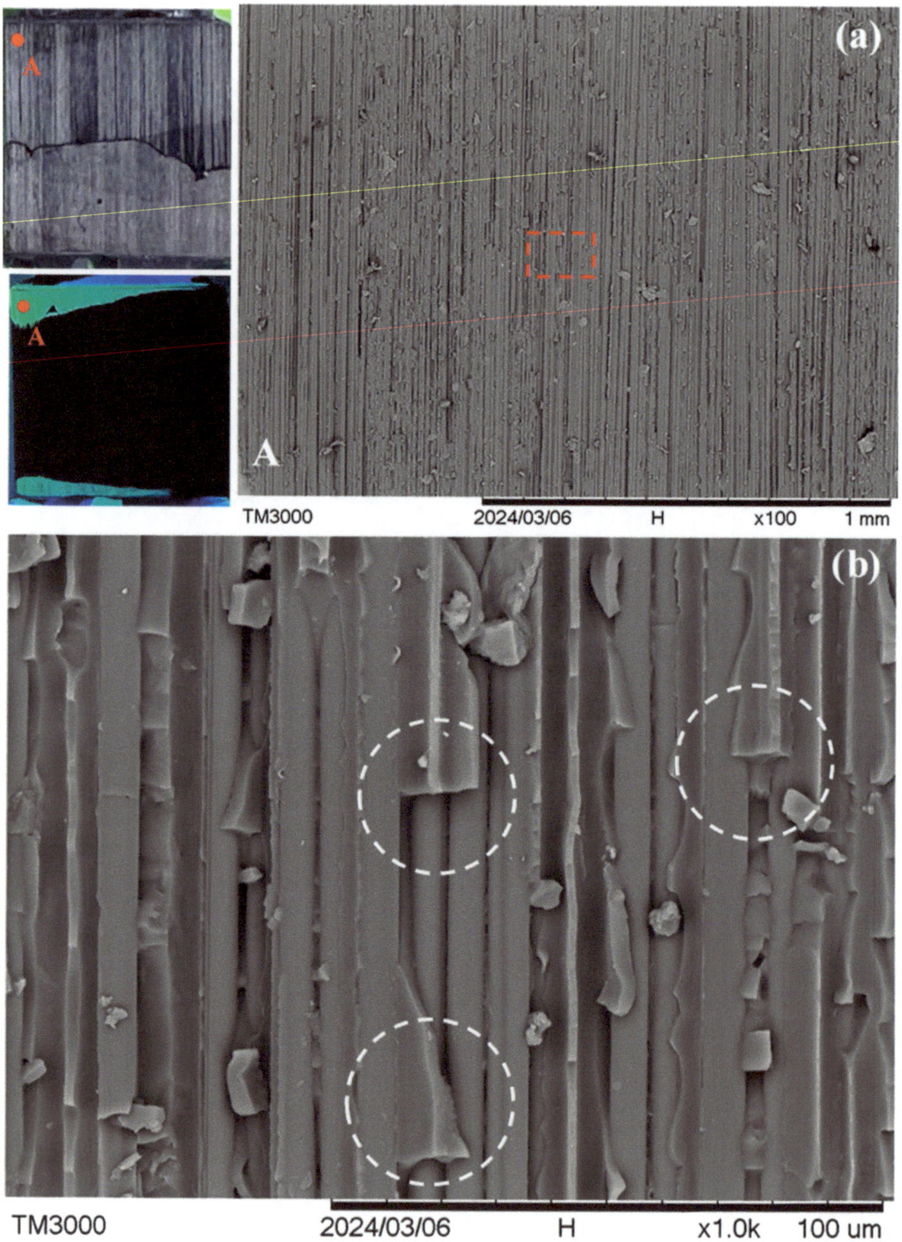

Figure 8. (**a**) A typical point A in visible and UV light views on the fractured joint in specimen *R* under low magnification; and (**b**) a magnified view of the dotted rectangle area in (**a**). Broken circles indicate intra- or inter-laminar debonding.

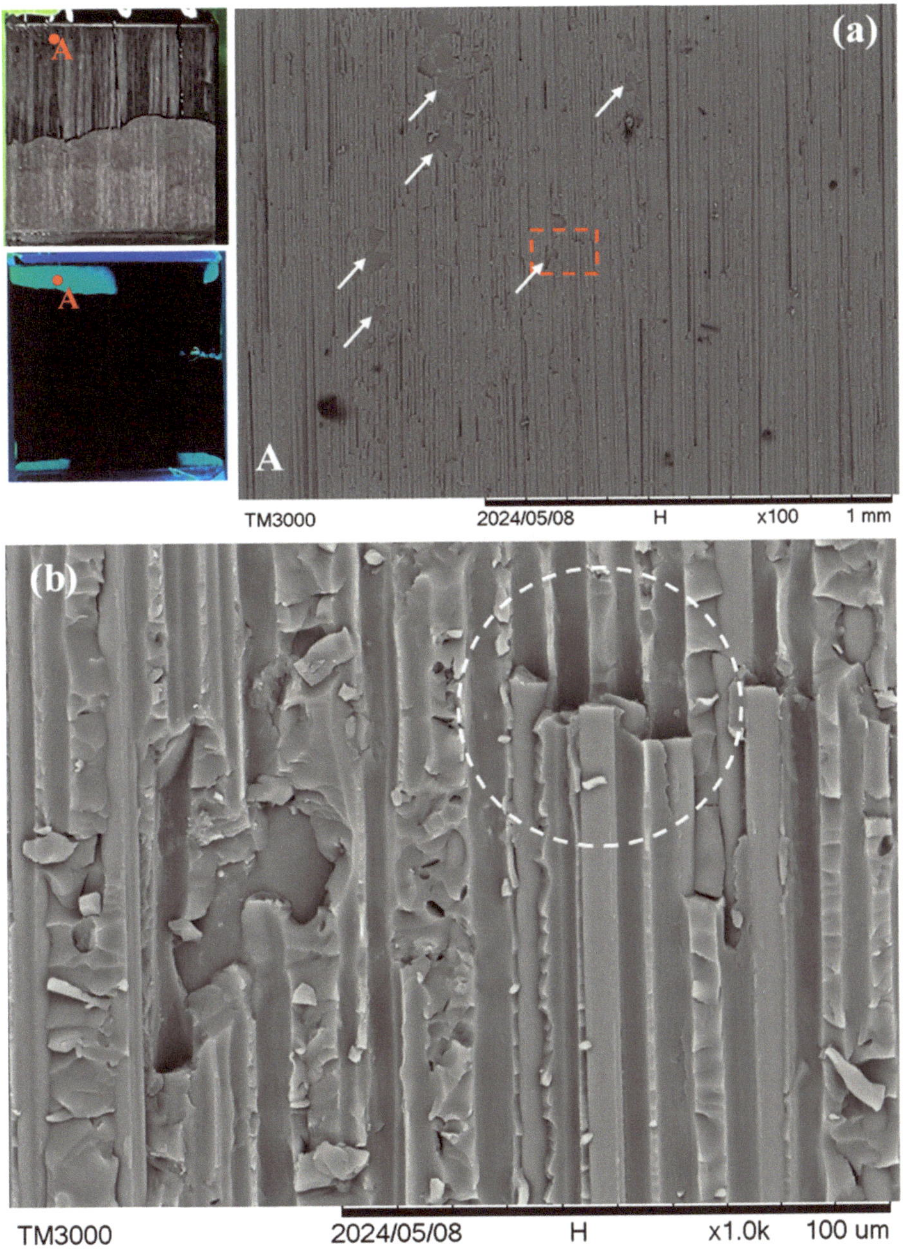

Figure 9. (**a**) A typical point A in visible and UV light views on the fractured joint in specimen *S* under low magnification, and white arrows show interfacial debonding sites; and (**b**) a magnified view of the dotted rectangle area in (**a**). Broken circles indicate intra- or inter-laminar debonding.

The lap-joint specimen is conductive because the adherends and the adhesive are both conductive. The adhesive is conductive as carbon nanotubes have been dispersed in the resin. Although the epoxy resin in the composite adherend is non-conducting, the high volume fraction of graphite fibers is packed very closely together and inevitably, there are direct contact and tunneling points along the fiber length. In fact, contact between fibers was encouraged by the compressive stress of ~10 atmospheric pressure that was maintained throughout the whole course of composite laminate fabrication. This would press the fibers tightly together and squeeze out part of the resin between the fibers when the former was still flowable. As a result, the composite adherend is conductive and in fact has a very low resistance as mentioned before. Sandblasting the surface to be joined lightly removed some of the resin and exposed the fiber in the outmost layer at some points. At these sites, the fibers can come into contact with, or very close to, the carbon nanotubes in the adhesive, completing the conductive pathway across the specimen. From the extensive interfacial debonding revealed in Figures 6 and 7, the debonded interface areas are relatively smooth and free from fiber imprint marking. This suggests that the contact between the exposed fibers and the adhesive is probably not very extensive nor tight at the joint region. A large fraction of the interfacial area may therefore have played little role in electrical conduction. As a result, intra- or inter-laminar debonding in the adherend will cause a more prominent resistance or voltage drop increase than interfacial debonding between the adhesive and adherend. This explains why the specimens inside the ellipse in Figure 4 give a significant increase in the voltage drop with a small amount of debonded area, as specimens in this category mainly exhibited intra- or inter-laminar debonding, as shown in Figures 8 and 9. On the other hand, specimens close to the dotted line in Figure 4 exhibited extensive interfacial debonding between the adhesive and adherend. As a large fraction of the interface may not be involved in electrical conduction, their debonding will not affect the resistance. In this case, much more debonded area is required to induce a significant increase in the resistance or voltage drop. The prevalence of two different micromechanisms in different specimens may probably be attributed to comparable adhesive/adherend interfacial strength and the inter- or intra-laminar debonding strength.

The current study demonstrated that when competing micromechanisms are involved in the failure of an adhesive joint, the resistance and the degree of joint degradation are not uniquely correlated. This is unfavorable if conductivity is intended to be used for SHM in practical applications. The failure of an adhesive joint may involve cohesive failures in the adhesive or in any one of the adherends, as well as adhesive failure at the adhesive/adherend interface. When any two or more of these failure modes are of comparable strengths, non-uniqueness of the resistance and damage status will probably occur, as exemplified in this work. Thus, a sufficient number of specimens tested under correctly simulated service loading conditions and environment is needed in order to establish a correlation between resistance and structural integrity for use in practical SHM.

The above observation may be further tested by using the resistance change to monitor fatigue damage progression when a single micromechanism predominates. A possible material system is the metal–metal adhesive joint.

4. Conclusions

Attempts have been made to monitor the fatigue damage of single-lap-joint composite specimens using conductivity change and debonded area. Conductivity change is reflected by the percentage voltage change when a constant current passes through the specimen. An instantaneous debonded area was recorded with a fluorescent liquid penetrant treatment. It was found that

1. The percentage voltage change monotonically increased from 0% to 49% with fatigue loading cycles. The percentage voltage changes measured at 500 N are higher than that at 0 N, with the difference initially at 0.11% and increasing to 4.48% as it approached the end of the fatigue life.

2. The debonded area is roughly directly proportional to the fatigue life expended. A debonded area of 200 mm^2, out of the total joint area of 625 mm^2, roughly corresponds to 70% expended life.
3. The percentage voltage change does not have a unique correlation with the debonded area. Two different trends existed. In one trend, the percentage voltage change correlated linearly with the debonded area. Specimens in this trend are fractured partly by interfacial debonding and partly by the peel-off of fiber from the outermost lamina of the composite adherend. In another trend, considerable voltage changes occurred with small debonded areas. The major fatigue fracture mechanism for specimens in this latter trend is intra- or inter-laminar debonding inside the adherend.
4. As there is no unique correlation between the debonded area and the percentage voltage change, the latter is not a valid quantifying parameter for the fatigue life expended in the current composite lap-joint specimens.

Supplementary Materials: The following supporting information can be downloaded at: https://www.mdpi.com/article/10.3390/polym16162374/s1, Table S1 The manufacturer provided specifications of Swancor 2261-A/BS epoxy resin (Swancor Industrial Corp., Nantou, Taiwan). Figure S1. More microscopic and magnified views of the dotted rectangle joint fracture views of specimen *P*; extensive white arrows show interfacial debonding sites. Figure S2: More microscopic and magnified views of the dotted rectangle joint fracture views of specimen *Q*; extensive white arrows show interfacial debonding sites. Figure S3: More microscopic and magnified views of the dotted rectangle joint fracture views of specimen *R*; sparing white arrows show interfacial debonding sites and broken circles indicate intra- or inter-laminar debonding. Figure S4: More microscopic and magnified views of the dotted rectangle joint fracture views of specimen *S*; sparing white arrows show interfacial debonding sites and broken circles indicate intra- or inter-laminar debonding. Refs. [66–68] are listed in References of main text.

Author Contributions: Conceptualization, C.-S.S.; methodology, C.-S.S.; software, S.-H.H.; validation, S.-H.H. and C.-S.S.; formal analysis, C.-S.S. and S.-H.H.; investigation, S.-H.H. and C.-S.S.; resources, C.-S.S.; data curation, C.-S.S.; writing—first draft preparation, S.-H.H.; writing—subsequent drafts, review, and editing, C.-S.S.; visualization, C.-S.S. and S.-H.H.; supervision, C.-S.S.; project administration, C.-S.S.; funding acquisition, C.-S.S. All authors have read and agreed to the published version of the manuscript.

Funding: This research was funded by the National Science and Technology Council, ROC, through the project MOST 110–2221-E-002-152 and NSTC 112-2221-E-002-226-MY2, and the National Taiwan University, ROC, through the project NTU-JP-111L7242.

Institutional Review Board Statement: Not applicable.

Data Availability Statement: The original contributions presented in the study are included in the article/Supplementary Materials; further inquiries can be directed to the corresponding author.

Acknowledgments: We thank the Ministry of Science and Technology, ROC and the National Taiwan University, ROC, for supporting this work.

Conflicts of Interest: The authors declare no conflicts of interest. The funders had no role in the design of the study; in the collection, analyses, or interpretation of data; in the writing of the manuscript; or in the decision to publish the results.

References

1. Casavola, C.; Palano, F.; De Cillis, F.; Tati, A.; Terzi, R.; Luprano, V. Analysis of CFRP Joints by Means of T-Pull Mechanical Test and Ultrasonic Defects Detection. *Materials* **2018**, *11*, 620. [CrossRef] [PubMed]
2. Jasiuniene, E.; Mazeika, L.; Samaitis, V.; Cicenas, V.; Mattsson, D. Ultrasonic non-destructive testing of complex titanium/carbon fibre composite joints. *Ultrasonics* **2019**, *95*, 13–21. [CrossRef]
3. Yilmaz, B.; Ba, A.; Jasiūnienė, E.; Bui, H.-K.; Berthiau, G. Evaluation of Bonding Quality with Advanced Nondestructive Testing (NDT) and Data Fusion. *Sensors* **2020**, *20*, 5127. [CrossRef]
4. Yu, X.; Fan, Z.; Puliyakote, S.; Castaings, M. Remote monitoring of bond line defects between a composite panel and a stiffener using distributed piezoelectric sensors. *Smart Mater. Struct.* **2018**, *27*, 035014. [CrossRef]

5. Ochôa, P.; Villegas, I.F.; Groves, R.M.; Benedictus, R. Diagnostic of manufacturing defects in ultrasonically welded thermoplastic composite joints using ultrasonic guided waves. *NDT E Int.* **2019**, *107*, 102126. [CrossRef]
6. Mohd Tahir, M.F.; Echtermeyer, A.T. Phased Array Ultrasonic Testing on Thick Glass Fiber Reinforced Thermoplastic Composite Pipe Implementing the Classical Time-Corrected Gain Method. *J. Nondestr. Eval.* **2024**, *43*, 74. [CrossRef]
7. de Almeida, P.D.; Pereira, G.R. Phased array inspection of glass fiber reinforced polymers pipeline joints. *J. Mater. Res. Technol.* **2019**, *8*, 4736–4740. [CrossRef]
8. Nagy, P.B. Ultrasonic detection of kissing bonds at adhesive interfaces. *J. Adhes. Sci. Technol.* **1991**, *5*, 619–630. [CrossRef]
9. Yılmaz, B.; Jasiūnienė, E. Advanced ultrasonic NDT for weak bond detection in composite-adhesive bonded structures. *Int. J. Adhes. Adhes.* **2020**, *102*, 102675. [CrossRef]
10. Liu, T.; Pei, C.; Cheng, X.; Zhou, H.; Xiao, P.; Chen, Z. Adhesive debonding inspection with a small EMAT in resonant mode. *NDT E Int.* **2018**, *98*, 110–116. [CrossRef]
11. Roth, W.; Giurgiutiu, V. Structural health monitoring of an adhesive disbond through electromechanical impedance spectroscopy. *Int. J. Adhes. Adhes.* **2017**, *73*, 109–117. [CrossRef]
12. Medeiros, R.d.; Souza, G.; Marques, D.; Flor, F.; Tita, V. Vibration-based structural monitoring of bi-clamped metal-composite bonded joint: Experimental and numerical analyses. *J. Adhes.* **2021**, *97*, 891–917. [CrossRef]
13. Dugnani, R.; Chang, F.-K. Analytical model of lap-joint adhesive with embedded piezoelectric transducer for weak bond detection. *J. Intell. Mater. Syst. Struct.* **2017**, *28*, 124–140. [CrossRef]
14. Dugnani, R.; Zhuang, Y.; Kopsaftopoulos, F.; Chang, F.-K. Adhesive bond-line degradation detection via a cross-correlation electromechanical impedance–based approach. *Struct. Health Monit.* **2016**, *15*, 650–667. [CrossRef]
15. Grammatikos, S.; Kordatos, E.; Matikas, T.; Paipetis, A. Real-time debonding monitoring of composite repaired materials via electrical, acoustic, and thermographic methods. *J. Mater. Eng. Perform.* **2014**, *23*, 169–180. [CrossRef]
16. Martens, U.; Schröder, K.-U. Evaluation of infrared thermography methods for analysing the damage behaviour of adhesively bonded repair solutions. *Compos. Struct.* **2020**, *240*, 111991. [CrossRef]
17. Shin, P.H.; Webb, S.C.; Peters, K.J. Pulsed phase thermography imaging of fatigue-loaded composite adhesively bonded joints. *NDT E Int.* **2016**, *79*, 7–16. [CrossRef]
18. Kryukov, I.; Böhm, S. Prospects and limitations of eddy current shearography for non-destructive testing of adhesively bonded structural joints. *J. Adhes.* **2019**, *95*, 874–886. [CrossRef]
19. Kryukov, I.; Thiede, H.; Böhm, S. Quality assurance for structural adhesively bonded joints by eddy current shearography. *Weld World* **2017**, *61*, 581–588. [CrossRef]
20. Davis, M.J.; McGregor, A. Assessing adhesive bond failures: Mixed-mode bond failures explained. In Proceedings of the ISASI Australian Safety Seminar, Canberra, Australia, 4–6 June 2010; Available online: http://www.adhesionassociates.com/papers/56%20Assessing%20Adhesive%20Bond%20Failures%20-%20Mixed-Mode%20Bond%20Failures%20Explained.pdf (accessed on 1 June 2024).
21. Baker, A.A.; Wang, J. *Chapter Six—Adhesively Bonded Repair/Reinforcement of Metallic Airframe Components: Materials, Processes, Design and Proposed Through-Life Management*; Jones, R., Baker, A.A., Matthews, N., Champagne, V., Eds.; Aircraft Sustainment and Repair, Butterworth-Heinemann: Oxford, UK, 2018; pp. 191–252. [CrossRef]
22. Sadeghi, M.Z.; Weiland, J.; Preisler, A.; Zimmermann, J.; Schiebahn, A.; Reisgen, U.; Schröder, K.U. Damage detection in adhesively bonded single lap joints by using backface strain: Proposing a new position for backface strain gauges. *Int. J. Adhes. Adhes.* **2020**, *97*, 102494. [CrossRef]
23. Sadeghi, M.Z.; Weiland, J.; Zimmermann, J.; Schiebahn, A.; Reisgen, U.; Schröder, K.U. Experimental and FE investigations on the influential parameters in positioning and measurement of strain gauges in adhesively bonded single lap joints. *Procedia Struct. Integr.* **2020**, *28*, 1590–1600. [CrossRef]
24. Augustin, T.; Karsten, J.; Kötter, B.; Fiedler, B. Health monitoring of scarfed CFRP joints under cyclic loading via electrical resistance measurements using carbon nanotube modified adhesive films. *Compos. Part A Appl. Sci. Manuf.* **2018**, *105*, 150–155. [CrossRef]
25. Dengg, A.; Kralovec, C.; Schagerl, M. Damage monitoring of pinned hybrid composite-titanium joints using direct current electrical resistance measurement. *Compos. Struct.* **2024**, *334*, 117972. [CrossRef]
26. Grefe, H.; Weiser, D.; Kandula, M.W.; Dilger, K. Deformation measurement within adhesive bonds of aluminium and CFRP using advanced fibre optic sensors. *Manuf. Rev.* **2020**, *7*, 14. [CrossRef]
27. Young, S.; Penumadu, D.; Foster, D.; Maeser, H.; Balijepalli, B.; Reese, J.; Bank, D.; Dahl, J.; Blanchard, P. Smart Adhesive Joint with High-Definition Fiber-Optic Sensing for Automotive Applications. *Sensors* **2020**, *20*, 614. [CrossRef] [PubMed]
28. Jaiswal, P.R.; Kumar, R.I.; Saeedifar, M.; Saleh, M.; Luyckx, G.; De Waele, W. Deformation and damage evolution of a full-scale adhesive joint between a steel bracket and a sandwich panel for naval application. *Proc. Inst. Mech. Engr. Part C J. Mech. Eng. Sci.* **2021**, *235*, 571–584. [CrossRef]
29. Zeng, H.; Yan, R.; Xu, L.; Gui, S. Application study on fiber Bragg grating sensors in damage monitoring of sandwich composite joints. *J. Sandw. Struct. Mater.* **2020**, *22*, 1542–1563. [CrossRef]
30. Webb, S.; Shin, P.; Peters, K.; Zikry, M.; Stan, N.; Chadderdon, S.; Selfridge, R.; Schultz, S. Characterization of fatigue damage in adhesively bonded lap joints through dynamic, full-spectral interrogation of fiber Bragg grating sensors: 1. Experiments. *Smart Mater. Struct.* **2013**, *23*, 025016. [CrossRef]

31. Karpenko, O.; Khomenko, A.; Koricho, E.; Haq, M.; Udpa, L. Monitoring of fatigue damage in composite lap-joints using guided waves and FBG sensors. In Proceedings of the AIP Conference Proceedings 1706, 42nd Annual Review of Progress in Quantitative Nondestructive Evaluation, Minneapolis, MN, USA, 26–31 July 2015; Chimenti, D.E., Bond, L.J., Eds.; AIP Publishing: Melville, NY, USA, 2016; p. 120005. [CrossRef]
32. Shin, C.-S.; Lin, T.-C. Adhesive Joint Integrity Monitoring Using the Full Spectral Response of Fiber Bragg Grating Sensors. *Polymers* **2021**, *13*, 2954. [CrossRef]
33. Shin, C.-S.; Lin, T.-C. Hygrothermal Damage Monitoring of Composite Adhesive Joint Using the Full Spectral Response of Fiber Bragg Grating Sensors. *Polymers* **2022**, *14*, 368. [CrossRef]
34. Andrew, J.J.; Arumugam, V.; Ramesh, C. Acoustic emission characterization of local bending behavior for adhesively bonded hybrid external patch repaired glass/epoxy composite laminates. *Struct. Health Monit.* **2019**, *18*, 739–756. [CrossRef]
35. Shin, C.S.; Chiang, C.C. Deformation monitoring by using optical fiber grating sensor. *J. Chin. Inst. Eng.* **2005**, *28*, 985–992. [CrossRef]
36. Wen, S.; Wang, S.; Chung, D. Piezoresistivity in continuous carbon fiber polymer-matrix and cement-matrix composites. *J. Mater. Sci.* **2000**, *35*, 3669–3675. [CrossRef]
37. Zhang, J.; Zhuang, R.; Liu, J.; Mäder, E.; Heinrich, G.; Gao, S. Functional interphases with multi-walled carbon nanotubes in glass fibre/epoxy composites. *Carbon* **2010**, *48*, 2273–2281. [CrossRef]
38. Park, J.-M.; Kwon, D.-J.; Wang, Z.-J.; DeVries, K.L. Review of self-sensing of damage and interfacial evaluation using electrical resistance measurements in nano/micro carbon materials-reinforced composites. *Adv. Compos. Mater.* **2015**, *24*, 197–219. [CrossRef]
39. Wang, Z.-J.; Kwon, D.-J.; Gu, G.-Y.; Kim, H.-S.; Kim, D.-S.; Lee, C.-S.; DeVries, K.L.; Park, J.-M. Mechanical and interfacial evaluation of CNT/polypropylene composites and monitoring of damage using electrical resistance measurements. *Compos. Sci. Technol.* **2013**, *81*, 69–75. [CrossRef]
40. Bilotti, E.; Zhang, H.; Deng, H.; Zhang, R.; Fu, Q.; Peijs, T. Controlling the dynamic percolation of carbon nanotube based conductive polymer composites by addition of secondary nanofillers: The effect on electrical conductivity and tuneable sensing behaviour. *Compos. Sci. Technol.* **2013**, *74*, 85–90. [CrossRef]
41. Hao, B.; Ma, Q.; Yang, S.; Mäder, E.; Ma, P.-C. Comparative study on monitoring structural damage in fiber-reinforced polymers using glass fibers with carbon nanotubes and graphene coating. *Compos. Sci. Technol.* **2016**, *129*, 38–45. [CrossRef]
42. Pramanik, P.; Khastgir, D.; De, S.; Saha, T. Pressure-sensitive electrically conductive nitrile rubber composites filled with particulate carbon black and short carbon fibre. *J. Mater. Sci.* **1990**, *25*, 3848–3853. [CrossRef]
43. Sánchez-Romate, X.F.; García, C.; Rams, J.; Sánchez, M.; Ureña, A. Structural health monitoring of a CFRP structural bonded repair by using a carbon nanotube modified adhesive film. *Compos. Struct.* **2021**, *270*, 114091. [CrossRef]
44. Sánchez-Romate, X.F.; Coca, Á.; Jiménez-Suárez, A.; Sánchez, M.; Ureña, A. Crack sensing mechanisms of Mode-II and skin-stringer joints between dissimilar materials by using carbon nanotubes. *Compos. Sci. Technol.* **2021**, *201*, 108553. [CrossRef]
45. Sánchez-Romate, X.F.; Sbarufatti, C.; Sánchez, M.; Bernasconi, A.; Scaccabarozzi, D.; Libonati, F.; Cinquemani, S.; Güemes, A.; Ureña, A. Fatigue crack growth identification in bonded joints by using carbon nanotube doped adhesive films. *Smart Mater. Struct.* **2020**, *29*, 035032. [CrossRef]
46. Li, W.; Frederick, H.; Palardy, G. Multifunctional films for thermoplastic composite joints: Ultrasonic welding and damage detection under tension loading. *Compos. Part A Appl. Sci. Manuf.* **2021**, *141*, 106221. [CrossRef]
47. Mactabi, R.; Rosca, I.D.; Hoa, S.V. Monitoring the integrity of adhesive joints during fatigue loading using carbon nanotubes. *Compos. Sci. Technol.* **2013**, *78*, 1–9. [CrossRef]
48. Kim, C.-H.; Choi, J.-H.; Kweon, J.-H. Defect detection in adhesive joints using the impedance method. *Compos. Struct.* **2015**, *120*, 183–188. [CrossRef]
49. Kang, M.-H.; Choi, J.-H.; Kweon, J.-H. Fatigue life evaluation and crack detection of the adhesive joint with carbon nanotubes. *Compos. Struct.* **2014**, *108*, 417–422. [CrossRef]
50. Ladani, R.B.; Wu, S.; Zhang, J.; Ghorbani, K.; Kinloch, A.J.; Mouritz, A.P.; Wang, C.H. Using carbon nanofibre sensors for in-situ detection and monitoring of disbonds in bonded composite joints. *Procedia Eng.* **2017**, *188*, 362–368. [CrossRef]
51. Bregar, T.; An, D.; Gharavian, S.; Burda, M.; Durazo-Cardenas, I.; Thakur, V.K.; Ayre, D.; Słoma, M.; Hardiman, M.; McCarthy, C. Carbon nanotube embedded adhesives for real-time monitoring of adhesion failure in high performance adhesively bonded joints. *Sci. Rep.* **2020**, *10*, 16833. [CrossRef]
52. Sam-Daliri, O.; Faller, L.-M.; Farahani, M.; Zangl, H. Structural health monitoring of adhesive joints under pure mode I loading using the electrical impedance measurement. *Eng. Fract. Mech.* **2021**, *245*, 107585. [CrossRef]
53. Li, W.; Palardy, G. Mechanical and electrical properties of MWCNT/PP films and structural health monitoring of GF/PP joints. In Proceedings of the SPE Automotive Composites Conference & Exhibition, Novi, MI, USA, 2–4 November 2021; Available online: https://speautomotive.com/wp-content/uploads/2021/10/MechanicalElectrical-Properties-of-MWCNTPP-Films-for-Structural-Health-Monitoring-of-GFPP-Joints.pdf (accessed on 1 June 2024).
54. Kumar, S.; Falzon, B.G.; Hawkins, S.C. Ultrasensitive embedded sensor for composite joints based on a highly aligned carbon nanotube web. *Carbon* **2019**, *149*, 380–389. [CrossRef]
55. Bekas, D.G.; Sharif-Khodaei, Z.; Baltzis, D.; Aliabadi, M.F.; Paipetis, A.S. Quality assessment and damage detection in nanomodified adhesively-bonded composite joints using inkjet-printed interdigital sensors. *Compos. Struct.* **2019**, *211*, 557–563. [CrossRef]

56. Lee, B.M.; Loh, K.J. A 2D percolation-based model for characterizing the piezoresistivity of carbon nanotube-based films. *J. Mater. Sci.* **2015**, *50*, 2973–2983. [CrossRef]
57. Li, C.; Chou, T. Modeling of damage sensing in fiber composites using carbon nanotube networks. *Comp. Sci. Technol.* **2008**, *68*, 3373–3379. [CrossRef]
58. Hu, N.; Karube, Y.; Arai, M.; Watanabe, T.; Yan, C.; Li, Y.; Liu, Y.; Fukunaga, H. Investigation on sensitivity of a polymer/carbon nanotube composite strain sensor. *Carbon* **2010**, *48*, 680–687. [CrossRef]
59. Hu, N.; Karube, Y.; Yan, C.; Masuda, Z.; Fukunaga, H. Tunneling effect in a polymer/carbon nanotube nanocomposite strain sensor. *Acta Mater.* **2008**, *56*, 2929–2936. [CrossRef]
60. Stampfer, C.; Helbling, T.; Jungen, A.; Hierold, C. Piezoresistance of Single-Walled Carbon Nanotubes. In Proceedings of the Transducers & Eurosensors '07, The 14th International Conference on Solid-State Sensors, Actuators and Microsystems, Lyon, France, 10–14 June 2007; pp. 1565–1568. [CrossRef]
61. Yang, L.; Anantram, M.P.; Han, J.; Lu, J. Bandgap change of carbon nanotubes: Effect of small uniaxial and torsional strain. *Phys. Rev. B* **1999**, *60*, 13874–13878. [CrossRef]
62. *D5868-01*; Standard Test Method for Lap Shear Adhesion for Fiber Reinforced Plastic (FRP) Bonding. ASTM International: West Conshohocken, PA, USA, 2014.
63. Schneider, C.A.; Rasband, W.S.; Eliceiri, K.W. NIH Image to ImageJ: 25 years of image analysis. *Nat. Methods* **2012**, *9*, 671–675. [CrossRef]
64. Gonçalves, J.P.M.; de Moura, M.F.S.F.; de Castro, P.M.S.T. A three-dimensional finite element model for stress analysis of adhesive joints. *Int. J. Adhes. Adhes.* **2002**, *22*, 357–365. [CrossRef]
65. Shin, C.-S.; Chen, L.-W. Damage Monitoring of Composite Adhesive Joint Integrity Using Conductivity and Fiber Bragg Grating. *Polymers* **2023**, *15*, 1575. [CrossRef]
66. *ASTM D638-22*; Standard Test Method for Tensile Properties of Plastics. ASTM International: West Conshohocken, PA, USA, 2022.
67. *ASTM D790-17*; Standard Test Methods for Flexural Properties of Unreinforced and Reinforced Plastics and Electrical Insulating Materials. ASTM International: West Conshohocken, PA, USA, 2017.
68. *ASTM D3418-21*; Standard Test Method for Transition Temperatures and Enthalpies of Fusion and Crystallization of Polymers by Differential Scanning Calorimetry. ASTM International: West Conshohocken, PA, USA, 2021.

polymers

MDPI

Article

Damage and Failure Monitoring of Aerospace Insulation Layers Based on Embedded Fiber Bragg Grating Sensors

Guang Yan [1], Boli Wan [1], Heng Huang [1] and Wuyi Li [2,*]

[1] College of Instrument Science and Opto-Electronics Engineering, Beijing Information Science & Technology University, Beijing 100101, China; yanguang79@bistu.edu.cn (G.Y.)
[2] Research Institute of Aero-Engine, Beijing University of Aeronautics and Astronautics, Beijing 102206, China
* Correspondence: 11194@buaa.edu.cn

Abstract: Carbon fiber-reinforced polymer (CFRP) composites are widely used in aviation thermal insulation layers due to their high strength-to-weight ratio and excellent high-temperature performance. However, challenges remain regarding their structural integrity and durability under extreme conditions. This study first employed finite element simulation to model the damage evolution of CFRP laminated plates under axial tensile loads and their thermal decomposition behavior in high-temperature environments, providing a theoretical reference. Subsequently, experimental research was conducted on CFRP laminated plates embedded with fiber Bragg grating (FBG) sensors. In the tensile tests, FBG sensors accurately monitored the entire process from elastic deformation to damage propagation and eventual failure. In the high-temperature tests, despite a 75% reduction in tensile strength, FBG sensors effectively monitored damage evolution. Conclusively, the results demonstrate that FBG sensors possess reliable monitoring capabilities under complex conditions, making them a promising solution for the long-term structural health monitoring of aviation thermal insulation materials and paving the way for future developments in this area.

Keywords: carbon fiber-reinforced composites; fiber Bragg grating; finite element analysis; damage evolution; structural health monitoring

Citation: Yan, G.; Wan, B.; Huang, H.; Li, W. Damage and Failure Monitoring of Aerospace Insulation Layers Based on Embedded Fiber Bragg Grating Sensors. *Polymers* **2024**, *16*, 3543. https://doi.org/10.3390/polym16243543

Academic Editor: Costas Charitidis

Received: 26 October 2024
Revised: 21 November 2024
Accepted: 2 December 2024
Published: 19 December 2024

1. Introduction

Aviation thermal insulation layers are crucial in the field of aerospace, protecting aircraft structures from extreme environments such as high temperatures and pressures. These layers effectively block heat and prevent the overheating of internal structures and critical equipment, ensuring safe and efficient flights during complex missions and the secure return of spacecraft. However, during flight, external loads such as vibrations, impacts, aerodynamic pressure, and thermal expansion/contraction cycles can cause material fatigue, stress concentration, and microcracking. High-temperature environments, especially during high-speed flight or atmospheric re-entry, accelerate material aging, leading to thermal stress concentration and thermal fatigue [1–3]. This damage gradually expands within the insulation layer, degrading its thermal insulation performance and structural integrity, ultimately compromising the safety of the aircraft.

Therefore, the structural health monitoring (SHM) of aviation thermal insulation layers to detect and repair damage in a timely manner is a pressing issue. Carbon fiber-reinforced polymer (CFRP) composites have been widely used in aviation thermal insulation layers due to their high strength, light weight, and corrosion resistance [4–6]. This paper focuses on CFRP laminated plates to study the structural health during the service life of insulation layers.

In recent years, fiber optic sensing technology, as a non-invasive, real-time monitoring method, has been increasingly applied in the aerospace field for damage detection [7,8]. Among them, fiber Bragg grating (FBG) sensors have become a research hotspot in damage monitoring due to their high sensitivity, resolution, and resistance to electromagnetic interference [9–11]. Combining FBG sensors with CFRP to create smart composites has shown strong potential in applications such as wind turbine blades and aerospace, where sensor size and weight are critical.

S. Takeda et al. [12] combined finite element analysis (FEA) with FBG sensors, embedding small-diameter FBG sensors into CFRP composite structures. By analyzing the changes in the reflected spectrum, they monitored damage such as delamination and matrix cracking under tensile load. The results showed that as the number of matrix cracks increased, the reflected spectrum width of the FBG sensor also increased, and the results were validated by FEA. Shin et al. [13] embedded FBG sensors into single-layer lap joints of composites and conducted tensile and fatigue loading experiments. The damage initiation and propagation were reflected by the peak splitting and broadening of the full spectrum response of the FBG. Zhu et al. [14] used FBG sensors to detect internal damage in composite samples under low-frequency cycling. The results showed that when the FBG sensor was 15 mm away from the damage and the damage size was below 1.5 mm, the damage could not be accurately identified. Under low-frequency cycling, when the FBG sensor was 15 mm away from the damage and the damage size exceeded 2 mm, the damage could be identified. Yin et al. [15] combined finite element analysis with FBG sensors and proposed a strain-based structural damage identification model. The experimental results showed that the model could accurately identify hole damage positions with an accuracy rate of over 98%, and crack direction recognition accuracy was over 91%. ZHAN et al. [16] encapsulated FBG in a capillary ceramic tube to create a strain-free temperature sensor, embedding it into composites to monitor the temperature and strain during the autoclave curing process. By comparing the temperature monitoring results of ordinary gratings, high-temperature gratings, and thermocouples, they verified the accuracy of temperature monitoring, demonstrating that FBG can monitor the out-of-autoclave curing process of composites and obtain internal temperature and strain data, improving the quality of thermoset composite products. Ma et al. [17] proposed a tilted fiber Bragg grating sensor to monitor the curing of CFRP composite materials. The experimental results show that the sensor is very sensitive to the curing residual strain of CFRP and can be used in applications involving composite curing processes.

These studies show that embedded fiber Bragg grating sensor technology has made significant progress in monitoring the damage and curing process of carbon fiber-reinforced polymer structures. However, there remains a gap in understanding the performance of FBG sensors in complex working conditions. This paper takes the CFRP laminates used as aviation thermal insulation materials as the research object, uses embedded FBG sensors for monitoring, and studies the evolution of the material from intact state to damage failure under the influence of external loads and high-temperature environments. The response characteristics of FBG sensors under complex working conditions are explored, and the mechanical response of CFRP laminates in complex stress environments is observed, including stress–strain relationships, changes in strength and stiffness, and fatigue and damage evolution of materials.

2. Principle

2.1. Fiber Bragg Grating Sensor Principle

A fiber Bragg grating is created by exposing the core of a single-mode optical fiber transversely to strong ultraviolet light with a periodic pattern. The exposure to UV light permanently increases the refractive index of the fiber core, creating a fixed refractive index modulation based on the exposure pattern, known as the grating.

At each periodic refractive index change, a small amount of light is reflected. When the grating period is approximately half the wavelength of the incident light, all reflected light coherently combines into a single beam of highly reflective light. This is known as the Bragg condition. The wavelength of the reflected light is referred to as the Bragg wavelength. Other wavelengths of light signals are almost unaffected by the Bragg grating and continue to transmit through the fiber. The Bragg wavelength condition is expressed as

$$\lambda_B = 2n_{\text{eff}}\Lambda \tag{1}$$

where λ_B is the Bragg wavelength, n_{eff} is the effective refractive index of the fiber core, and Λ is the grating period.

When the FBG sensor is affected by strain or temperature, the Bragg wavelength shifts as shown in Equation (2):

$$\frac{\Delta\lambda_B}{\lambda_B} = (1 - P_e)\Delta\varepsilon + (\alpha + \xi)\Delta T \tag{2}$$

where $\Delta\lambda_B$ is the Bragg wavelength shift, $\Delta\varepsilon$ is the strain variation, P_e is the effective photoelastic coefficient, α is the thermal expansion coefficient, and ξ is the thermo-optic coefficient. According to Equation (2), if the experiment is conducted at a constant temperature or temperature compensation is applied, the relationship between the Bragg wavelength shift and the strain variation is expressed as

$$\frac{\Delta\lambda_B}{\lambda_B} = (1 - P_e)\Delta\varepsilon \tag{3}$$

Since P_e is a constant, it can be deduced from Equation (3) that under conditions without temperature influence, the Bragg wavelength shift and strain variation are linearly related. Thus, FBG can serve as a strain sensing unit. When both temperature and mechanical strain cause a shift in the FBG sensor's Bragg wavelength, to accurately measure strain using the FBG sensor, temperature compensation is necessary to decouple the strain value.

2.2. Failure Criteria for CFRP Laminates

This study adopts the 3D Hashin failure criterion [18] and develops the corresponding Abaqus subroutine to determine whether damage occurs in composite laminates and to identify the damage modes. The expressions for the 3D Hashin failure criterion are as follows:

1. Fiber tensile failure ($\sigma_{11} > 0$)

$$F_{ft} = \left(\frac{\sigma_{11}}{X_t}\right)^2 + \left(\frac{\tau_{12}}{S_{12}}\right)^2 + \left(\frac{\tau_{13}}{S_{13}}\right)^2 \geq 1 \tag{4}$$

2. Fiber compression failure ($\sigma_{11} < 0$)

$$F_{fc} = \left(\frac{\sigma_{11}}{X_c}\right)^2 \geq 1 \tag{5}$$

3. Matrix tensile failure ($\sigma_{22} > 0$)

$$F_{mt} = \left(\frac{\sigma_{22}}{Y_t}\right)^2 + \left(\frac{\tau_{12}}{S_{12}}\right)^2 + \left(\frac{\tau_{13}}{S_{13}}\right)^2 \geq 1 \tag{6}$$

4. Substrate compression failure ($\sigma_{22} < 0$)

$$F_{mc} = \left(\frac{\sigma_{22}}{Y_c}\right)^2 + \left(\frac{\tau_{12}}{S_{12}}\right)^2 + \left(\frac{\tau_{23}}{S_{23}}\right)^2 \geq 1 \tag{7}$$

5. Tensile delamination failure ($\sigma_{33} > 0$):

$$F_{dlt} = \left(\frac{\sigma_{33}}{Z_t}\right)^2 + \left(\frac{\sigma_{13}}{S_{13}}\right)^2 + \left(\frac{\sigma_{23}}{S_{23}}\right)^2 \geq 1 \tag{8}$$

6. Compressive delamination failure ($\sigma_{33} < 0$):

$$F_{dlc} = \left(\frac{\sigma_{33}}{Z_c}\right)^2 + \left(\frac{\sigma_{13}}{S_{13}}\right)^2 + \left(\frac{\sigma_{23}}{S_{23}}\right)^2 \geq 1 \tag{9}$$

where σ_{11}, σ_{22}, and σ_{33} are positive stresses; τ_{12}, τ_{13}, and τ_{23} are shear stresses; X_t and X_c are the longitudinal tensile and compressive strengths of the single-layer plate; Y_t and Y_c are the transverse tensile and compressive strengths of the single-layer plate; Z_t and Z_c are the tensile and compressive strengths perpendicular to the direction of the single-layer plate; and S_{12}, S_{13}, and S_{23} are the longitudinal and the two transverse shear strengths, respectively.

After satisfying the failure criteria, the damage variables representing the remaining elastic modulus of the damaged material are used to describe the constitutive relationships of the composite material. The overall damage states of the fibers and matrix are represented by the damage variables d_f and d_m, respectively [18].

$$d_{\mathrm{f}} = 1 - (1 - d_{\mathrm{ft}})(1 - d_{\mathrm{fc}}) \tag{10}$$

$$d_{\mathrm{m}} = 1 - (1 - d_{\mathrm{mt}})(1 - d_{\mathrm{mc}}) \tag{11}$$

When $F_{ft} \geq 1$, $d_{ft} = 1$; $F_{fc} \geq 1$, $d_{fc} = 1$; when $F_{mt} \geq 1$, $d_{mt} = 1$; $F_{mc} \geq 1$, $d_{mc} = 1$.

3. Finite Element Simulation Analysis

3.1. Finite Element Simulation of CFRP Laminated Plate Strength

To explore the structural changes in aviation thermal insulation materials under axial loads, a 3D Hashin failure criterion is used to simulate the tensile failure process of CFRP laminated plates. In ABAQUS 2023 software, a 3D solid structure with dimensions of 250 × 25 × 2.5 mm is created. The selected element type is C3D8R, and the mesh is generated from the top to the bottom in the thickness direction. The lamination sequence is alternated at 0°/90°, with each single-layer plate thickness set to 0.125 mm, and a total of 20 layers are used for the specimen, with an overall thickness of 2.5 mm. The lamination sequence of the test specimen is shown in Figure 1. The material parameters used are shown in Table 1. The mesh division result is shown in Figure 2.

Figure 1. Lamination sequence of CFRP laminated plate.

Table 1. Material parameters of carbon fiber composite laminates.

Parameter	Value
ρ (kg/m^3)	1700
E_1/GPa	132
E_2, E_3/GPa	9.2
G_{12}, G_{13}/GPa	4.68
G_{23}/GPa	3.35
μ_{12}, μ_{13}	0.312
μ_{23}	0.32
X_T/MPa	1622
X_C/MPa	1270
Y_T, Z_T/MPa	46.2
Y_C/Z_C/MPa	179
S_{12}, S_{13}, S_{23}/MPa	69

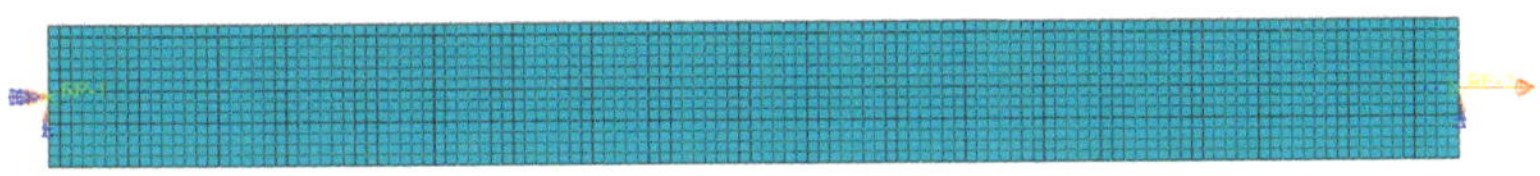

Figure 2. Schematic diagram of boundary conditions of the specimen.

The material properties are defined using the UserMaterial option in the ABAQUS material definition module. The material density, elastic parameters, and 3D Hashin damage criterion parameters are assigned. A displacement load of 10 mm is applied at the right end of the CFRP laminated plate. In the output manager, configuring and extracting the stress, strain, and damage variables for each layer of the CFRP laminate is essential for analyzing the stress response and damage evolution of the material under loading conditions.

3.2. Tensile Load Damage Study of the Model

After the calculations, the overall stress distribution of the CFRP model is shown in Figure 3. The middle section of the elements satisfies the damage criterion defined in the VUMAT subroutine, and the fully damaged elements are deleted.

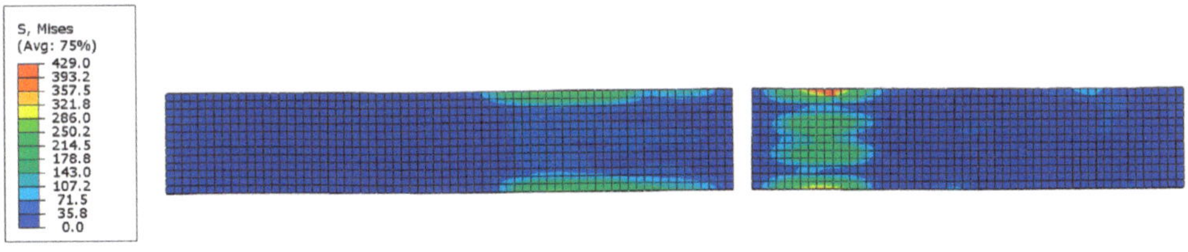

Figure 3. Overall stress distribution of the model.

The load–displacement curve during the tensile process of the CFRP laminated plate is shown in Figure 4. The curve exhibits a clear linear characteristic, and after reaching the ultimate load, the material immediately loses its load-bearing capacity and undergoes brittle fracture, resulting in overall failure. The displacement at the peak load is 3.15 mm, and the ultimate load is 11,342 N.

Since the CFRP laminated plate has structural symmetry, we select the average strain changes between the first 10 layers to analyze and understand the response characteristics of the plate during the loading process.

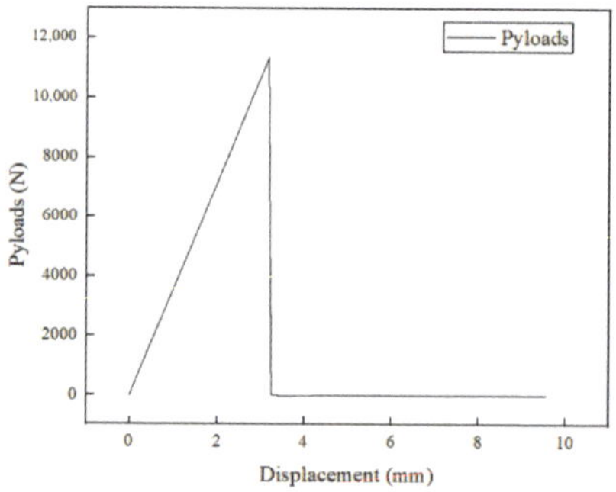

Figure 4. Load–displacement curve of the tensile fracture process of the model.

Figure 5a shows that the strain in the 0° lamination layer increases linearly with the displacement load and decreases sharply after reaching the limit. The strain–displacement curve of the 0° lamination layer is consistent with the overall load–displacement curve, indicating that the 0° lamination fibers dominate the mechanical response of the material during tensile loading. Due to the high strength and modulus of carbon fibers, the material undergoes brittle fracture when the fiber strength limit is reached.

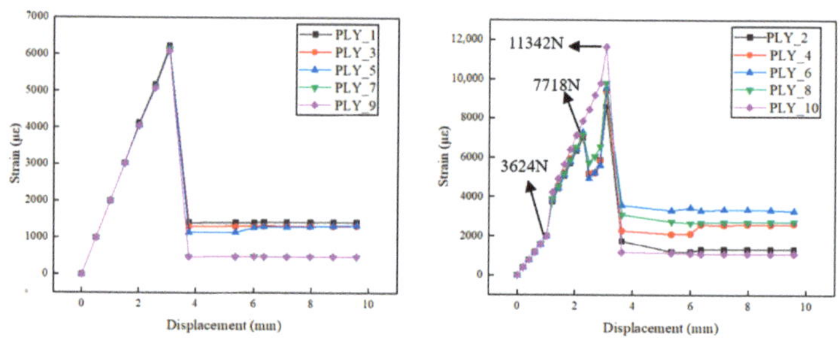

(a) (b)

Figure 5. Interlayer strain variation during tensile loading: (**a**) strain in 0° lamination layer; (**b**) strain in 90° lamination layer.

Figure 5b shows the strain changes in the 90° laminations. The figure indicates that although the strain in each layer increases with the load, the relationship gradually deviates from linearity. This suggests that with the increasing load, internal damage begins in the 90° fiber layer, leading to nonlinear strain variations. Below, we analyze the damage variables to explore the relationship between interlayer strain changes and damage evolution in the CFRP structure.

In this study, the damage propagation of the same layers in the orthogonal laminate was similar. Therefore, the first layer (0° ply) and the second layer (90° ply) were selected to examine the damage. The main damage type in the 0° sub-layer was fiber tensile damage, while the 90° sub-layer primarily exhibited matrix tensile damage. The damage variation in the simulation process is shown in Figures 6 and 7:

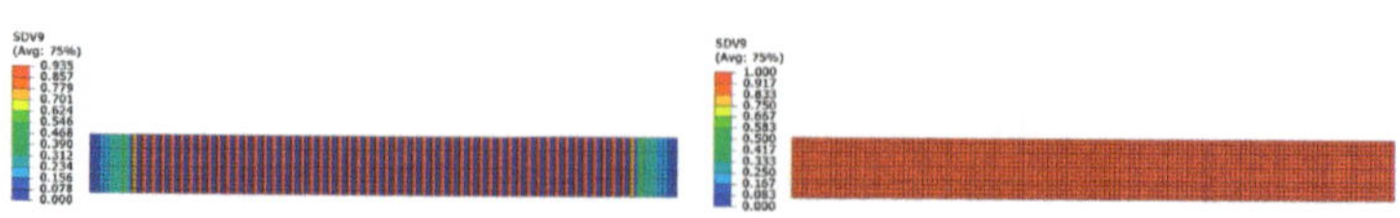

(a) (b)

Figure 6. Matrix tensile damage in the 90° sub-layer: (**a**) F = 3624 N; (**b**) F = 7718 N.

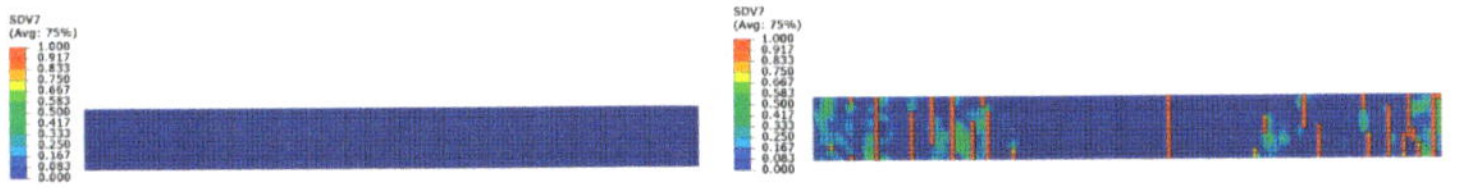

(**a**) (**b**)

Figure 7. Fiber tensile damage in the 0° sub-layer: (**a**) F = 7718 N; (**b**) F = 11,432 N.

At the first stage, within the load range of 3624 N, the strain in all layers exhibited a linear upward trend with the increasing load, indicating that the material's internal condition remained good.

In the second stage, between 3624 N and 7718 N, the strain increased sharply with the load at approximately 3624 N, and the rate of strain change with load also increased. The reason is that the matrix has a relatively low tensile strength. Microcracks and plastic deformation accumulated in the matrix, causing matrix tensile failure in the middle region of the sub-layer. The damage led to a reduction in matrix tensile capacity, changing the local stiffness of the material and resulting in strain concentration, which accelerated the strain increase with load.

In the third stage, between 7718 N and 11,342 N, the strain in the 2nd, 4th, 6th, and 8th layers initially decreased and then increased with load, while the strain in the 10th layer continued to increase with load. At around 7718 N, the matrix in the 2nd, 4th, 6th, and 8th layers completely failed, with the damage index reaching 1, causing the strain to decrease rapidly. Stress and strain redistributed, and since the matrix in the 10th layer had not fully failed, the strain continued to increase with the load.

Finally, at around 11,432 N, the fibers in the 0° ply layer experienced tensile failure after bearing a large amount of tensile load. The fiber breakage caused the entire structure of the laminated plate to fail, resulting in the material cracking and losing its load-bearing capacity.

By conducting a tensile failure simulation of CFRP laminated plates, the evolution process of internal structural damage was revealed, and a corresponding relationship between internal damage variation and interlayer strain variation was established. By monitoring the internal strain variation in the laminated plate, internal damage can be effectively monitored, and preliminary judgments about damage types can be made. The high accuracy and sensitivity of FBG sensors enable the precise capture of strain variation in the material. Therefore, an experiment was designed to embed FBG sensors into locations in the CFRP laminated plate prone to damage, allowing for effective monitoring and evaluation of internal structural damage.

3.3. Finite Element Simulation Analysis of High-Temperature Decomposition

Given that aviation thermal insulation materials are often exposed to high-temperature environments, a high-temperature thermal decomposition simulation analysis was conducted using ABAQUS to further explore the internal structural changes under these conditions. Through numerical simulation, the thermal decomposition behavior and damage evolution process of the material under high temperatures were studied in detail to provide reference for subsequent experiments.

A 3D solid model with dimensions of 250 × 25 × 2.5 mm was established in ABAQUS. The entity is divided into 20 pieces along the thickness direction, and 0°/90° alternating layer laying is achieved by assigning material directions. After the laying is completed, the interlayer stacking of the test piece is shown in Figure 8, which is the same as Figure 1. The mesh was generated by first applying global seeds of 0.005, and then assigning local seeds of 10 in the thickness direction. The element type used was the thermal transmission DC3D8 element. The surface heat exchange coefficient was set to 15, the surface radiation emissivity was set to 1, and the environmental temperature was set to 650 °C. A predefined temperature field of 25 °C was applied to the entire model.

Figure 8. Lamination sequence of CFRP laminated plate.

Based on the 1D transient thermal model for composites proposed by Henderson et al. [19], and the equation for predicting the ultimate strength of CFRP plates calibrated by Ke Wang et al. [20], the UMATHT subroutine was developed in FORTRAN for secondary development in ABAQUS. The calculation flow of the UMATHT subroutine is shown in Figure 9. The thermal property parameters used are the thermal decomposition rate, anisotropic thermal conductivity, and specific heat capacity of the solid material.

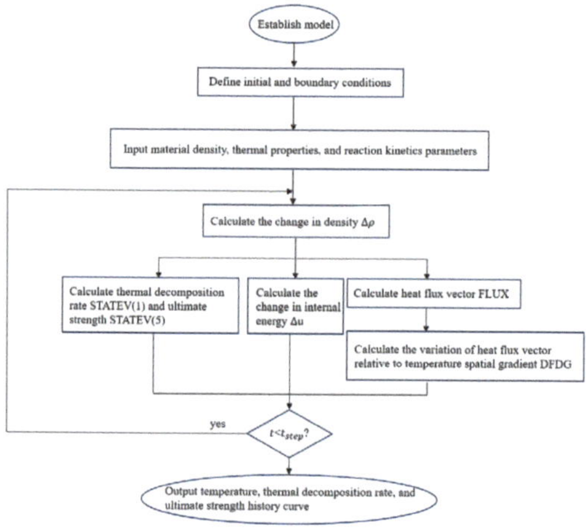

Figure 9. Subroutine calculation flow chart.

The thermal decomposition rate is expressed as follows:

$$F = \frac{\rho_v - \rho}{\rho_v - \rho_d} \tag{12}$$

where ρ_v represents the density of the original material, ρ_d represents the density of the carbonized material, and ρ represents the instantaneous material density. A thermal decomposition rate of 0 indicates that the material is in its original state and has not undergone pyrolysis, while a rate of 1 indicates that the material is fully carbonized and completely pyrolyzed.

The expression for anisotropic thermal conductivity is as follows:

$$k_i = (1 - F)k_{vi} + Fk_{di}(i = x, y, z) \tag{13}$$

where k_{vi} and $k_{di}(i = x, y, z)$ are the thermal conductivity of the raw material and the residual material after pyrolysis reaction, respectively, and F is the thermal decomposition rate.

The specific heat capacity of solid materials can be expressed as

$$C_p = (1 - F)C_{pv} + FC_{pd} \tag{14}$$

where C_{pv} is the specific heat capacity of the raw material, C_{pd} is the specific heat capacity of the residual material after pyrolysis reaction, and F is the thermal decomposition rate.

Chen and Young [21] proposed a model to describe the change in ultimate strength of materials exposed to high temperatures:

$$\frac{f_{u,T}}{f_{u,normal}} = A - \frac{(T - B)^n}{C} \tag{15}$$

$f_{u,T}$ represents the ultimate strength of the material at temperature T, and $f_{u,normal}$ represents the ultimate strength of the material under normal conditinos. In Ke Wang [20], the coefficients A, B, C, and n were calibrated to align with experimental results for carbon fiber composites at a high temperature of 706 °C, which is close to the experimental temperatures in this study. Therefore, this model is utilized in the development of subroutines to predict changes in ultimate strength.

3.4. Simulation Results and Analysis

Figure 10 shows the thermal decomposition rate curves along the thickness direction of the material at the initial moment, 50 s, and the final moment under an environmental temperature of 650 °C. Here, Z = 0 mm represents the heated surface of the material, while Z = −2.5 mm represents the backside surface. The thermal decomposition rate reflects the degree of thermal decomposition reactions in the material, where a higher thermal decomposition rate indicates a more complete decomposition reaction, leading to a greater change in material density and a more significant reduction in the resin matrix. As the simulation progresses, the longer the heating time, the higher the material's thermal decomposition rate. Analysis indicates that the heated surface of the material is affected by external heat flux, causing the temperature of the heated surface to rise the fastest and reach the pyrolysis temperature the earliest, resulting in the most complete pyrolysis. At the same moment, the closer to the Z = 0 mm surface, the higher the thermal decomposition rate.

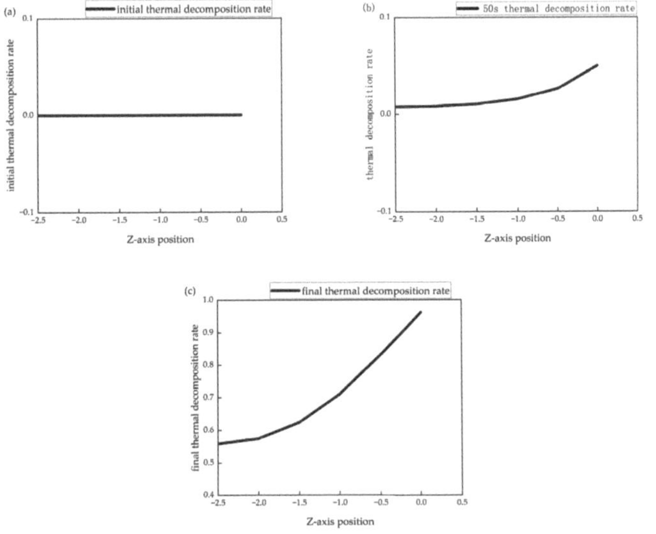

Figure 10. Thermal decomposition rate of model: (**a**) initial thermal decomposition rate; (**b**) 50 s thermal decomposition rate; (**c**) final thermal decomposition rate.

From the heated surface of the laminated plate, six surfaces were evenly selected along the negative Z-axis, with 0 mm representing the heated surface and −2.5 mm representing the unheated back surface of the material. The thermal decomposition rate data during the heating process for these six surfaces is shown in Figure 11. As observed in Figure 11, during the first 40 s of heating, since the material temperature did not yet reach the critical thermal degradation temperature, no thermal decomposition reaction occurred. After 40 s, as the temperature further increased, the heated surface of the laminated plate first reached the critical thermal decomposition temperature and began the thermal decomposition reaction. Subsequently, the other surfaces gradually reached the critical thermal decomposition temperature and started to degrade. By the end of the simulation, the thermal decomposition rate at the heated surface of the laminate reached 0.96. This rate indicates that the material density at the heated surface is very close to that of fully decomposed material, suggesting almost complete thermal decomposition. Conversely, the thermal decomposition rate was found to be the lowest at 0.56 at the z = −2.5 mm surface.

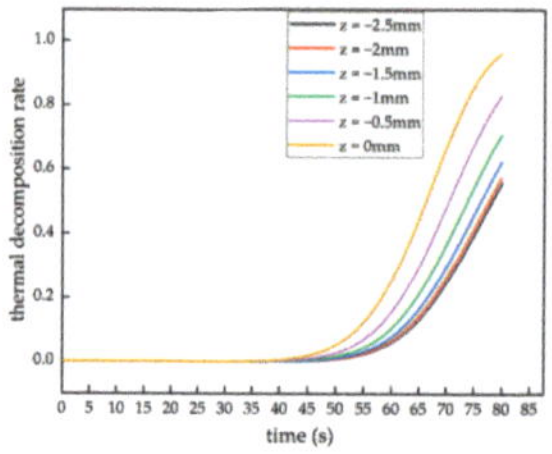

Figure 11. Thermal decomposition rate changes with time at different locations.

Figure 12 shows the variation in the ultimate strength of the model with the material thickness direction at the initial moment, 50 s and the final moment under the ambient temperature of 650 °C. At the beginning of the simulation, the laminated plate had not yet been affected by the heat flow, and the ultimate strength of each surface was at its initial value. As the simulation progressed, at 50 s, the ultimate strength of the heated surface decreased to 50% of its original value, while the surface at −2.5 mm had decreased to 51.6% of its original value. By the end of the simulation, the ultimate strength of the heated surface had decreased to 40%, while the surface at −2.5 mm had decreased to 46%. The rate of ultimate strength reduction varied at different locations, with surfaces farther from the heated side requiring more time to reach the same level of strength reduction. Figure 13 shows the ultimate strength changes over time at different locations in the CFRP laminated plate under a 650 °C single-sided heating condition.

By combining Figures 11 and 13, we can explore the reasons for the performance degradation of carbon fiber-reinforced composite materials (CFRPs) under high temperatures. During the first 40 s of the simulation, the thermal degradation reaction in the laminated plate matrix had not yet started. Therefore, the performance degradation at this stage was not caused by matrix degradation, but rather due to the softening of the epoxy resin matrix when the temperature gradually reached its glass transition temperature. The softening of the epoxy resin led to a gradual loss of its original strength and stiffness. After 40 s, the thermal degradation reaction in the laminated plate matrix began, and the carbon fibers lost support from the matrix, rendering them unable to effectively transfer stress. This further caused a reduction in the material's ultimate strength. As mentioned in Wang's study [20], when the temperature increased above 600 °C, carbon fibers began to oxidize. Although carbon fibers are resistant to high temperatures, they undergo chemical reactions at this temperature, generating carbon monoxide (CO) and carbon dioxide (CO_2), leading to the gradual consumption of carbon fibers and the loss of their load-bearing capacity. This further decreased the ultimate strength of the material.

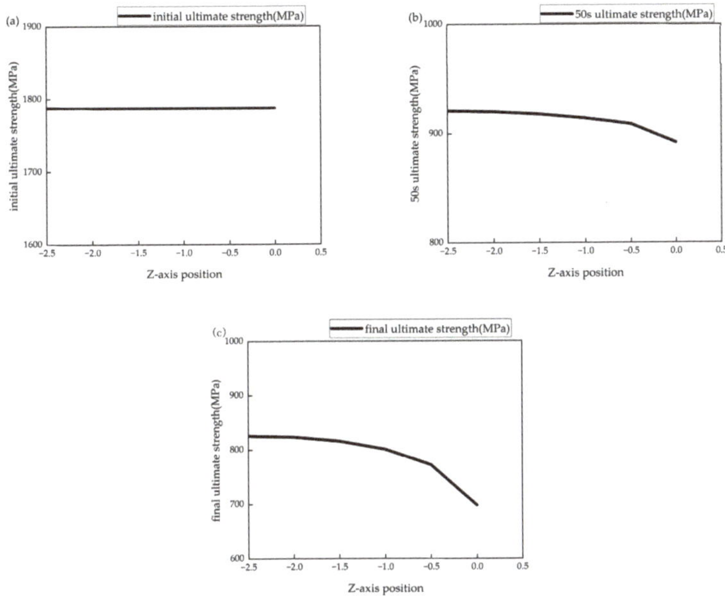

Figure 12. Ultimate strength of model: (**a**) initial ultimate strength; (**b**) 50 s ultimate strength; (**c**) end ultimate strength.

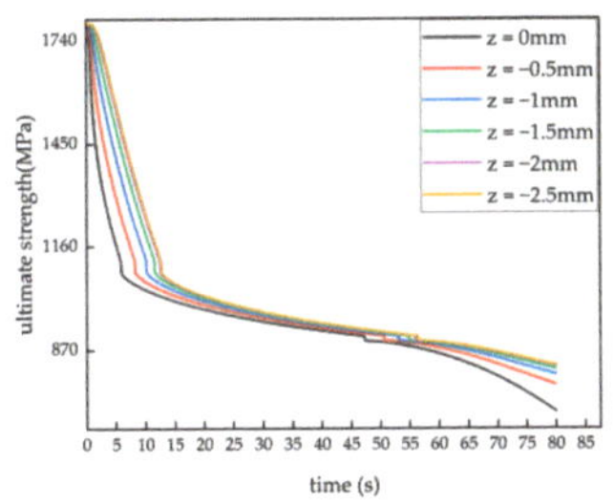

Figure 13. Ultimate strength changes over time at different locations.

4. Experiment and Analysis

Based on the materials used in aviation thermal insulation layers, this study used T300/M03 carbon fiber prepreg to fabricate CFRP laminated plates, with a ply sequence of $[0/90]_{20s}$, and each ply had a thickness of 0.125 mm, with a total thickness of 2.5 mm. The FBG sensor used is a single bare optical fiber. It should be noted that the temperature can reach 650 °C in the subsequent test, and the fiber coating layer cannot withstand this high temperature. Therefore, the coating layer should be removed before burying the fiber. Before curing, place the bare optical fiber in the corresponding layer and cure it together with the composite material. The curing process uses RTM technology and high-temperature curing. The specimens were fabricated according to the recommended tensile specimen geometry in ASTM D3039. FBG sensors were embedded inside the CFRP specimens, with sufficient protection for the internal and exposed ends of the sensors to ensure the feasibility and reliability of embedding sensors into the composite materials. To minimize the structural impact of the embedded fiber, the total thickness of composite laminates was designed to be more than ten times the diameter of the bare fiber, and the fiber was drawn out from the side of the composite material. The structure and dimensions of the CFRP laminated plates are shown in Figure 14. The red rectangle in Figure 14 represents the FBG sensor.

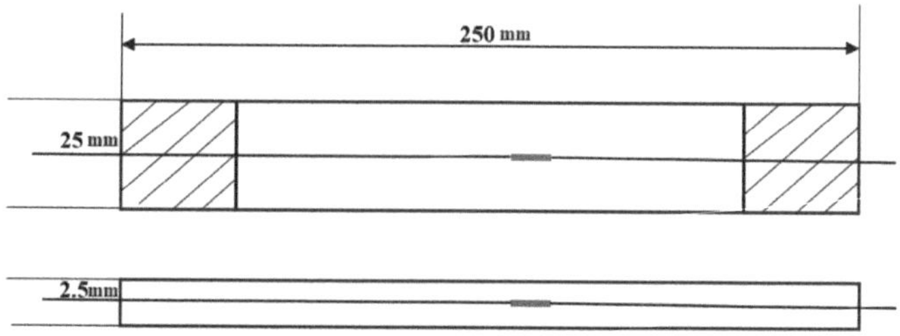

Figure 14. Dimensions of the specimen.

In this experiment, two CFRP laminates embedded with FBG sensors and one CFRP laminate without FBG sensors were designed and prepared, referred to as specimens 1, 2, and 3, respectively. The two CFRP laminates embedded with FBG sensors share the same materials, dimensions, ply angles, number of layers, and fiber embedding techniques. The dimensions and shapes of specimens 1 and 2 are shown in Figure 14.

Specimen 3, while not embedded with FBG sensors, has identical conditions to specimens 1 and 2 in all other aspects. According to Zhou et al. [22], the tensile strength of composite laminates embedded with fiber Bragg gratings in symmetric internal layers is reduced by 5.36% compared to those without embedded gratings, indicating minimal impact. To verify whether embedding FBGs in this experiment results in a similar effect on the strength of laminates, a tensile fracture test was designed for specimen 3.

In order to simulate the complex load environment during the use of CFRP laminates as aviation insulation layers and explore the ability of FBG sensors to monitor the internal state of aviation insulation layers in real-time, stably, and for a long time, tensile fracture tests were conducted on specimen 1, and high-temperature sintering and tensile fracture tests were conducted on specimen 2. To ensure the standardization and comparability of the experiments, all tensile tests were performed according to ASTM D3039 [23].

4.1. Tensile Fracture Test for Specimen 1

This experiment aimed to simulate the working conditions of CFRP laminated plates under axial loads without temperature influence, mimicking their performance on the ground. By performing long-term axial loading experiments, the internal structural changes in aviation thermal insulation materials were studied, and the real-time monitoring capability of FBG sensors under these conditions was further validated.

A static test was conducted on the INSTRON-8801(manufacturer: INSTRON, equipment source: Shanghai, China) testing machine in a room temperature dry condition of 18 °C. The prepared CFRP specimen 1 was placed in the grips of the Instron 8801 testing machine and securely fixed to ensure the alignment of the grips with the central axis, thereby avoiding uneven stress distribution due to eccentric tension, which could increase the risk of premature specimen failure. The experimental setup is shown in Figure 15. During the test, a demodulator was used to decode the wavelength signals from the sensor, with a demodulation frequency of 1 Hz.

First, a calibration test for specimen 1 was conducted. The axial load was increased in steps of 500 N, with each load step held for 30 s to reach a stable state. The load was increased to 3 kN and then unloaded in steps, repeating the process three times. The strain values measured by the extensometer and the fiber sensor were recorded in real-time during the calibration test of specimen 1. Then, a tensile fracture test was performed, with the loading rate controlled at 1.0 mm/min by displacement. After the test, the load and deformation data were recorded, and the failure mode of the specimen was observed.

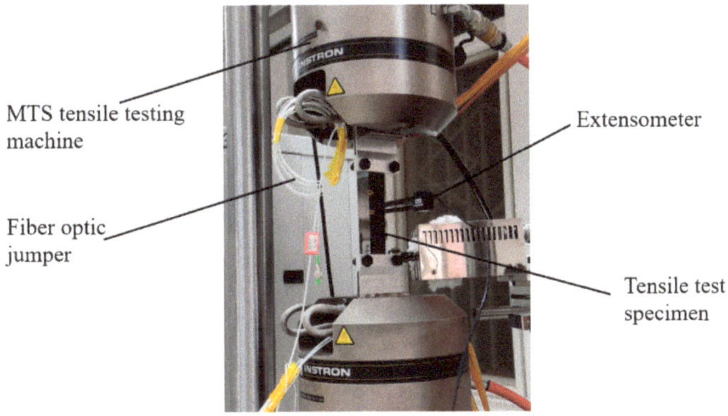

Figure 15. Tensile testing system diagram.

4.1.1. Calibration Data Processing for Specimen 1

The wavelength variation data of the FBG sensor during the calibration test for CFRP specimen 1 is shown in Figure 16.

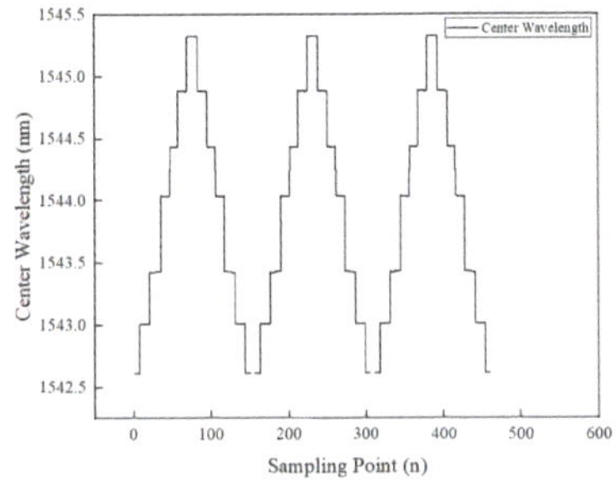

Figure 16. Calibration data for specimen 1.

As shown in Figure 16, the FBG wavelength data from the three repeated tests were almost identical, demonstrating the excellent stability and reliability of the FBG sensor. The average strain values from the extensometer and the fiber sensor at each load level were calculated from the first cycle and correlated with the corresponding load for analysis. The load–strain statistics for each load level and the fitted curves for the extensometer and fiber sensor are shown in Figure 17a,b.

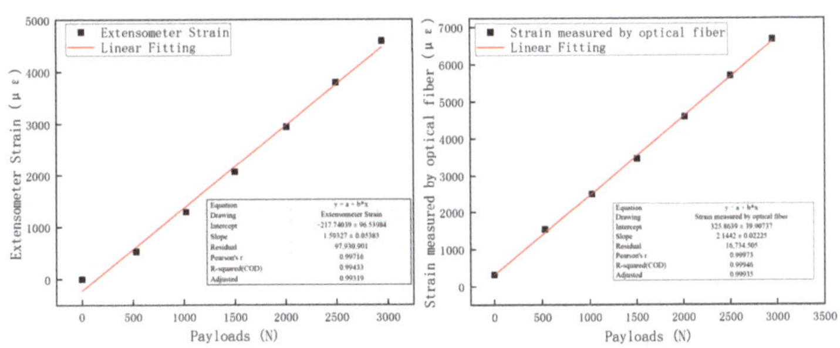

(**a**) (**b**)

Figure 17. Load–strain fitting curves: (**a**) extensometer; (**b**) optical fiber.

As shown in Figure 17, within the 3 kN range, the external strain measured by the extensometer exhibits a linear relationship with the axial load, with a linearity of 0.997. This indicates that the surface of the CFRP specimen undergoes elastic deformation within this range. Similarly, the internal strain of the specimen measured by the optical fiber also shows a linear relationship with the axial load, with a linearity of 0.999. The fitted equation is y = 2.14x + 325.86. This suggests that within the 3 kN range, the internal structure of CFRP specimen 1 remains intact, exhibiting elastic deformation behavior. The sensitivity of strain to the external load is 2.14 με/N. Table 2 is the Load–strain statistics.

Table 2. Load–strain statistics.

Payloads/N	0	530	1021.800	1500.500	2009.390	2494.136	2946.701
Extensometer strain/με	0	529	1292.325	2066.714	2940.667	3790.483	4590.015
Strain measured by optical fiber/με	321.824	1538.721	2504.193	3459.608	4596.049	5692.262	6687.905

It is worth noting that the extensometer is placed on the surface of the specimen, so it measures the surface strain, while the optical fiber is embedded inside the specimen and measures the internal strain. Due to the laminate structure of the specimen, there is an attenuation of stress transfer between layers. Therefore, it is normal for the strains measured by the extensometer and the optical fiber to differ, and a nonlinear relationship between them is expected. At this stage, it is demonstrated that the embedded FBG sensor can monitor the internal strain of the CFRP specimen in real-time and establish correspondence between strain and load under normal conditions, providing a reference for subsequent experiments.

4.1.2. Tensile Fracture Test Data Processing for Specimen 1

The load–displacement curve obtained from the tensile test and the finite element simulation results are compared in Figure 18.

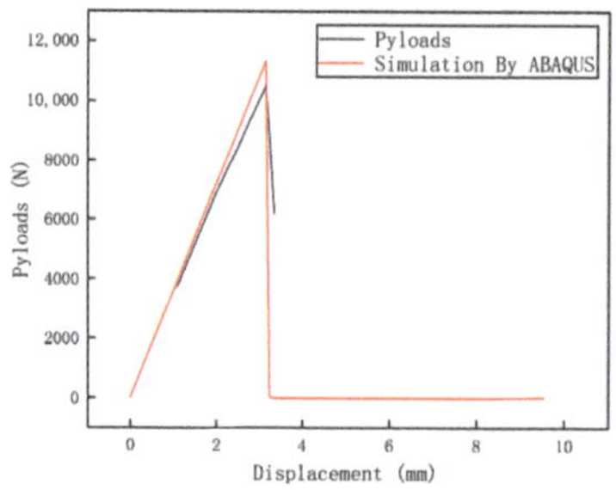

Figure 18. Comparison between numerical calculation and experimental results.

As shown in Figure 18, the ultimate load predicted by the finite element simulation was 11,432 N, which is 902 N higher than the experimental result of 10,530 N, yielding a relative error of 8.5%. This indicates that the finite element prediction of ultimate strength is reliable.

The wavelength shift in the FBG sensor during the tensile test is shown in Figure 19.

As shown in Figure 19, within the 4 kN range, the wavelength increased linearly with the load, indicating elastic deformation, and the internal structure of the CFRP remained in good condition. This demonstrates that the FBG sensor could normally monitor the internal structure of the CFRP.

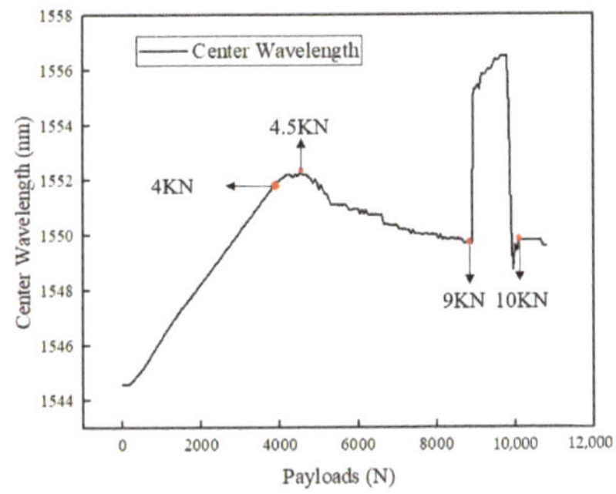

Figure 19. Wavelength of the FBG sensor during tensile testing.

Between 4 kN and 4.5 kN, the wavelength exhibited an overall fluctuating upward trend with the increase in load, and the linear relationship disappeared. Based on the finite element analysis, it is inferred that at this stage, the matrix in the 90° plies began to experience tensile damage, and microcracks started to form, leading to uneven strain distribution inside the material. The FBG sensor accurately captured this non-uniformity and the initial damage in the material.

From 4.5 kN to 9 kN, the wavelength of the FBG sensor began to decrease as the load increased, indicating that internal damage was gradually accumulating, and that the structure was progressively deteriorating. Damage in the matrix near the FBG sensor caused delamination between the plies, leading to local strain release and a nonlinear strain distribution. The wavelength changes recorded by FBG sensors reflect the impact of these local damages on the overall structure, and the fluctuating strain curve also indicates that the damage is expanding, affecting the area near the FBG sensor.

Around 9 kN, the FBG waveform exhibited chirping, in the case of only bearing axial loads, the influence of transversal strain is not sufficient to cause chirping in the waveform. Therefore, it is believed that fibers in the 0° ply near the FBG sensor began to experience tensile failure, generating shear forces that compressed the FBG sensor's grating area. The FBG waveform chirping is shown in Figure 20. The demodulation software displayed the original waveform splitting into two, which caused a significant increase in the recorded wavelength value. This reflects the drastic changes that occur when the FBG sensor gate area is squeezed.

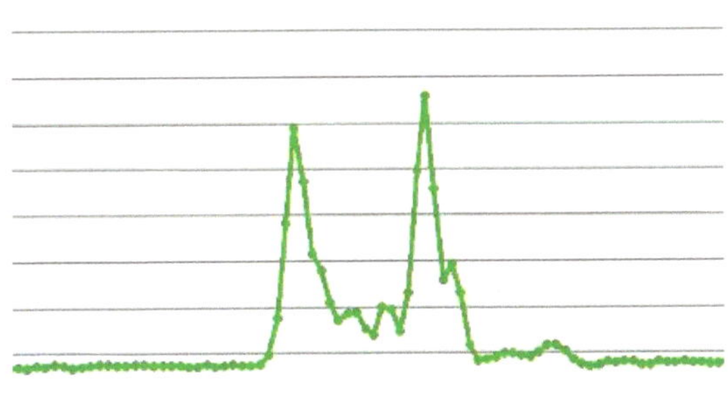

Figure 20. Chirping phenomenon of the FBG waveform.

Finally, at around 10 kN, the material reached its load-bearing limit. The 0° ply fibers experienced massive tensile failure, leading to the total failure of the laminated plate, with the material cracking and losing its entire load-bearing capacity.

The failure of the specimen is shown in Figure 21, where a sound was heard as the sample neared fracture. Fibers in the 0° layer were pulled out and fractured in large quantities. The fracture occurred mainly in the middle of the specimen, and the primary failure modes were fiber pull-out and matrix damage, consistent with the results of the finite element simulation.

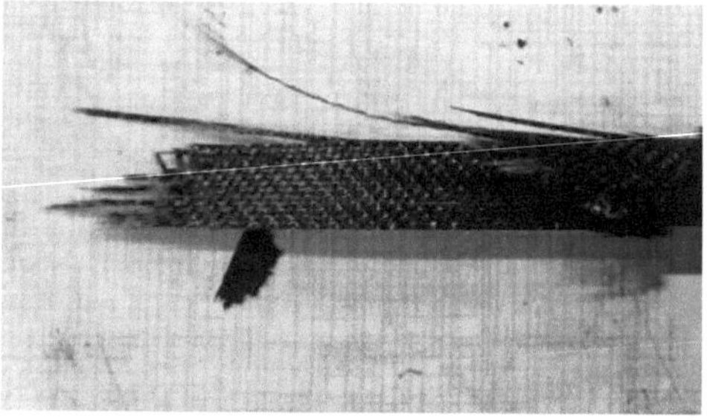

Figure 21. Fracture of the specimen.

In summary, from the initial elastic deformation stage to damage accumulation and expansion, and finally to complete fracture failure, the FBG sensor was able to precisely capture strain changes caused by internal micro-damage in the material. This demonstrates that the FBG sensor, through the real-time monitoring of wavelength shifts, can effectively detect internal damage in aviation thermal insulation materials and accurately reflect microstructural changes under tensile load. The sensor sensitively detected damage initiation and the evolution process, providing a preliminary assessment of the type of damage. This provides reliable data support for structural health monitoring and early damage warning in aviation thermal insulation materials.

4.1.3. Comparison of Ultimate Strength Between Specimen 3 and Specimen 1

Specimen 3, which does not contain embedded FBG sensors, does not require calibration testing, and was directly subjected to tensile failure testing. The experiment was conducted using an INSTRON-8801 testing machine under ambient conditions at 18 °C in a dry state. The prepared CFRP specimen 3 was placed in the grips of the Instron 8801 testing machine and firmly secured to ensure the alignment of the grips with the central axis, thereby preventing uneven stress distribution caused by eccentric loading, which could increase the risk of premature failure.

Figure 22 presents the test results for the ultimate loads of specimens 1 and 3. From the results, it can be observed that the ultimate tensile load of specimen 1 was 10,530 N, whereas that of specimen 3 was 10,862 N. The tensile ultimate load of the composite laminate embedded with fiber Bragg gratings in the center-symmetric layer decreased by 3.06% compared to the laminate without embedded gratings.

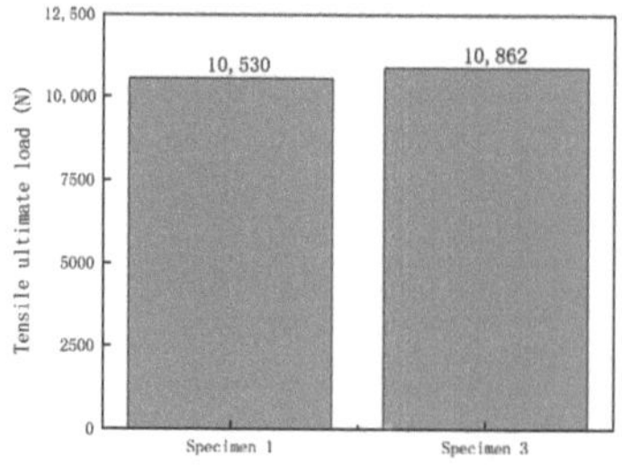

Figure 22. The impact of embedded fiber Bragg gratings on the tensile strength of laminate specimens.

4.2. High-Temperature Sintering Experiment of Specimen 2

Aerospace thermal insulation materials typically serve in extreme high-temperature environments, where prolonged exposure to such conditions can adversely affect the mechanical properties, structural stability, and lifespan of the material. This experiment aims to study whether the fiber Bragg grating embedded in aerospace thermal insulation materials can continue to provide stable monitoring in these environments. A split-type SX2-5–12 muffle furnace is used to perform high-temperature sintering on test specimen 2. Starting from room temperature, the initial set temperature is 400 °C. Once the temperature stabilizes, the set temperature is raised, and this process is repeated 10 times. The temperature is increased to 650 °C before stopping the experiment. The FBG sensor is used to dynamically monitor the strain changes inside the material in real time throughout the process. The center wavelength changes in the FBG during sintering are recorded using a PC. The experimental setup is shown in Figure 23.

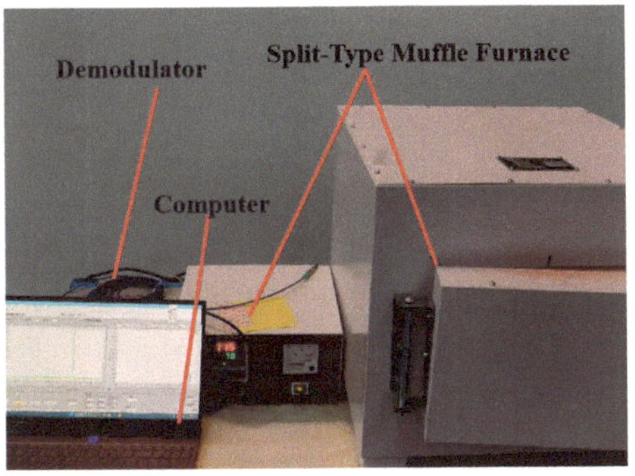

Figure 23. High-temperature sintering experimental setup.

High-Temperature Sintering Experimental Data Processing

The wavelength variation in the FBG sensor during the high-temperature experiment is shown in Figure 24. From Figure 24, it can be seen that during the process of heating up to 650 °C, FBG stably monitored the changes throughout the entire process, indicating that FBG can work stably in extremely high-temperature environments.

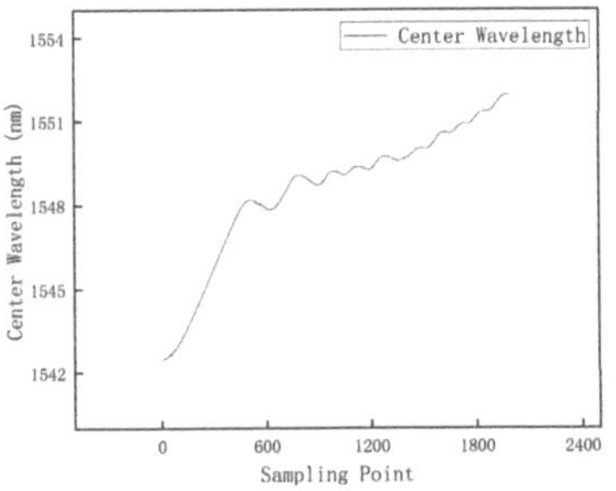

Figure 24. High-temperature sintering wavelength variation diagram for specimen 2.

4.3. Tensile Fracture Test for Specimen 2 After High-Temperature Treatment

The high-temperature sintering experiment verified that the FBG can still function normally for monitoring in high-temperature environments. To investigate whether the FBG can continue to stably monitor the internal structural state of the material after high-temperature treatment, a tensile failure test was conducted on cooled test specimen 2. The experiment used the same tensile test system as test specimen 1, and specimen 2 was mounted on the fixture. Since the mechanical properties of the carbon fiber composite laminate after high-temperature burning are unclear, a stepwise stable load was applied to the specimen until material failure and fracture occurred. After organizing the data measured by the optical fiber, the strain data measured from the high-temperature burning test specimen was obtained, as shown in Figure 25. Compared with the tensile test of test specimen 1, it was found that after high-temperature treatment, the tensile properties of the CFRP laminate decreased significantly. It can withstand 8KN of support under normal conditions, but brittle fracture occurred at 2KN after high-temperature treatment. Growth intensity decreased by 75%.

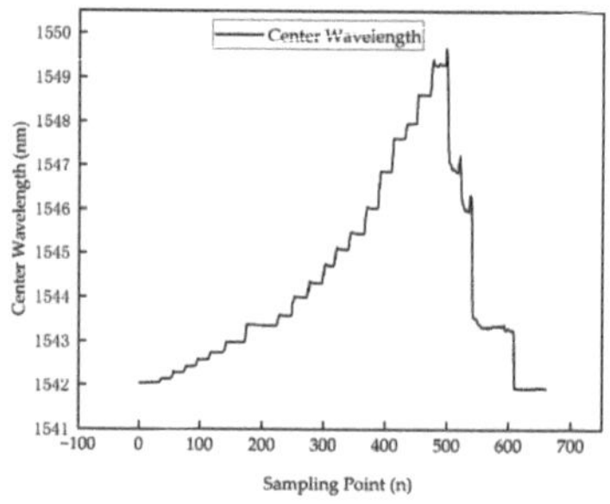

Figure 25. Tensile fracture experiment for specimen 2.

As shown in Figure 25, despite undergoing high-temperature treatment, the FBG sensor structure and performance were not irreversibly damaged, and the sensor was still able to monitor the mechanical performance of the specimen, continuing to measure strain in real-time. This demonstrates the stability and reliability of the FBG sensor in high-temperature environments and its ability to monitor the structural health of CFRP materials under complex conditions.

The scatter plot in Figure 26 was drawn by selecting the average wavelength values of the FBG sensor at each stable load step.

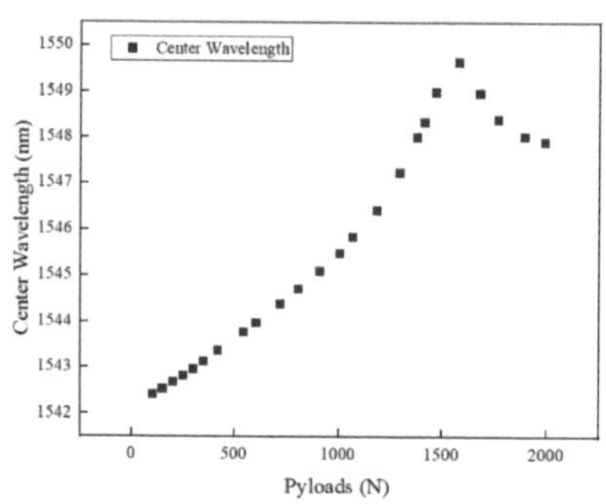

Figure 26. Wavelength changes in specimen 2 during tensile fracture testing.

As shown in Figure 26, within the initial 1 kN load range, despite the high-temperature treatment, the strain measured by the FBG sensor still showed a good linear relationship with the load. This indicates that at low loads, the overall stiffness and integrity of the material had not completely degraded, and the CFRP material still exhibited elastic deformation.

When the load exceeded 1 kN, the strain–load relationship began to deviate from linearity, and the FBG sensor displayed strain values that increased more significantly than the load. This nonlinear response suggests that internal damage had gradually accumulated, particularly as the resin's evaporation during high-temperature treatment weakened the bond between fibers and the matrix. This stress concentration at the interface further triggered matrix cracking, microcrack propagation, and interlayer delamination damage mechanisms. At this stage, the stress distribution in the material became uneven, and the internal structural damage slowly expanded.

When the load exceeded 1.5 kN, despite the continued increase in external load, the FBG wavelength began to decrease. This indicates that internal damage accumulation had accelerated, and the micro-damage caused by high-temperature treatment further spread, reducing the interfacial stress transfer efficiency between fibers and the matrix. Fiber bundles may have begun to slip or debond. The local stiffness gradually decreased, explaining the observed strain drop.

At around 2 kN, specimen 2 reached its ultimate failure, with damage accumulation leading to brittle fracture. Compared to specimen 1, which was not exposed to high-temperature treatment, specimen 2 exhibited early failure at a much lower load due to the evaporation of the resin, which caused significant microcracks and interface damage, making the material unable to bear higher loads. The FBG sensor accurately captured the process of damage accumulation and expansion, further confirming its sensitivity and effectiveness in monitoring internal strain changes and damage evolution.

The tensile test results for specimen 2 after high-temperature treatment showed that although significant internal changes occurred, the FBG sensor structure remained functional, and it could still monitor the strain of the material in real time. This highlights the stability and reliability of FBG sensors in high-temperature conditions.

5. Conclusions

This study employed finite element simulations and experiments on CFRP laminated plates embedded with FBG sensors to verify the effectiveness of FBG sensors in monitoring damage progression under complex stress and high-temperature conditions. The main conclusions are as follows:

- In CFRP laminated plates without high-temperature treatment, FBG sensors accurately captured the entire process from elastic deformation to final failure and provided preliminary identification of damage types. The damage evolution process, dynamically tracked through wavelength changes, was highly consistent with the finite element simulation results, validating the reliability of FBG sensors in complex stress environments.
- In CFRP laminated plates subjected to 650 °C high-temperature treatment, although the mechanical properties significantly deteriorated, with a 75% reduction in tensile strength, FBG sensors continued to effectively monitor internal strain changes, recording the entire progression from damage accumulation to final failure. This demonstrates the stability and durability of FBG sensors in high-temperature environments.

Although FBG sensors have stability and reliability in complex stress and high-temperature environments, there are still certain limitations as composite material structural health monitoring devices: firstly, calibration is required before use, secondly, temperature compensation is needed, and finally, there may be creep, and the long-term stability of the sensor needs to be verified.

Through this study, the long-term application potential of FBG sensors in monitoring aviation thermal insulation layers has been validated, providing important technical support for improving the safety and reliability of aircraft. Future research can focus on ident ifying the nonlinear relationship between internal strain and surface strain in laminated panels.

Author Contributions: Conceptualization, G.Y.; Methodology, G.Y.; Software, B.W. and H.H.; Validation, B.W.; Formal analysis, G.Y., B.W. and W.L.; Investigation, G.Y.; Resources, W.L.; Data curation, B.W., H.H. and W.L.; Writing—original draft, G.Y.; Writing—review & editing, W.L.; Visualization, H.H.; Supervision, W.L.; Project administration, W.L.; Funding acquisition, G.Y. All authors have read and agreed to the published version of the manuscript.

Funding: This research received no external funding.

Institutional Review Board Statement: Not applicable.

Data Availability Statement: Restrictions apply to the datasets. The dataset presented in this article is not readily available as the data are part of an ongoing follow-up study. For requests to access the dataset, please contact email: yanguang79@bistu.edu.cn.

Conflicts of Interest: The authors declare no conflict of interest.

References

1. Congdon, E.; Mehoke, D.; Buchta, M.; Nagle, D.; Zhang, D. Development of a high-temperature optical coating for thermal managemnt on Solar Probe Plus. In Proceedings of the 10th AIAA/ASME Joint Thermphysics Heat Transfer Conference, Chicago, IL, USA, 28 June–1 July 2010.
2. Reynolds, E.L.; Driesman, A.; Kinnison, J.; Lockwood, M.K. Solar probe plus mission overview. In Proceedings of the AIAA Guidance, Navigation and Control (GNC) Conference, Boston, MA, USA, 19–22 August 2013.
3. Dombrovsky, L.A.; Reviznikov, D.L.; Kryukov, A.P.; Levashov, V.Y. Self-generated clouds of micron-sized particles as a promsing way of a Solar Probe shielding from intense thermal radiation of the Sun. *J. Quant. Sprectroscopy Radiat. Transf.* **2017**, *200*, 234–243. [CrossRef]
4. Pramanik, A.; Basak, A.K.; Dong, Y.; Sarker, P.K.; Uddin, M.S.; Littlefair, G.; Dixit, A.R.; Chattopadhyaya, S. Joining of carbon fibre reinforced polymer (CFRP) composites and aluminium alloys—A review. *J. Compos. Part A Appl. Sci. Manuf.* **2017**, *101*, 1–29. [CrossRef]
5. Wang, F.F. Study on the application of carbon fiber composite materials in high-speed trains. *Mater. Sci. Forum* **2017**, *893*, 31–34. [CrossRef]
6. Rajak, D.K.; Wagh, P.H.; Linul, E. Manufacturing technologies of carbon/glass fiber-reinforced polymer composites and their properties: A review. *J. Polym.* **2021**, *13*, 3721. [CrossRef] [PubMed]
7. Scheel, I.; Udd, E. Fiber gratings strain sensor systems for composites and adhesive joints. In Proceedings of the SPIE Smart Structures and Materials + Nondestructive Evaluation and Health Monitoring, Portland, OR, USA, 25–29 March 2017; Sensors and Smart Structures Technologies for Civil, Mechanical, and Aerospace Systems; SPIE: Bellingham, WA, USA, 2017; Volume 10168, pp. 448–455.
8. Kahandawa, G.C.; Epaarachchi, J.; Wang, H.; Lau, K.T. Use of FBG sensors for SHM in aerospace structures. *J. Photonic Sens.* **2012**, *2*, 203–214. [CrossRef]
9. Zhang, W.; Zhang, H.; Jin, L.; Liu, B.; Zhao, J.; Tu, Q.; Kai, G.; Dong, X. An embedded FBG sensor for simultaneous measurement of stress and temperature. *J. IEEE Photonics Technol. Lett.* **2005**, *18*, 154–156.
10. Leng, J.; Asundi, A. Structural health monitoring of smart composite materials by using EFPI and FBG sensors. *J. Sens. Actuators A Phys.* **2003**, *103*, 330–340. [CrossRef]
11. Torres, B.; Payá-Zaforteza, I.; Calderón, P.A.; Adam, J.M. Analysis of the strain transfer in a new FBG sensor for structural health monitoring. *J. Eng. Struct.* **2011**, *33*, 539–548. [CrossRef]
12. Takeda, S.; Yokozeki, T.; Ogasawara, T. Damage monitoring of polymer-lined carbon fibre-reinforced plastic using small-diameter fibre Bragg grating sensors. *J. Reinf. Plast. Compos.* **2015**, *34*, 454–462. [CrossRef]
13. Shin, C.-S.; Lin, T.-C. Adhesive Joint Integrity Monitoring Using the Full Spectral Response of Fiber Bragg Grating Sensors. *J. Polym.* **2021**, *13*, 2954. [CrossRef] [PubMed]
14. Zhu, Q.; Xu, C.; Yang, G. Experimental research on damage detecting in composite materials with FBG sensors under low frequency cycling. *J. Int. J. Fatigue* **2017**, *101*, 61–66. [CrossRef]
15. Yin, W.; Hu, Y.; Ding, G.; Fan, Z.; Jiang, N. Damage identification of CFRP laminate based on finite element analysis and FBG sensor network. *J. Opt. Fiber Technol.* **2022**, *73*, 103057. [CrossRef]
16. Zhan, Y.; Feng, C.; Shen, Z.; Xie, N.; Liu, H.; Xiong, F.; Wang, S.; Sun, Z.; Yu, M. Fiber Bragg grating monitoring for composites in the out-of-autoclave curing process. *J. Opt. Technol.* **2018**, *85*, 371–376. [CrossRef]
17. Ma, K.-P.; Wu, C.-W.; Tsai, Y.-T.; Hsu, Y.-C.; Chiang, C.-C. Internal residual strain measurements in carbon fiber-reinforced polymer laminates curing process using embedded tilted fiber bragg grating sensor. *J. Polym.* **2020**, *12*, 1479. [CrossRef] [PubMed]
18. Hashin, Z. Failure criteria for unidirectional fiber composites. *J. Appl. Mech.* **1980**, *47*, 329–334. [CrossRef]
19. Henderson, J.B.; Wiebelt, J.A.; Tant, M.R. A model for the thermal response of polymer composite materials with experimental verification. *J. Compos. Mater.* **1985**, *19*, 579–595. [CrossRef]
20. Wang, K.; Young, B.; Smith, S.T. Mechanical properties of pultruded carbon fibre-reinforced polymer (CFRP) plates at elevated temperatures. *J. Eng. Struct.* **2011**, *33*, 2154–2161. [CrossRef]
21. Chen, J.; Young, B. Stress–strain curves for stainless steel at elevated temperatures. *Eng. Struct.* **2006**, *28*, 229–239. [CrossRef]

22. Zhou, Y.-J.; Song, H.; Liu, G.; Wang, W.-J.; Li, X.Q. Tensile strain of composite laminates using embedded fiber bragg grating sensors. *J. Mater. Eng.* **2012**, *9*, 58–61+65.
23. *ASTM D3039/D3039M-17*; Standard Test Method for Tensile Properties of Polymer Matrix Composite Materials. ASTM International: West Conshohocken, PA, USA, 2017.

polymers

MDPI

Article

Interlaminar Shear Strength Change and Storage Life Prediction of Carbon Fiber/Epoxy Composites with Hygrothermal Accelerated Aging

Jinjuan Fan [1,2,*], Qin Zhang [1,2], Xinwen Chen [1,2] and Yuhuai He [1,2]

[1] AECC Beijing Institute of Aeronautical Materials, Beijing 100095, China
[2] Beijing Key Laboratory of Aeronautical Materials Testing and Evaluation, Beijing 100095, China
* Correspondence: fanjinjuan621@163.com

Abstract: In order to investigate the durability of fiber-reinforced polymer composites in hygrothermal environments, hygrothermal accelerate aging tests, for 360 days at 70 °C, RH70%; 70 °C, RH85%; 85 °C, RH70%; and 85 °C, RH85% and natural storage for 2 years in Guangzhou, China, were carried out for composite laminates. Then, the moisture absorption and interlaminar shear strength were measured. The hygrothermal damage mechanism of the composite was studied by Fourier transform infrared (FTIR), differential scanning calorimetry (DSC), thermogravimetric analysis (TGA), and field emission scanning electron microscopy (FSEM). A dual stress storage life prediction model and the equivalent relationship between natural storage and hygrothermal acceleration were established. The results show that the order of moisture absorption rates, moisture absorption contents, and the severity effect order on the interlaminar shear strength is RH85%; 85 °C > 70 °C; RH85% > 85 °C; RH70% > 70 °C; and RH70%. The time to achieve an effective moisture absorption balance is opposite to this. The moisture absorption rate meets Fick's law before the effective moisture absorption balance, and then shows a linear trend. The interlayer shear strength still decreases exponentially with aging, which is mainly caused by the resin plasticization and interface weakening. Hygrothermal accelerated aging for 13.4831 days at 85 °C; RH85% is equivalent to that for one-year actual storage in Guangzhou. According to the failure criterion of shear strength decreasing to 77%, the storage life of T700/epoxy in Guangzhou is 14.4661 years.

Keywords: carbon fiber-reinforced resin-based composite; hygrothermal accelerated aging; interlaminar shear strength; failure mechanism; storage life prediction

Citation: Fan, J.; Zhang, Q.; Chen, X.; He, Y. Interlaminar Shear Strength Change and Storage Life Prediction of Carbon Fiber/Epoxy Composites with Hygrothermal Accelerated Aging. *Polymers* **2024**, *16*, 1109. https://doi.org/10.3390/polym16081109

Academic Editor: Antonino Alessi

Received: 3 March 2024
Revised: 4 April 2024
Accepted: 9 April 2024
Published: 16 April 2024

1. Introduction

Carbon fiber-reinforced polymer composites have been widely used in the aerospace field because of their advantages of high specific strength and specific modulus, good designability, good fatigue resistance, and easy large-scale integral molding. The main reinforcing material currently used in polymer composites is T300. T700 and T800 carbon fibers with high strength and high toughness are the new generation of reinforcing materials [1,2].

According to the requirements of aerospace system applications, composite structural components should be suitable for long-term storage and ready for use at any time. Thus, polymer composite structures are inevitably exposed to environments with, for example, UV, high temperatures, high humidity, and high salt spray, which would lead to material changes, reduce their performance, and affect their durability [3–5]. Wang [6] studied the simulated marine environment aging behavior of a T300-reinforced polymer composite and found that the deterioration severity order of aging methods is hydrothermal aging, salt spray, and salt water immersion. Most research indicates that hygrothermal is one of the main environmental factors leading to performance decrease in polymer composites [6–10].

The effects of a hygrothermal environment on polymer composites were investigated by accelerated testing due to the fact that natural aging tests would take a long time [11–15].

The accelerated testing method subjects the specimen to a controlled and simpler, but more severe, test condition than the service environment and accelerates the material property degradation. Simultaneously, environmental effect factors can be separated into multiple influencing factors such as temperature, humidity, ultraviolet, salt mist, etc. Single-factor acceleration tests or coupled-factor acceleration tests can be conducted to obtain the key environmental factors that affect material performance. This method, combined with predictive models, can effectively save experimental time and costs.

Yang [16], Peret [17], Almeida [18], Sui [19], and Kesentini [20] found that the moisture absorption characteristics of carbon fiber-reinforced polymer composites, under water immersion at temperatures of 25 °C, 70 °C, and 85 °C and hygrothermal environments of 80 °C, RH85%, and 80 °C, RH 90%, conformed to Fick's law. Wu Rui [21] found that the later moisture absorption behavior of polymer composites deviated from Fick's law. Wang et al. [6,16,17] found that the tensile performance was less affected, the interlayer shear performance decreased rapidly, and the resin matrix underwent aging degradation in hygrothermal environments. Niu Yifan et al. [22,23] found that the swelling and plasticization of the resin after moisture absorption led to a significant decrease in the shear strength, and that a brittle fracture formed by the coupling of shear and compression; the interface debonding and resin hydrolysis led to a decrease in bending strength, and the fracture form was compressive brittle fracture. Liu [24] analyzed the failure mechanism of resin-based composite materials in hygrothermal environments, and found that the molecular chain structure of the resin would change after moisture absorption. DSC testing showed the glass transition temperature decrease in the resin.

Usually, shear and compression properties, which were significantly affected by the hygrothermal environment, were used as life-characterizing parameters. The degradation of physical and chemical properties is also used for life prediction; for example, Wani [25] used dynamic mechanical analysis to predict storage life prediction. There are two main types of life prediction models: one is a dynamic model or empirical model with a certain physical meaning, including the single factorial Arrhenius Model and Eyring Model, as well as the multiple-factor generalized Arrhenius Model, Eyring Model, Pecking Model, etc. [1]. These models have been used to predict the creep life [26], fatigue life [27], and storage life [28] of T300 fiber-reinforced composites. Olesja [2] et al. summarized the usage conditions of various models, which were related to the environmental factors, performance degradation patterns, and failure mechanisms of materials. It is generally required that the failure mechanism does not change throughout the entire experimental process. The other model is a numerical simulation model, which fits equations based on data trends and has no physical meaning. This type of model has a relatively high degree of fit, but their application conditions are often limited.

At present, there is more research on performance changes and life predictions in hygrothermal environments for T300 fiber-reinforced polymer composites. However, research on T700 fiber-reinforced composites is few. Wang [29] investigated the hygrothermal behavior of T700 and T300 BMI and found that the tensile/compressive strength of two composites decreased in humid and hot environments, while the glass transition temperature became lower with the rate of water absorption. YU [30] found that, for T700 and T300 carbon fiber-strengthened epoxy resin composites aged at 70 °C, RH85% for 70 days, the longitudinal and transverse compressive strength and shear strength slightly reduced. The property retention rate of T700 carbon fiber-reinforced composites was slightly better. The surfaces of T300 fiber and T700 fiber exhibit significantly different characteristics, with T300 appearing as bark and having distinct grooves of varying depths. On the other hand, T700 presents a smooth surface [31]. This may lead to different resistances of T300 and T700 to humidity and heat.

In addition, the hygrothermal aging time of T700 polymer composites is relatively short, the long-term durability in humid and hot environments has not been fully reflected, and the storage life of T700 polymer composites is not focused. Current prediction models are mainly single-factor models and cannot simultaneously investigate the effects of temperature and humidity on composite properties.

In this article, the property degradation and hygrothermal failure mechanism of T700 fiber-reinforced polymer composites were investigated by long-term hygrothermal aging tests. A dual factorial life prediction model of temperature and humidity was established on the basis of property degradation and failure mechanism analysis. The natural storage life of T700 fiber-reinforced polymer composite in Guangzhou, China, where the temperature and humidity are high, was prediction.

2. Experimental Specimens and Experimental Methods

2.1. Specimens

First, 200 mm × 200 mm carbon fiber-reinforced polymer matrix composite unidirectional laminates were prepared using T700 grade fibers with a fiber volume content of 60%. The polymer matrix was epoxy resin with an epoxy value of 85 eq/100 g, and a schematic diagram is shown in Figure 1. Composite laminates were made by wet winding method. The winding angle was 0°. Curing pressure was 0.6 MPa and curing process was 130 °C/4 h + 160 °C/2 h. The size of composite laminates was 300 mm × 300 mm.

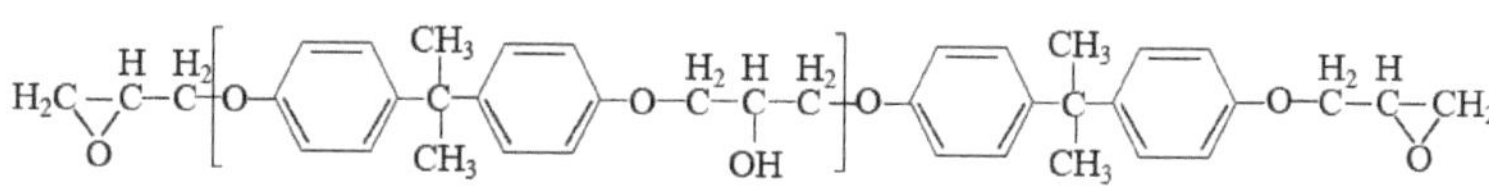

Figure 1. Schematic diagram of epoxy resin structure.

Before hygrothermal aging and natural storage, composite laminate edges were sealed with HM108 sealant. After hygrothermal aging, laminates were machined by the precision cutting machine. In order to eliminate the edge effect of defects and moisture absorption, the laminates edges were removed by 25 mm.

The size of the moisture absorption specimen was 50 mm × 50 mm × 2 mm and the size of the interlayer shear specimen was 20 mm × 10 mm × 2 mm, which met the specimen requirements in the standard HB7401 [32]. The length and width accuracy of all specimens was ±0.02 mm and the thickness accuracy was ±0.05 mm, which met the specimen requirements in the standard JC-773 [33].

The samples were divided into two parts. One part was used for the aging test, while the other part was used for natural storage tests in Guangzhou, China.

2.2. Hygrothermal Accelerated Aging Test

Hygrothermal accelerated aging tests were conducted according to standard HB7401 [32] by constant temperature and humidity testing equipment with a temperature accuracy of ±0.2 °C and a humidity accuracy of ±RH2%. The hygrothermal accelerated aging conditions were 70 °C, RH70%, 70 °C, RH85%, 85 °C, RH70%, and 85 °C, RH85%, respectively. The test period was 360 days.

The masses of moisture absorption specimens were measured using a precision analytical balance with an accuracy of 0.0001 g. For the first four days, the specimen masses were measured every day, and then measured every three days. After reaching an effective moisture equilibrium, the masses of the specimens were measured every 10 days.

The moisture content ΔM was calculated using the following formula:

$$\Delta M = \frac{W_i - W_0}{W_0} \times 100\%, \quad (1)$$

where

W_i—current specimen mass, g;
W_0—initial specimen mass, g.

When the moisture content change in the specimens is less than 0.02% for two consecutive times within 7 ± 0.5 days, it is considered that the specimens have reached an effective moisture equilibrium.

2.3. Interlayer Shear Test

After the moisture absorption test for 30 days, 60 days, 120 days, 180 days, 240 days, 300 days, and 360 days, 5 specimens were taken out and subjected to interlayer shear tests according to the standard JC/T773. A schematic diagram of the interlayer shear test is shown in Figure 2. The hardness of the loading head was 60–62 HRC, and there should be no burrs on the loading head surface. During the experiment, the loading head was equidistant from the supports on both sides, with an accuracy of ±0.3 mm. The movement speed of the loading head was 1 mm/min. Loading continued until the load dropped by 30% and the test was stopped. The interlayer shear strength was calculated by the following formula:

$$\tau = 0.75 \times \frac{F}{bh}, \tag{2}$$

where

τ—interlayer shear strength, MPa;
F—the maximum load, N;
b—specimen width, mm;
h—specimen thickness, mm.

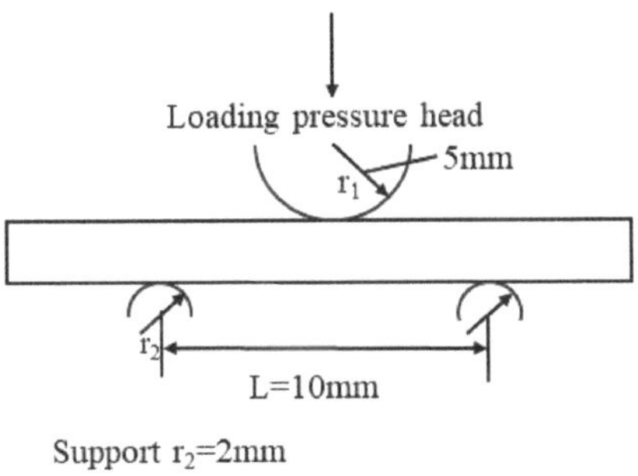

Figure 2. Schematic diagram of interlayer shear test.

2.4. Mechanism Analysis of Hygrothermal Accelerated Aging

To determine the changes in the type and content of characteristic functional groups of T700/epoxy resin before and after the hygrothermal accelerated aging test, infrared analysis was conducted on the unaged and aged specimens using the Spectrum 100 infrared spectrometer from PerkinElmer Company, Waltham, MA, USA.

The crystallization temperature and glass transition temperature of the resin matrix before and after hygrothermal accelerated aging were analyzed using the DSC25 differential scanning calorimeter from TA Company, New Castle, DE, USA, according to ASTM D3418 [34]. The test conditions were a nitrogen atmosphere, the heating rate was 10 °C/min, and the testing temperature range was −80 °C~300 °C.

The initial thermal decomposition temperatures of composite materials before and after aging were analyzed using the TGA2050 thermogravimetric analyzer from TA Company. The test conditions were a nitrogen atmosphere, the heating rate was 10 °C/min, and the testing temperature range was 30 °C~800 °C.

The specimens after interlaminar shear tests were embedded with epoxy resin in both vertical and parallel fiber directions. After curing at room temperature for 2 h, the specimens were ground using sandpaper with particle sizes of 200 #, 600 #, and 1200 #, then burnished using silicon dioxide polishing solution (particle size of 0.5 microns) and the metallographic samples were prepared. A thin layer of gold (about nm) was sprayed on the metallographic sample surfaces to increase the conductivity of the material using a KYKY-SBC-12 ion-sputtering instrument from KYKY Technology Company, Beijing, China. The conditions for spraying gold were as follows: voltage of 25 kV, vacuum degree of 6 Pa, and time of 45 s. The delamination, crack propagation path, interface changes between the fiber and resin, and resin change were analyzed using a field emission scanning electron microscope (FESEM).

2.5. Storage Life Prediction

On the basis of the failure mechanism analysis of hygrothermal accelerated aging, and the variation law analysis of interlayer shear strength, a temperature–humidity dual stress storage life prediction model was established.

Taking the Guangzhou region of China as an example, where high temperatures and humidity have a significant impact on the performance of composite materials (as shown in Table 1), the storage life of the composite under the actual storage conditions was calculated with the failure criterion of interlaminar shear strength degradation to 77% of the original value [35]. The reliability of the model was analyzed by comparing the prediction value and the actual measured value of specimens stored in Guangzhou for 2 years.

Table 1. Temperature and humidity conditions of actual storage in Guangzhou, China.

Storage Time	Average Temperature and Humidity
Spring (From March to May)	21 °C ± 0.5 °C, RH65% ± RH 2%
Summer (From June to August)	29 °C ± 0.5 °C, RH 75% ± RH 2%
Autumn (From September to November)	25 °C ± 0.5 °C, RH60% ± RH 2%
Winter (From December to February)	18 °C ± 0.5 °C, RH58% ± RH 2%

3. Results Analysis and Discussion

3.1. Moisture Content

Figure 3a shows that the moisture contents of the specimens increased, the times to reach effective moisture equilibrium decreased, and the moisture contents increased with increasing temperature and humidity under the accelerated conditions of 70 °C, RH70%; 70 °C, RH85%; 85 °C, RH70%; and 85 °C, RH85%. For example, the time to reach effective moisture equilibrium was 1320 h and the moisture content was 0.46% under the condition of 70 °C and RH70%, and the time to reach effective moisture equilibrium was 840 h and the moisture absorption was 1.77% under the conditions of 85 °C and RH85%. The moisture contents data under four temperature and humidity conditions were fitted nonlinearly. It was found that the moisture content changes meet Fick's diffusion law before reaching effective moisture equilibrium, as shown in Figure 3b. The fitting correlation coefficients R2 were all higher than 0.988, showing a higher fitting degree. After reaching the effective moisture equilibrium, the moisture contents of the specimens changed slightly and showed a linear trend.

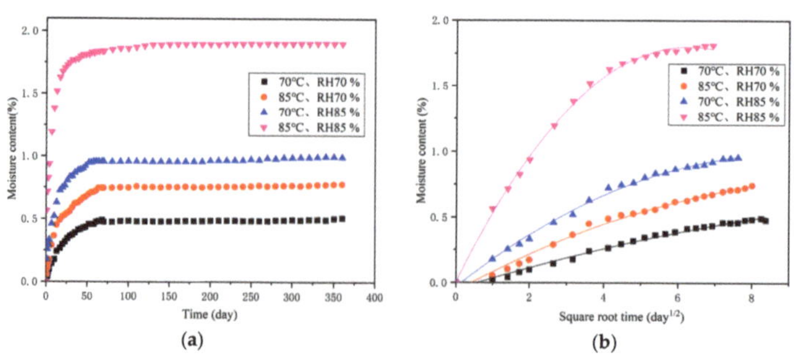

Figure 3. Moisture content curves with (**a**) the moisture content change with the accelerated aging time and (**b**) the initial stages of moisture content, which conform to Fick's law.

3.2. Interlayer Shear Strength under Different Hygrothermal Conditions

The changes in interlayer shear strength with hygrothermal accelerated aging time are shown in Figure 4. It can be seen that, with increasing aging time, the interlaminar shear strengths showed a decreasing trend, and that the higher the temperature and humidity, the faster the interlaminar shear strengths declined. The data dispersions also showed an overall decreasing trend. For example, the average value of original interlaminar shear strength was 73.8 MPa, with a standard deviation of 1.8; after hygrothermal accelerated aging for 360 days under 85 °C and 85% conditions, the interlaminar shear strength average value was 42.1 MPA (decreased by 43%), with a standard deviation of 0.8. Obviously, the influence of moisture absorption on interlayer shear strength gradually increased, relatively reducing the influence of differences in sample preparation.

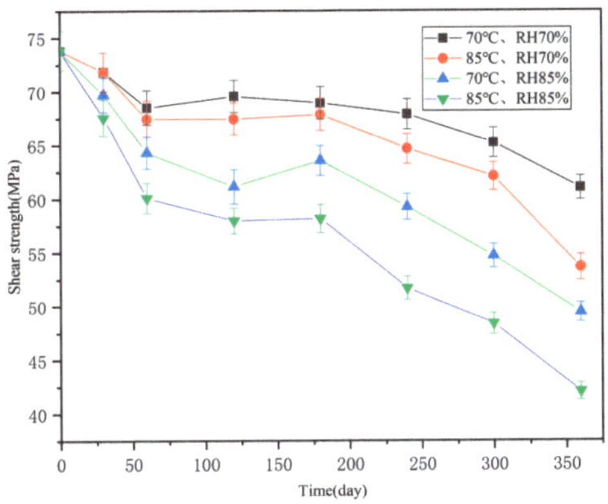

Figure 4. Interlaminar shear strength changes with accelerated aging time under different hygrothermal conditions.

The interlayer shear strengths of specimens stored in Guangzhou for 2 years are shown in Table 2. The average value was 68.6 MPa; compared to the original performance of 73.8 MPa, this decreased by 7%.

Table 2. Interlayer shear strength of specimens stored in Guangzhou for two years.

Interlayer Shear Strength (MPa)	Average Value (MPa)	Coefficient of Variation (%)
68.3		
69.0		
70.0	68.6	1.6
67.0		
68.5		

3.3. Failure Mechanism of Hygrothermal Accelerated Aging

3.3.1. Fourier Transform Infrared Analysis

Figure 5 shows the infrared spectra of the T700/epoxy composites before and after hygrothermal accelerated aging. The peak at 3339 cm^{-1} is the stretching vibration absorption peak of −OH. The peaks at 2932 cm^{-1} and 2850 cm^{-1} are the stretching vibration absorption peaks of saturated C-H.

The peak at 1730 cm^{-1} is the stretching vibration peak of C=O in the imine ring; at 1605 cm^{-1} and 1507 cm^{-1}, there are stretching vibration peaks of the benzene ring skeleton C=C. The C-O stretching vibration peak is located at 1180 cm^{-1}; 1031 cm^{-1} is the stretching vibration peak of the ether bond C-O-C.

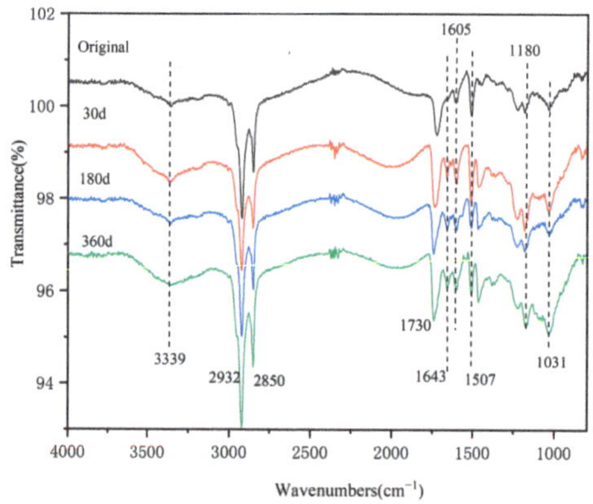

Figure 5. Infrared spectra of the specimens before and after hygrothermal accelerated aging.

After hygrothermal accelerated aging for 30 days to 360 days, the trend of infrared peak changes is the same. Only a small absorption peak is formed at 1643 cm^{-1}, which is the stretching vibration absorption peak of the decomposition product–benzene ring produced by the slight hydrolysis of the resin [6]. This indicates that the resin slightly hydrolyzes after hygrothermal aging. The degree of resin hydrolysis does not increase when prolongating the aging time.

3.3.2. DSC Analysis of Resin Matrix

Figure 6 shows the DSC test curves of the resin matrix aged under 85 °C and RH85% conditions for different times. The glass transition temperature of epoxy resin is about 120 °C. An exothermic peak appears around 102 °C, which is a secondary reaction between the amino groups in the epoxy resin and curing agent.

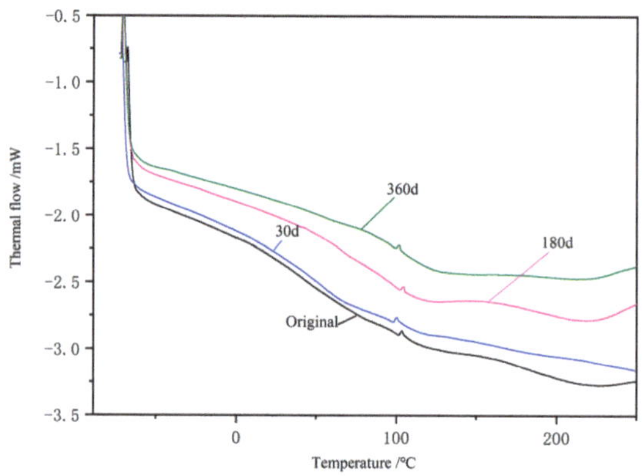

Figure 6. DSC results of resin matrix aged at 85 °C and RH85% for different times.

It can be seen that there is no significant difference in the DSC curves between the original specimen and the specimens aged for different times. During the aging process, there was no significant chemical structural change in the matrix resin.

3.3.3. TGA Results of Composites

The TGA results of composites are shown in Figure 7. The results show that the initial thermal decomposition temperature of composite materials is about 300 °C. There was no significant change in the initial thermal decomposition temperature of the composite material before and after aging. There is a slight difference in the remaining mass of composites aged for different times, which is related to the different resin and fiber contents of the test samples. The TGA results further demonstrate that there is no significant chemical change in the resin matrix.

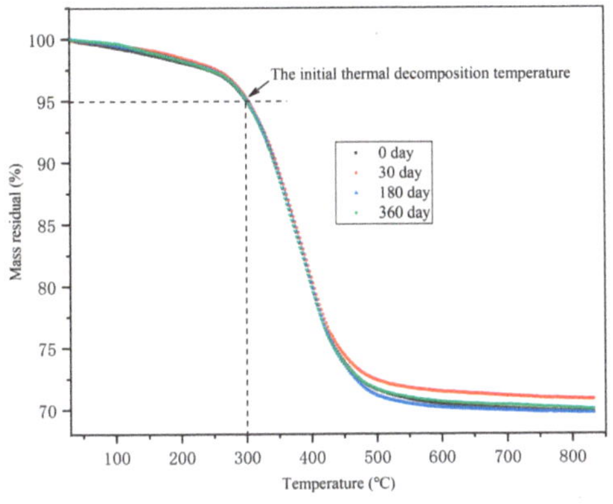

Figure 7. TGA results of resin matrix aged at 85 °C and RH85% for different times.

3.3.4. Interlaminar Cracks and Interface Analysis

Figure 8 shows that the number of delamination cracks observed from the specimen side increased with the hygrothermal accelerated aging time. Hygrothermal environments weaken the interface between the fiber and matrix, which is the main reason for the decrease in its shear strength.

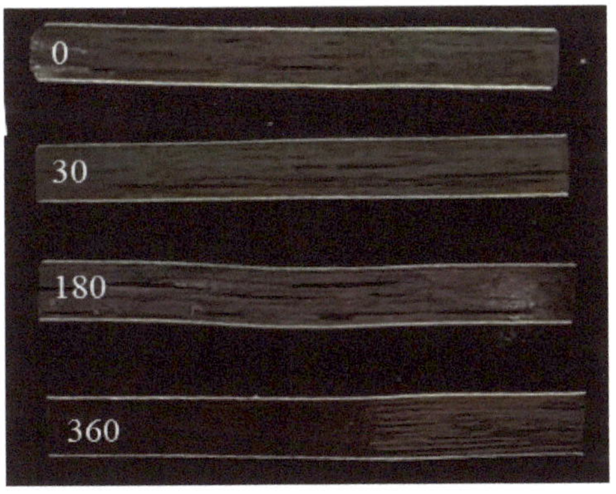

Figure 8. Interlayer cracks before and after hygrothermal accelerated aging.

According to the FESEM analysis, the interlayer cracking of the unaged specimen was mainly located at the interface between fiber bundles and rich resin, which is a typical brittle cracking. There was no obvious cracking inside the matrix and fiber bundles, and the interface between the fibers and the matrix was well bonded in the non-rich resin area, as shown in Figure 9a. The uneven distribution of resin is also one of the reasons for the interlaminar shear strength dispersion. After hygrothermal accelerated aging for 30 days, cracks still mainly occurred at the interface between rich resin and fiber bundles, but the number and length of cracks increased significantly, as shown in Figure 9b. After hygrothermal accelerated aging for 180 days, cracks appeared at the interface between the resin and fibers, as well as within the fiber bundles, and the interface between all resin and fibers significantly weakened, as shown in Figure 9c,d.

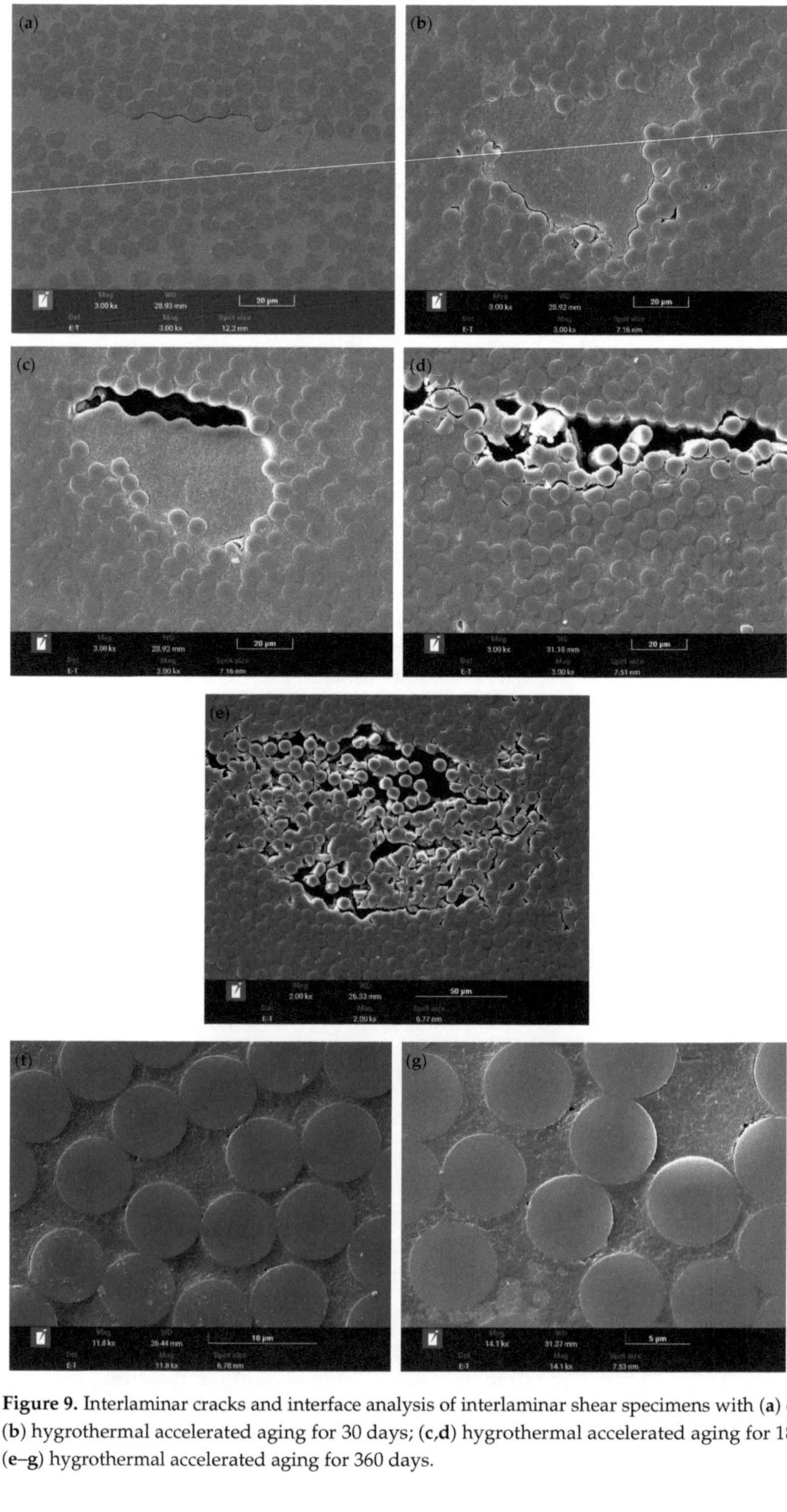

Figure 9. Interlaminar cracks and interface analysis of interlaminar shear specimens with (**a**) original; (**b**) hygrothermal accelerated aging for 30 days; (**c**,**d**) hygrothermal accelerated aging for 180 days; (**e**–**g**) hygrothermal accelerated aging for 360 days.

The composites aged for 360 days showed more obvious cracking within the fiber bundles, as shown in Figure 9e, and pores caused by moisture absorption appeared at the interface between the fiber and matrix. There were no obvious pore characteristics inside the resin, as shown in Figure 9f,g.

3.3.5. Fracture Morphology Analysis

The cracks were artificially opened, and the microscopic characteristics of the fracture surface were observed by FESEM. For the T700/epoxy composites without moisture and after hygrothermal accelerated aging for 30 days, the resin fracture characteristics were tear characteristics, which are typical brittle cracking, and the interfaces between the resin and fibers were well bonded, as shown in the red box in Figure 10a,b. With increasing hygrothermal accelerated aging time, resins were significantly elongated, which is typical ductile fractures. At the same time, adhesive resins on the fiber surface decreased the fiber surfaces became smooth, and the interface bond between the fibers and the resin matrix weakened, as shown in the red box in Figure 10c,d. And the resin fracture characteristics gradually became ductile with increasing aging time, which indicates that hygrothermal aging causes plasticization of the resin matrix.

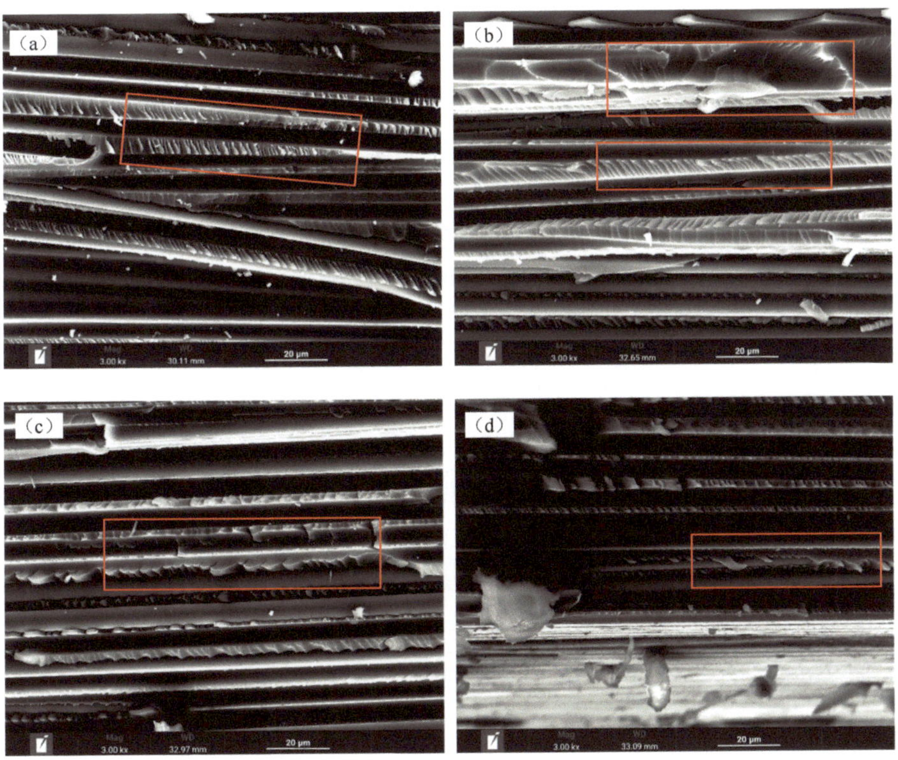

Figure 10. Fracture characteristics of interlaminar shear specimens with (**a**) original; (**b**) hygrothermal accelerated aging for 30 days; (**c**) hygrothermal accelerated aging for 180 days; (**d**) hygrothermal accelerated aging for 360 days.

From the above test results, it can be seen that the infrared and DSC tests of the resin matrix indicated that the resin matrix underwent slight hydrolysis after hygrothermal accelerated aging, but the degree of hydrolysis did not increase with the increase in hygrothermal accelerated aging time. The glass transition temperature of the resin matrix did not change significantly before and after hygrothermal aging. The chemical changes in the resin were not obvious. The fracture surfaces of the cracks showed that, with the increase in hygrothermal aging time, the resin matrix cracking gradually transitioned from brittle to plastic, and the resin matrix underwent plasticization.

An observation of the cross-section of the interlayer shear specimens revealed that cracks only appeared at the interface between the rich resin and fiber bundles in the non-aged and early hygrothermal accelerated aged specimens. With increasing hygrothermal accelerated aging time, cracks also appeared inside the fiber bundles, and pores caused by moisture absorption occurred at the interface between the fibers and the matrix. The main reasons for the decrease in the interlaminar shear strength were the gradual plasticization of resin moisture and the weakening of the interface between fibers and resin. The failure mechanism of the interlaminar shear did not obviously change throughout the whole hygrothermal accelerated aging process.

3.4. Storage Life Estimation

The life estimation model is based on the analysis of property degradation and the failure mechanism. If the property degradation law or failure mechanism of the interlaminar shear change during the hygrothermal accelerated aging test, different life prediction models would be used at different change stages. In this article, the degradation mechanism of the resin matrix remained unchanged during the whole hygrothermal accelerated aging test. Therefore, only one life prediction model was used throughout the whole hygrothermal aging process.

In general, the mechanical property degradation of polymer composites experiencing long-term hygrothermal effects decreases exponentially, which can be expressed by the following equation:

$$y_t = y_0 exp(-kt), \tag{3}$$

where:=

y_t—the mechanical property of composite materials at time t;

y_0—the original mechanical property;

k—material property degradation rate;

t—hygrothermal accelerated aging time.

Under different hygrothermal conditions, the reaction rate of composite materials was different, and the time used for the performance to decrease to the same value was different.

The interlayer shear strengths of T700/epoxy, aged under the conditions of 70 °C, RH70%; 70 °C, RH85%; 85 °C, RH70%; and 85 °C, RH85%, were fitted using Origin software 2018. The degradation equations were obtained, as shown in Equations (4)–(7) and Figure 11.

$$70\ ^\circ\mathrm{C}, \mathrm{RH70\%} : \mathrm{y} = 72.86142\mathrm{e} - 0.0004\mathrm{t}(\text{fitting degree R2} = 0.84), \tag{4}$$

$$85\ ^\circ\mathrm{C}, \mathrm{RH70\%} : \mathrm{y} = 73.33141\mathrm{e} - 0.00066\mathrm{t}(\text{fitting degree R2} = 0.84), \tag{5}$$

$$70\ ^\circ\mathrm{C}, \mathrm{RH85\%} : \mathrm{y} = 71.49943\mathrm{e} - 0.00092\mathrm{t}(\text{fitting degree R2} = 0.90), \tag{6}$$

$$85\ ^\circ\mathrm{C}, \mathrm{RH85\%} : \mathrm{y} = 70.36816\mathrm{e} - 0.00134\mathrm{t}(\text{fitting degree R2} = 0.93), \tag{7}$$

The fitting degrees R2 of the above four equations are, respectively, 0.84, 0.84, 0.90, and 0.93.

The interlaminar shear strength degradation rates k of T700/epoxy resin under different hygrothermal accelerated aging conditions were obtained from the four equations above, as shown in Table 3.

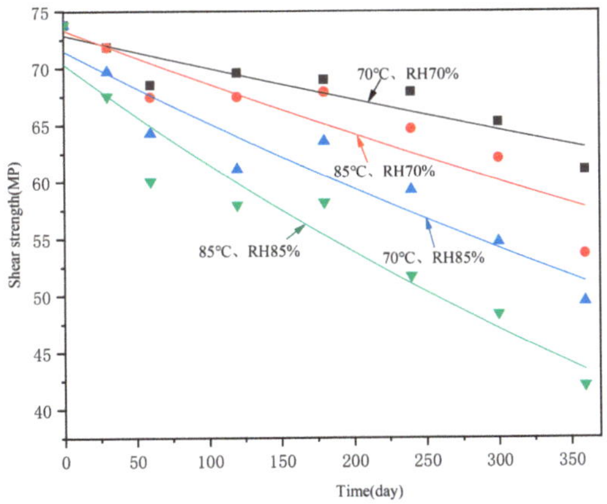

Figure 11. Regression curve of interlaminar shear strength with hygrothermal accelerated aging time.

Table 3. Interlaminar shear strength degradation rate *k* under different hygrothermal accelerated aging conditions.

Hygrothermal Accelerated Aging Conditions	Degradation Rate *k*
70 °C, RH70%	0.0004
85 °C, RH70%	0.00066
70 °C, RH85%	0.00092
85 °C, RH85%	0.00134

Due to the accelerated test temperature being lower than the glass transition temperature of the matrix (90 °C), and there was no significant change in the glass transition temperature during the accelerated tests; the relationship between the interlaminar shear strength degradation rate of composite materials under two stress levels for humidity and temperature can be represented by the generalized Eyring's Model [2]:

$$k(T,H) = \frac{A}{T}e^{-\frac{Ea}{KT}}e^{H(C+\frac{D}{KT})} \tag{8}$$

where

A, C, and D are undetermined constants;

Ea—activation energy;

K—Boltzmann constant, 8.61738×10^{-5};

T—absolute temperature, K;

H—relative humidity, %.

For the convenience of calculation, Equation (8) is converted to a logarithmic form:

$$ln(kT) = lnA + Ea/(KT) + CH + DH/(KT) \tag{9}$$

Let $a = \ln A$, $b = Ea/K$, $c = C$, and $d = D/K$; Formula (9) can be simplified as

$$ln(kT) = a + b/T + cH + dH/T \tag{10}$$

where

a, *b*, *c*, and *d* are the undetermined constants.

Substituting the reaction rates k at the conditions of 70 °C, RH70%; 85 °C, RH70%; 70 °C, RH85%; and 85 °C, RH85% (in Table 3) into Equation (10), the following equations can be obtained:

$$-1.986316 = a + 0.0029155b + 0.70c + 0.002041d \tag{11}$$

$$-1.442738 = a + 0.0027933b + 0.70c + 0.001955d \tag{12}$$

$$-1.153406 = a + 0.0029155b + 0.85c + 0.002478d \tag{13}$$

$$-0.734553 = a + 0.0027933b + 0.85c + 0.002374d \tag{14}$$

Four undetermined coefficients, a, b, c, and d, are obtained by solving Equations (11) to (14); thus, a two-factor reaction rate model regarding temperature and humidity for T700/epoxy resin is obtained:

$$ln(kT) = 20.9916 - 9214.6538/T - 14.2922H + 6806.8297H/T \tag{15}$$

Then, the interlaminar shear strength degradation rate at different temperatures and humidity levels k can be calculated from the following equation:

$$k = \exp\,(20.9916 - 9214.6538/T - 14.2922H + 6806.8297H/T)/T \tag{16}$$

The degradation rate under storage conditions of each season in Guangzhou, China can be obtained, as shown in Table 4.

Table 4. Degradation rate of interlayer shear strength stored in Guangzhou, China.

Storage Conditions	Degradation Rate k
Spring (From March to May)	0.000034
Summer (From June to August)	0.000121
Autumn (From September to November)	0.000028
Winter (From December to February)	0.000015

The durability prediction methodology is based on the time shift concept—the ratio of the time which the performance composites take to decrease to a certain value at different humidities and temperatures [4].

According to Equation (3), the time required for the performance degradation to y_1 of composite materials can be expressed as

$$t = ln(y_1/y_0)/(-k)$$

Thus, the time shift factor (*TSF*) can be calculated by the following equation:

$$TSF = \frac{t_1}{t_2} = \frac{ln(y_1/y_0)/(-k_1)}{\ln(y_1/y_0)/(-k_2)} = \frac{k_2}{k_1} \tag{17}$$

Thus, it can be concluded that the *TSF* of the laboratory at 85 °C, RH85% under the storage conditions of spring, summer, autumn, and winter in Guangzhou, China, are 39.4118, 11.0744, 47.8571, and 89.3333, respectively. In other words, the entire spring, summer, autumn, and winter storage period in Guangzhou, China is equivalent to 2.3153 days, 8.2397 days, 1.9067 days, and 1.0214 days stored under laboratory conditions of 85 °C and RH85%. Assuming that the interlaminar shear strength degradation is a cumulative damage process during the whole storage process, the storage time of T700/epoxy resin composite in Guangzhou, China for one year is equivalent to 13.4831 days under the laboratory conditions of 85 °C and RH85%. Thus, the equivalent relationship between actual storage and accelerated testing is obtained.

According to the safety factor guidelines of the Federal Aviation Administration for fiber-reinforced resin based composite laminates [35], the failure criterion is that the strength of composite materials decreases to 77% of the original value, meaning that

$$F_{failure} = 0.77F_{unaged} \quad (18)$$

when $\frac{y_t}{y_0} = 0.77$, the storage time of the T700/resin matrix composite at 85 °C and RH85% is 195.0483 days, and the storage life in Guangzhou, China is 14.4661 years.

The interlayer shear strength of T700/epoxy resin measured after 2 years of actual storage is 68.6 MPa, as shown in Table 2. Based on the equivalent relationship between actual storage and laboratory storage at 85 °C and RH85%, the interlaminar shear strength calculated for two years storage in Guangzhou, China is 67.87 MPa. The calculated values are close to the measured values, and the model is effective. The model still needs to be validated with more actual stored data.

4. Conclusions

The durability of carbon fiber-reinforced polymer composites was investigated by hygrothermal accelerated aging tests for 360 days and a natural storage test. Based on the analysis of property changes and the hygrothermal aging mechanism, a dual factor storage life prediction model was established. The main conclusions obtained are as follows:

(1) The order of moisture absorption rates, moisture absorption contents, and the severity effect order on the interlaminar shear strength is RH85%; 85 °C > 70 °C, RH85% > 85 °C, RH70% > 70 °C, RH70%. The times to achieve effective moisture absorption balance were opposite to this.
(2) The moisture absorption rate satisfies Fick's law before the effective moisture absorption balance, then shows a linear trend. The interlayer shear strength still decreases exponentially with aging time, which is mainly caused by the resin plasticization and interface weakening.
(3) With the increase in hygrothermal accelerated aging time, the resin matrix fracture gradually changes from a brittle fracture to a plastic fracture. The main reasons for the interlayer shear strength decreasing are resin moisture plasticization and interface weakening.
(4) A dual stress storage life prediction model and the equivalent relationship between natural storage and hygrothermal acceleration were established. According to the model, based on the failure criterion of the interlayer shear strength decreasing to 77% of the initial value, the storage life of T700/epoxy resin in Guangzhou, China was calculated to be 14.4661 years. The effectiveness of the model was verified by actual storage data for 2 years.

Author Contributions: Formal analysis, Q.Z.; design of the work, X.C. and Y.H.; Writing—original draft, J.F. All authors have read and agreed to the published version of the manuscript.

Funding: This research received no external funding.

Institutional Review Board Statement: Not applicable.

Data Availability Statement: Data are contained within the article.

Conflicts of Interest: The authors declare no conflict of interest.

References

1. Xing, L.; Li, Y.; Chen, X. The status and role of advanced composite materials in the development of aviation equipment. *Acta Mater. Compos. Sin.* **2022**, *39*, 8–15. [CrossRef]
2. Nakada, M.; Miyano, Y. Accelerated testing for long-term durability of FRP laminates for Marine Use. *Compos. Sci. Technol.* **2009**, *69*, 805–813. [CrossRef]

3. Frigione, M.; Rodriguez-Prieto, A. Can Accelerated Aging Procedures Predict the Long Term Behavior of Polymers Exposed to Different Environments? *Polymers* **2021**, *13*, 2688. [CrossRef] [PubMed]
4. Starkova, O.; Gagani, A.I.; Karl, C.W.; Rocha, I.B.; Burlakovs, J.; Krauklis, A.E. Modelling of environmental ageing of polymers and polymer composites-durability prediction methods. *Polymer* **2022**, *14*, 907. [CrossRef] [PubMed]
5. Zhang, Y.; Wang, J.; Wei, J.; Liu, M.; Li, X.; Ding, A. Long-term mechanical properties of carbon fiber reinforced vinyl resin composites in hygrothermal environment. *Acta Mater. Compos. Sin.* **2023**, *40*, 1406–1416. [CrossRef]
6. Wang, D.X.; Sun, Y.; Xie, K.Y.; Li, H.; Wang, X.B.; Duan, J.; Shao, M. Long term aging and failure behaviors of carbon fiber reinforced polymer composites in simulated marine environments. *Acta Mater. Compos. Sin.* **2022**, *39*, 1353–1362. [CrossRef]
7. Bauer, H.; Ilie, N. Effects of aging and irradiation time on the properties of a highly translucent resin-based composite. *Dent. Mater. J.* **2013**, *32*, 592–599. [CrossRef] [PubMed]
8. Rao, R.M.V.G.K.; Kumari, H.V.S.; Raju, K.S. Moisture Diffusion Behavior of T300-914C Laminates. *J. Reinf. Plast. Compos.* **1995**, *14*, 513–522. [CrossRef]
9. Burks, B. The effect of atmospheric aging on a hybrid polymer matrix composites material properties. *Diss. Theses—Gradworks* **2012**, *72*, 1803–1811. [CrossRef]
10. Park, S.Y.; Choi, W.J.; Choi, C.H.; Choi, H.S. An experimental study into aging unidirectional carbon fiber epoxy composite under thermal cycling and moisture absorption. *Compos. Struct.* **2019**, *207*, 81–92. [CrossRef]
11. Xu, L.; Jia, Y.; Zhou, S.; Li, H. Effect of Aging Environment on the Properties of T800 Carbon Fiber/Epoxy Resin Composites. *Polym. Mater. Sci. Eng.* **2021**, *37*, 7. [CrossRef]
12. Wang, W.; Wei, C.; Tian, J. Research on hygrothermal ageing models of carbon fiber composites. *Compos. Sci. Eng.* **2023**, *11*, 44–48. [CrossRef]
13. Devine, M.; Bajpai, A.; Obande, W.; Conchur, M.O.; Bradaigh, D.R. Seawater ageing of thermoplastic acrylic hybrid matrix composites for marine applications. *Compos. Part B Eng.* **2023**, *15*, 263. [CrossRef]
14. Zhao, Y.Q.; Zhou, Y.; Huang, Z.M.; Batra, R.C. Experimental and micromechanical investigation of T300/7901 unidirectional composite strength. *Polym. Compos.* **2019**, *40*, 2639–2652. [CrossRef]
15. Das, S. Life cycle assessment of carbon fiber-reinforced polymer composites. *Int. J. Life Cycle Assess* **2011**, *16*, 268–282. [CrossRef]
16. Yang, X.; An, T.; Zou, T. Effect of hygrothermal environment on mechanical properties and damage mechanism of CFRP. *J. Mater. Eng.* **2019**, *47*, 84–91. [CrossRef]
17. Peret, T.; Clement, A.; Freour, S.; Jacquemin, F. Effect of mechanical states on water diffusion based on the free volume theory: Numerical study of polymers and laminates used in marine application. *Compos. Part B Eng.* **2017**, *118*, 54–66. [CrossRef]
18. Almeida, J.H.S.; Souza, S.D.; Botelho, E.C.; Amico, S.C. Carbon fiber-reinforced epoxy filament-wound composite laminates exposed to hygrothermal conditioning. *J. Mater. Sci.* **2016**, *51*, 4697–4708. [CrossRef]
19. Sui, X.; Xiong, S.; Zhu, L.; Li, Y.; Li, N. Hygrothermal properties of domestic T800 carbonfiber/expoxy composites. *J. Aeronaut. Mater.* **2019**, *39*, 88–93.
20. Kesentini, Z.; El Mahi, A.; Rebiere, J.L.; El Guerjouma, R.; Beyaoui, M.; Haddar, M. Static and Fatigue Tensile Behavior and Damage Mechanisms Analysis in Aged Flax Fiber/PLA Composite. *Int. J. Appl. Mech.* **2022**, *14*, 2250080. [CrossRef]
21. Wu, R.; Li, Y.; Yu, T. Comparative study on the hygrothermal durability of different fiber reinforced composites. *Acta Mater. Compos. Sin.* **2022**, *39*, 4406–4419. [CrossRef]
22. Niu, Y.; Li, Z.; Zhu, X. Flexural properties and life-time estimation of carbon fiber/epoxy composite under hygrothermal conditions. *Acta Mater. Compos. Sin.* **2020**, *37*, 104–112.
23. Niu, Y.; Li, Z.; Zhu, X. Interlaminar shear strength of carbon fiber/epoxy composites under hygrothermal conditions. *Polym. Mater. Sci. Eng.* **2021**, *37*, 113–120. [CrossRef]
24. Liu, S.; Cheng, X.; Bao, J. Hygrothermal Effects on Properties of Composite Materials. *Polym. Mater. Sci. Eng.* **2014**, *30*, 183–189.
25. Wani, V.; M, M.; Jain, S.; Singh, P.; Bhattacharya, B. Prediction of storage life of propellants having different burning rates using dynamic mechanical analysis. *Def. Sci. J.* **2012**, *62*, 290–294. [CrossRef]
26. Krauklis, A.E.; Akulichev, A.G.; Gagani, A.; Echtermeyer, A.T. Time-Temperature Plasticization Superpositon principle: Predicting creep of a Plasticized Epoxy. *Polymers* **2019**, *11*, 1848. [CrossRef] [PubMed]
27. Gagani, A.I.; Monsas, A.B.; Krauklis, A.E. The effect of temperature and water immersion on the interlaminar shear fatigue of glass fiber epoxy composites using the I-beam method. *Compos. Sci. Technol.* **2019**, *181*, 107703. [CrossRef]
28. Ishisaka, A.; Kawagoe, M. Examination of the time–water content superposition on the dynamic viscoelasticity of moistened polyamide 6 and epoxy. *J. Appl. Polym. Sci.* **2004**, *93*, 560–567. [CrossRef]
29. Yu, Z.G.; Yang, S.C.; Song, B.F. Comparison of Wet and Hot Aging Resistance of T700 and T300 Carbon Fiber Strengthened Epoxy Resin Composites. *Mater. Mech. Eng.* **2009**, *33*, 48–51.
30. Wang, S.; Zhang, W.F.; Chen, Y.; Wu, Y.F. Hygrothermal Behavior of T700 and T300 BMI Used for Advanced Polymeric Composite. *Adv. Mater. Res.* **2012**, *476–478*, 632–635. [CrossRef]
31. Li, C.; Cui, R.; Lei, C. Comparison of Structure and Properties in Different Polyacrylonitrile-based Carbon Fibers. *Mater. Rev.* **2011**, *16*, 51–55.
32. *HB 7401-2021*; Test Methods for Moisture Absorption of Polymer Matrix Composites for Aerospace Applications. Ministry of Industry and Information Technology of the People's Republic of China: Beijing, China, 2021. Available online: http://www.biaozhun8.cn/biaozhun51930/ (accessed on 1 July 2021).

33. *JC/T773-2010*; Fiber-Reinforced Plastics Composites-Determination of Apparent Interlaminar Shear Strength by Short-Beam Method. Ministry of Industry and Information Technology of the People's Republic of China: Beijing, China, 2010. Available online: http://www.standardcnjc.com/index/standard/detail/id/320.html (accessed on 1 March 2011).
34. *ASTM D 3418-15*; Standard Test Method for Transition Temperatures and Enthalpies of Fusion and Crystallization of Polymers by Differential Scanning Calorimetry. American Society for Testing and Materials: West Conshohocken, PA, USA, 2015. [CrossRef]
35. Yan, F.F. *Performance of Composite Material Dispersion and Safety Coefficient*; Nanjing University of Aeronautics and Astronautics: Nanjing, China, 2009. [CrossRef]

polymers

MDPI

Article

Synergistic Effects of Liquid Rubber and Thermoplastic Particles for Toughening Epoxy Resin

Zhaodi Wang [1], Yuanchang Lai [1], Peiwen Xu [1], Junchi Ma [2], Yahong Xu [1,*] and Xin Yang [1,*]

1 College of Materials Science and Engineering, Nanjing Tech University, Nanjing 211816, China; 201910006628@njtech.edu.cn (Z.W.); 202061203297@njtech.edu.cn (Y.L.); 202261203197@njtech.edu.cn (P.X.)
2 Yangtze River Delta Carbon Fiber and Composites Innovation Center, Changzhou 213000, China; majunchi@ccicyd.com
* Correspondence: 201910006672@njtech.edu.cn (Y.X.); xin12.25@hotmail.com (X.Y.)

Abstract: This study aims to investigate the toughening effects of rubber and thermoplastic particles on epoxy resin (EP), and to understand the mechanism underlying their synergistic effect. For this purpose, three EP systems were prepared using diglycidyl ether of bisphenol-A (DGEBA) epoxy resin (E-54) and 4,4-Diamino diphenyl methane (Ag-80) as matrix resin, 4,4-diaminodiphenyl sulfone (DDS) as a curing agent, and phenolphthalein poly (aryl ether ketone) particles (PEK-C) and carboxyl-terminated butyl liquid rubber (CTBN) as toughening agents. These systems are classified as an EP/PEK-C toughening system, EP/CTBN toughening system, and EP/PEK-C/CTBN synergistic toughening system. The curing behavior, thermal properties, mechanical properties, and phase structure of the synergistic-toughened EP systems were comprehensively investigated. The results showed that PEK-C did not react with EP, while CTBN reacted with EP to form a flexible block polymer. The impact toughness of EP toughened by PEK-C/CTBN was improved obviously without significantly increasing viscosity or decreasing thermal stability, flexural strength, and modulus, and the synergistic toughening effect was significantly higher than that of the single toughening system. The notable improvement in toughness is believed to be due to the synergistic energy dissipation effect of PEK-C/CTBN.

Keywords: resin; liquid rubber; thermoplastic particles; toughening agent; synergistic toughening

Citation: Wang, Z.; Lai, Y.; Xu, P.; Ma, J.; Xu, Y.; Yang, X. Synergistic Effects of Liquid Rubber and Thermoplastic Particles for Toughening Epoxy Resin. *Polymers* **2024**, *16*, 2775. https://doi.org/10.3390/polym16192775

Academic Editors: Chengji Zhao and Yizhuo Gu

Received: 23 August 2024
Revised: 26 September 2024
Accepted: 29 September 2024
Published: 30 September 2024

1. Introduction

Epoxy resin (EP) is an essential thermosetting resin due to its good processability, flexible formulation design, and low-volume shrinkage during curing. It has excellent adhesion and mechanical properties, high chemical stability, and low cost [1], making it a popular material in the chemical industry, electronics and electrical appliances, aerospace, transportation, and construction [2–5]. However, the cured EP molecular chain is a three-dimensional cross-linked network structure, which makes it difficult for the molecular chain to slide. As a result, the cured EP has a high cross-link density and internal stress, which leads to the existence of brittleness, insufficient impact performance, poor fatigue resistance and heat resistance in the cured materials [6–8], and limits the application in high-end fields. Therefore, EP must be toughened and modified.

The current EP toughening paths mainly include the following: changing the EP cross-linked network structure to toughen EP; and adding rubber elastomer, thermoplastic resin, liquid crystal polymer, interpenetrating network polymer, hyperbranched polymer, inorganic rigid particles, etc., to EP [1,9]. Among them, the study of the rubber toughened epoxy system was the earliest, which started in the 1960s [10]. The most commonly used toughening material are liquid rubbers, which can be both toughening agents and curing agents. During the curing process of EP, the active groups at both ends of the rubber molecular chain react with the epoxy system to form a three-dimensional network structure. It can effectively improve the inherent brittleness, impact toughness, and elongation of the cured epoxy system [11,12]. The reactive rubbers and elastomers which were used in the epoxy toughening system contain the following: carboxyl-terminated butyl liquid rubber (CTBN), hydroxyl-terminated liquid nitrile rubber (HTBN), polysulfide rubber, and polyurethane rubber [13–15]. G Tripathi et al. [16] prepared six toughening components by physically mixing different concentrations of CTBN with EP. The results showed that the tensile strength of the cured materials decreased from 11% to 46%, and the elongation at break and impact strength increased significantly, but the Tg decreased significantly. N Chikhi et al. [17] toughened EP with amine-terminated butadiene acrylonitrile (ATBN) and showed that by adding only 12.5 phr of ATBN, the impact strength of EP increased by 3 times and the fracture toughness by 1.5 times, while Tg and fracture stress decreased significantly. Wang et al. [18] investigated the mechanical and dielectric properties of HTBN-modified EP. The results showed that with the increase in HTBN content, the glass transition temperature (Tg) and the volume resistivity decrease, while the relative permittivity and dielectric loss tangent increase. The toughness of the composites was enhanced when the HTBN content was more than 15%. The results of the several researchers mentioned above show that even though rubber is effective in toughening EP, there are drawbacks that make the toughened material lose strength, modulus, and heat resistance, which undoubtedly greatly limits its application in toughening EP.

Due to the defects of rubber toughened epoxy systems, since the 1980s, more attention has been turned to the research of toughened epoxy systems using thermoplastic engineering plastics with high heat resistance and good mechanical properties [19,20]. The main thermoplastic resins used for toughening epoxy resins are polyethersulfone, pol-ysulfone, polyetherimide, polyimide, polycarbonate, polyphenylene ether, poly (ether ether ketone) (PEEK), etc. [21–25]. HS Jung et al. [26] toughened EP with PES and showed that the fracture toughness and Tg values of PES-toughened EP increased with the addition of PES, and the tensile and impact strengths of the 10 phr PES-toughened specimens increased by 14% and 106%, respectively. Zhou et al. [27] investigated the effects of phenolphthalein poly (aryl ether ketone) particles (PEK-C) on the mechanical, phase structure, and thermal properties of commercial EP. The results showed that the flexural strength, flexural modulus, and impact strength of the cured EP/PEK-C were increased by 13%, 11%, and 154%, respectively, compared with the pure resin, while the Tg and decomposition temperature were also significantly increased. PEK-C is a high-performance thermoplastic resin with high intermiscibility with epoxy resins and its own advantages of a high glass transition temperature, high modulus, and dimensional stability. According to the previous reports, it is clear that thermoplastic resins could toughen EP without decreasing the strength, modulus, and Tg. However, the toughness can only be effectively achieved when the added amount of thermoplastic resin is large. The thermoplastic resin could form a continuous phase with the epoxy resin spherical domain, or form a co-continuous phase with the epoxy resin. However, the solubility and fluidity of epoxy resin will reduce with the increase in the added amount of thermoplastic resin, which will limit the processing of prepreg.

Although the toughening of EP using single modifiers such as rubber and thermoplastic resins has been studied for decades, the preparation of hybrid ternary composites containing two different modifiers is a relatively new concept. So far, few studies on the synergistic toughening of EP with CTBN and PEK-C have been reported, and it is unknown whether CTBN and PEK-C have synergistic toughening effects on EP and what the synergistic toughening mechanism is. Therefore, in this study, PEK-C and CTBN were added into EP by the hot-melt method, and the toughness, thermal properties, mechanical properties, and phase structures of CTBN/PEK-C-modified EP were investigated. It is hoped to develop a toughening EP system with no reduction in strength and heat resistance, moderate viscosity, and good processing properties. Furthermore, the synergistic toughening mechanism of CTBN/PEK-C-modified EP was studied.

2. Experimental

2.1. Materials

Diglycidyl ether of bisphenol-A (DGEBA) epoxy resin (E-54) with an epoxy value (Eq/Kg) of 0.51–0.54 and molecular weight of 375.9 was purchased from Jiangsu Nantong Xingchen Synthetic Materials Co., Ltd. (Nantong, China). 4,4-Diamino diphenyl methane (AG-80), epoxy value (Eq/Kg) 0.8–0.87, molecular weight 442, was purchased from Shanghai Huayi Resin Co., Ltd. (Shanghai, China). The curing agent was 4,4-Diaminodiphenyl sulfone (DDS), purchased from Nanjing Chemical Reagent Co., Ltd. (Nanjing, China). Carboxyl-terminated butadiene nitrile liquid rubber (CTBN), grade CTBN1300, Mn = 3150, was supplied by Shenzhen Jiadida New Material Technology Co., Ltd. (Shenzhen, China). Phenolphthalein poly (aryl ether ketone) particles (PEK-C) with a particle size of 60 m, Mn = 36,700~43,000, were provided by Xuzhou Aviation Engineering Plastics Co., Ltd. (Xuzhou, China).

2.2. Preparation of Untoughened EP

Firstly, 40 Phr of E-54 and 30 Phr of AG-80 were mixed. After stirring at high speed for 10 min at 100 °C, the temperature was increased to 130 °C, and then 30 Phr of DDS was added. After the DDS was completely dissolved, the resin solution was transferred to a vacuum oven at 120 °C for 30 min, and finally poured into a preheated mold for curing. The curing system is 130 °C/1 h + 160 °C/1 h + 180 °C/2 h + 200 °C/2 h.

2.3. Preparation of EP/CTBN Composites

The mixture of 40 Phr of E-54 and 30 Phr of Ag-80 was stirred at 100 °C for 10 min and then heated to 170 °C, and 5 wt % CTBN was added. After 1 h, the mixture was cooled to 130 °C and 30 Phr DDS was added. After the DDS was completely dissolved in the resin, the mixture was transferred to a vacuum oven at 120 °C for degassing, and finally poured into a preheated mold for curing. The curing system is the same as above.

2.4. Preparation of EP/PEK-C Composites

The mixture of 40 Phr E-54 and 30 Phr Ag-80 was stirred at 100 °C for 10 min, heated to 170 °C, and then 15 wt % PEK-C was added. After PEK-C was completely dissolved in the resin, it was cooled to 130 °C and 30 Phr DDS was added. After the DDS was completely dissolved, the resin solution was transferred to a vacuum oven at 130 °C for degassing, and finally poured into a preheating mold for curing. The curing system is the same as above.

2.5. Preparation of EP/CTBN/PEK-C Composites

Firstly, 40 Phr of E-54 and 30 Phr of Ag-80 were mixed. After high-speed stirring at 100 °C for 10 min, the temperature was raised to 170 °C, and then 5 wt % CTBN was added. After 1 h, 15 wt % PEK-C was added and stirred until PEK-C was completely dissolved in the resin. Then, the temperature was reduced to 130 °C, and 30 Phr DDS was added. After the DDS was completely dissolved, the mixture was transferred to a vacuum oven at 120 °C for degassing. Finally, the mixture was poured into a preheated mold for curing. The curing system is the same as above.

2.6. Characterization

A Nicolet 6700 Fourier transform infrared spectrometer was used to analyze the structure of prepared samples. FTIR spectra were recorded on a Bruker's Tensor II spectrometer in the range of 500 to 4000 cm^{-1} at a resolution of 4 cm^{-1}.

According to the GB/T 2567-2008 standard [28], the flexural properties of the samples were carried out by an Instron 3367 universal material testing machine. The specimen size was 80 × 15 × 4 mm^3, and the number of test samples was no less than 5. According to the GB/T 2571-1995 standard [29], the impact performance of the samples was tested on the PIT501J cantilever impact testing machine. The specimen size was 80 × 10 × 4 mm^3, and the quoted data are the average of the results from five specimens.

Thermogravimetric analysis (TGA) was carried out under a nitrogen atmosphere with a Q50 thermogravimetric analyzer at a heating rate of 10 °C/min and a heating range of room temperature to 600 °C.

The dynamic mechanical properties of the cured material were determined by using a Q800 dynamic mechanical analyzer (TA Q800, New Castle, DE, USA) in a three-point mode under a nitrogen atmosphere. The test was carried out at a heating rate of 5 °C/min from room temperature to 350 °C with a fixed frequency of 1 Hz.

The rheological properties of each resin system were tested by a DHR-1 type advanced rotational rheometer. The viscosity of the resin was measured at 25 °C~200 °C using the plate viscometer mode. The heating rate was 3 °C/min and the shear rate was 6.28 rad/s.

The fracture morphology of the impact sample was observed by a Japan Hitachi TM300 scanning electron microscope. The fracture surface was sprayed with gold before observation. The PEK-C toughened sample was etched with THF for 72 h before gold spraying.

3. Results and Discussion

3.1. FT-IR Analysis of Different Toughened Systems

Figure 1 shows the FT-IR curve of untoughened EP, PEK-C powder, the EP/PEK-C system blend, and EP/PEK-C system-cured material. The peak at 1770 cm^{-1} in the PEK-C FT-IR curve is the C=O stretching vibration absorption peak on PEK-C phenolphthalein. From the FT-IR curve of the EP/PEK-C blends and EP/PEK-C-cured materials, it can be seen that the peak did not disappear. In addition, 3650 cm^{-1} is the characteristic peak of the stretching of the -OH bond on the carboxyl group in pekc, which can form a hydrogen bond with the hydroxyl group on the epoxy matrix. The peak corresponding to the polar hydroxyl group disappears completely in the infrared spectrum after curing, which is due to its participation in the cross-linking reaction of the epoxy system. The carboxyl group on the PEK-C side chain opens the ring and etherifies with the ring-opened epoxy group, eventually forming a cross-linked network. The two peaks at 830 cm^{-1} and 910 cm^{-1} on the FT-IR curve of EP are the characteristic absorption peaks of -CH(O)CH- [30]. The wavenumber and transmittance of the two peaks on the FT-IR curve of the EP/PEK-C blend did not change, but the two peaks on the FT-IR curve of the EP/PEK-C-cured material disappeared. The C-O-C absorption peak was formed at 1097 cm^{-1}, indicating that the EP and the curing agent DDS underwent a cross-linking curing reaction, and the reaction was complete. The cross-linking reaction process is as follows: (1) DDS is an aromatic polyamine curing agent, and the active hydrogen atom on the primary amine reacts with the epoxy group in EP to form secondary amines; (2) the active hydrogen in the secondary amine further reacts with the epoxy group to form tertiary amine; (3) the hydroxyl group in the reactant and the epoxy group undergo an etherification reaction, followed by chain expansion, branching, and cross-linking until the formation of a large molecule. The chemical reaction formula of the above three steps is shown in Scheme 1.

Figure 2 shows the FT-IR curves of EP, CTBN, EP/CTBN prepolymer, and EP/CTBN-cured material. In Figure 2, the peaks at 910 cm^{-1} and 830 cm^{-1} on the EP FT-IR curve are the characteristic absorption peaks of -CH(O)CH-. The peaks at 1712 cm^{-1} and 1738 cm^{-1} on the CTBN FT-IR curve are the characteristic absorption peaks of -COOH stretching vibration of CTBN [31]. The peak at 2237 cm^{-1} is the characteristic absorption peak of -C≡N stretching vibration. It can be seen from the FT-IR curve of EP/CTBN prepolymer that the two absorption peaks disappeared at 1712 cm^{-1} and 1738 cm^{-1}, and a new absorption peak appeared at 1740 cm^{-1}, which is the absorption peak of -COOR stretching vibration. In addition, the two absorption peaks at 910 cm^{-1} and 830 cm^{-1} became weaker, indicating that EP and CTBN underwent an esterification reaction during pre-polymerization [32,33] to generate block polymers and the chemical reaction formula [34] is shown in Scheme 2. As shown in the FT-IR curve of EP/CTBN-cured material, the two absorption peaks at 910 cm^{-1} and 830 cm^{-1} disappeared. The C-O-C characteristic absorption peak was formed at 1097 cm^{-1}, which indicated that the EP and the curing agent DDS had a cross-linking curing reaction (etherification reaction), and the reaction was completely. The chemical reaction formula is shown in Scheme 1(3). In addition, in the FT-IR curves of EP/CTBN prepolymer and EP/CTBN-cured material, the -C≡N absorption peak at 2237 cm^{-1} still exists, but the intensity becomes very weak. The reason is that the nitrile group participates in the modification of the network structure during the copolymerization or curing reaction. On the other hand, as the reaction proceeds, the volume fraction of the nitrile group decreases [35].

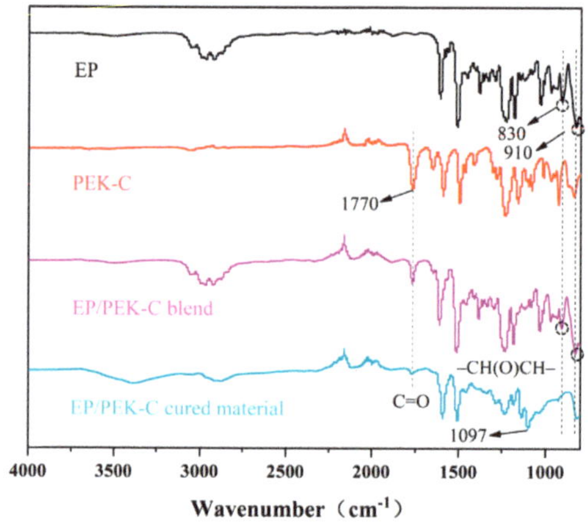

Figure 1. FT−IR spectrum of EP, PEK−C, EP/PEK−C blend, and EP/PEK−C-cured material.

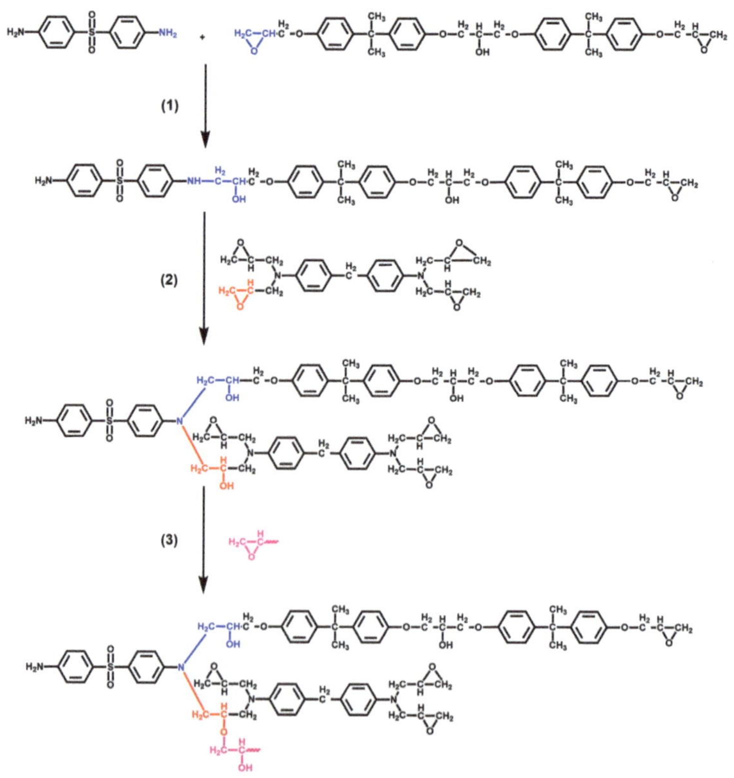

Scheme 1. The crosslinking reaction process of EP and DDS.

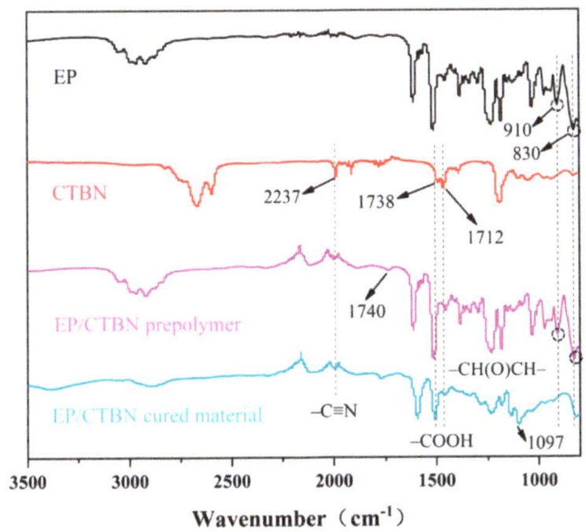

Figure 2. FT−IR spectrum of EP, CTBN, EP/CTBN prepolymer, and EP/CTBN−cured material.

Scheme 2. Chemical reaction formula of EP with CTBN.

3.2. Analysis of Mechanical Properties

3.2.1. Flexural Properties

Figure 3 shows the effect of three toughening agents on the flexural strength and flexural modulus of EP. Figure 3 shows the four resin systems' corresponding flexural strength, flexural modulus, and impact strength values. Among the three toughened systems, the flexural strength and flexural modulus of the EP/CTBN toughening system decreased the most. The flexural strength decreased from 143.93 MPa to 113.75 MPa, a decrease of 21%, and the flexural modulus decreased from 3.04 GPa to 2.6 GPa, a decrease of 15%. Followed by the EP/PEK-C/CTBN synergistic toughening system, the flexural strength decreased from 143.93 MPa to 130.36 MPa, a decrease of 9%. The flexural modulus decreased by 7% from 3.04 GPa to 2.83 GPa. The reason why the flexural strength and flexural modulus of the EP/CTBN toughening system decrease significantly, on the one hand, in the pre-polymerization stage, is because the carboxyl group of CTBN and the EP epoxy group undergo an esterification reaction, so that the cross-linked EP contains a large number of flexible nitrile segments. Although CTBN can cause reaction-induced phase separation during the high-temperature curing process and be separated from the EP/CTBN homogeneous solution, in most cases, the separation of the two phases is incomplete. A small part of CTBN will still be dissolved in the EP continuous phase [36], and the dissolved rubber flexible phase reduces the strength and modulus of EP/CTBN.

The lower decrease in the strength and modulus of the EP/PEK-C/CTBN synergistic system is because the introduction of PEK-C can enhance the entanglement between polymer chains, thereby strengthening the longitudinal and transverse connections between the long chains. During flexural tests, the entanglement between chains inhibits the relative sliding of molecular chains, thereby improving the flexural strength and modulus. Unlike the two toughening systems mentioned above, the flexural strength and flexural modulus of the EP/PEK-C toughening system are higher than those of the untoughened EP system. The flexural strength has increased from 143.93 MPa to 162.31 MPa, an increase of 13%, and the flexural modulus has increased from 3.04 GPa to 3.37 GPa, an increase of 11%. This increase is because the PEK-C molecular chain contains many benzene rings and polar ketone groups. The molecular chain exhibits greater rigidity, making its strength and modulus significantly greater than EP's and exhibiting an enhanced effect after adding EP. In addition, the presence of many polar groups strengthens the interaction between PEK-C and EP, which positively reinforces the system and improves the strength and modulus of the EP/PEK-C system. In addition, the cured epoxy resin-thermoplastic polymer system has phase segregation and formation of a heterogeneous structure, which also induces performance enhancement.

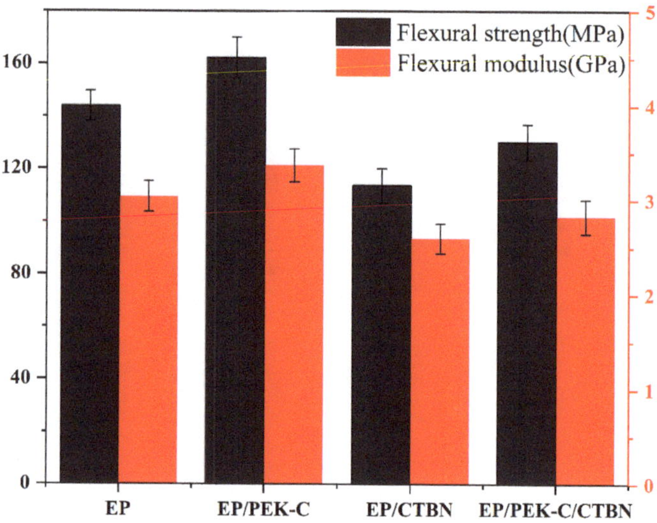

Figure 3. The effects of PEK-C and CTBN on the flexural strength and flexural modulus of EP.

3.2.2. Impact Performance

Figure 4 shows that the untoughened EP system has the lowest impact strength, indicating the worst toughness. When PEK-C is added, the toughness of the EP/PEK-C system is significantly improved, and the impact strength value increases from 9.51 kJ/m^2 to 24.16 kJ/m^2, an increase of 154%. PEK-C undergoes reaction-induced phase separation during the curing reaction and forms a mutually interpenetrating bicontinuous phase structure with EP [37], which causes cracks to inevitably pass through the tough PEK-C continuous phase, thereby triggering the yielding effect of the thermoplastic resin and significantly improving its toughness [38]. The toughness of the EP/CTBN system is further improved compared to the EP/PEK-C system, increasing from 9.51 kJ/m^2 to 32.64 kJ/m^2, an increase of 243% compared to the untoughened EP system. The reason for the improvement is that, on the one hand, EP and CTBN undergo esterification reactions to form block copolymers, which introduces a large number of nitrile flexible chain segments into EP [39]. However, the flexible segments will reduce the cross-linking density of the cured EP, reduce the binding of the molecular chain movement, make the matrix prone to plastic deformation, and absorb the fracture energy, thereby improving the toughness of the cured EP.

In contrast, as the curing reaction progresses, the viscosity of the EP/CTBN system gradually increases, inducing phase separation [40]. The rubber particles separate from EP to form a "sea-island" structure, where the sea phase is EP, and the island phase is rubber particles. When the material is damaged, the rubber particles disperse and absorb external impact stress, preventing the propagation of cracks in the EP and slowing down material fracture [41], strengthening and toughening the entire system. Figure 4 shows that the toughness of the EP/PEK-C/CTBN synergistic toughening system is significantly improved compared to the single toughening systems of EP/PEK-C and EP/CTBN, increasing from 9.51 kJ/m^2 to 42.37 kJ/m^2, an increase of 342% compared to the untoughened EP system. This suggests that PEK-C/CTBN has a significant synergistic toughening effect on EP related to its unique microstructure. The microstructure and the mechanism for improving toughness are described in detail in the following text.

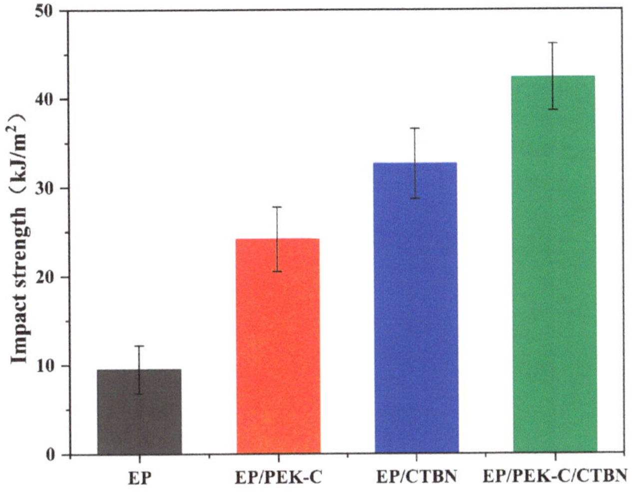

Figure 4. The impact strength value of each toughening system.

3.3. *Analysis of Heat Resistance Performance*

Figure 5 shows the relationship between the modified EP temperature and tanδ. The graph shows that the addition of PEK-C increases the Tg of EP from 254 °C to 264 °C. The increase is because PEK-C is a macromolecule polymer, and its own Tg is significantly higher than that of EP. Its addition to EP naturally increases the Tg of the whole toughening system. In contrast to the effect of PEK-C addition, the addition of CTBN significantly reduces the Tg of EP from 254 °C to 244 °C. The decrease is due to the flexible chains of the rubber and the rigid chains of EP forming a block copolymer. A portion of the added rubber is not completely separated and dissolved in EP, forming a plastic-to-glass transition [42]. In addition, Figure 5 shows that the Tg of the EP/PEK-C/CTBN toughening system is approximately the same as that of the untoughened EP system, differing only by 2 °C. This is because the Tg increase effect caused by the addition of PEK-C to EP is offset by the decreasing effect caused by the addition of CTBN to EP.

Figure 6 shows the TG curves of different toughening systems. The graph shows that all four systems begin to lose weight at approximately 136 °C. The temperature between 316 °C and 136 °C is the initial weight loss stage, during which weight loss is relatively slow due to the volatilization of small molecule volatiles. When the temperature interval is between 323 °C and 477 °C, the epoxy resin mainly decomposes and therefore the curve shows a rapid weight loss phase [43]. In this stage, the main chain of the resin casting body undergoes cracking, fusing, and carbonization decomposition, resulting in the maximum weight loss rate. After 478 °C, there is the final weight loss stage, during which the weight loss rate is much lower than that of the previous stage, and the reason for this weight loss should be the decomposition of certain composite resin components.

In addition, the graph shows that the temperatures corresponding to 50% weight loss for the four toughening systems, EP, EP/PEK-C, EP/CTBN, and EP/CTBN/PEK-C, are 431 °C, 465 °C, 412 °C, and 418 °C, respectively. The corresponding weight loss rates at 600 °C are 66%, 62%, 83%, and 72%, respectively. The addition of CTBN reduces the heat resistance of EP, while the addition of PEK-C improves the heat resistance of EP. In addition, the EP/CTBN toughening system begins to lose weight first, and its weight loss rate is the highest in the rapid weight loss stage due to the low molecular weight, short molecular chain, and low cohesive force of CTBN. The reason why the EP/PEK-C toughening system has the lowest weight loss rate in the rapid weight loss stage and the highest carbon residue rate at 600 °C is that the PEK-C molecular chain contains a large number of benzene rings and large phenolphthalein side groups, which greatly increases its own rigidity and heat resistance. After mixing with the EP system, the temperature resistance of the EP/PEK-C system is greatly improved.

Combining the DMA and TG curves leads to the conclusion that adding CTBN to EP decreases its Tg and heat resistance, whereas adding PEK-C to EP increases its Tg and heat resistance. The synergistic toughening of PEK-C/CTBN on EP essentially has no effect on EP's Tg or heat resistance.

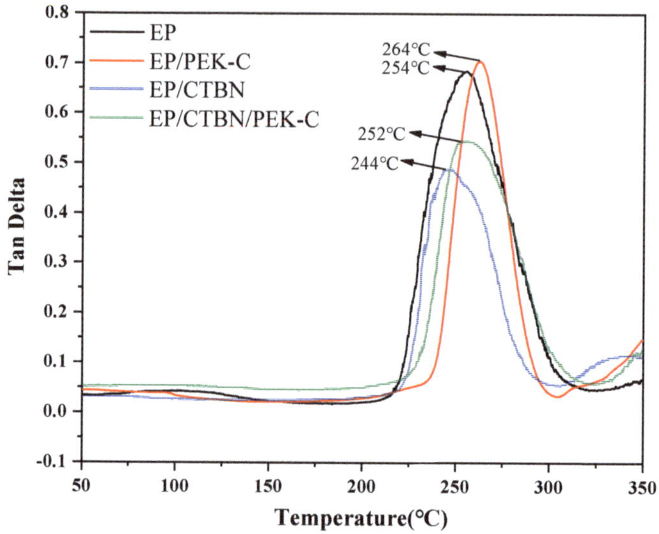

Figure 5. The relationship between modified EP temperature and tanδ.

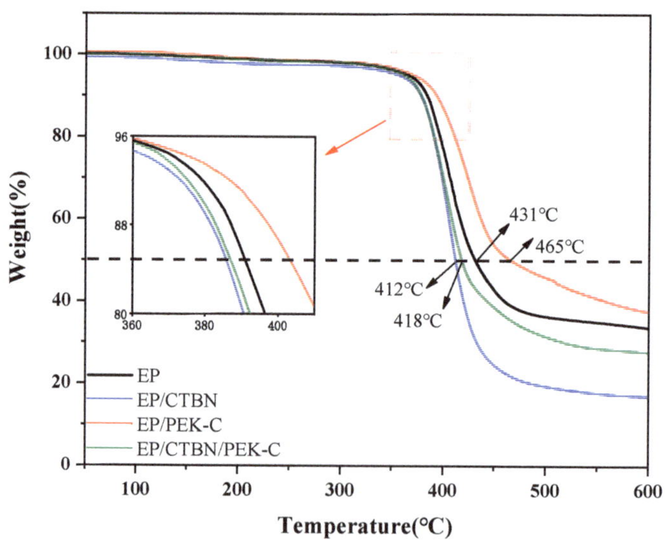

Figure 6. TG curves of different toughening systems.

3.4. Rheological Properties of Modified Resin

The viscosity under different toughening systems significantly impacts the processability of prepreg and composite molding. The dynamic rheological properties of three different toughening systems were investigated in this experiment, and the results are shown in Figure 7.

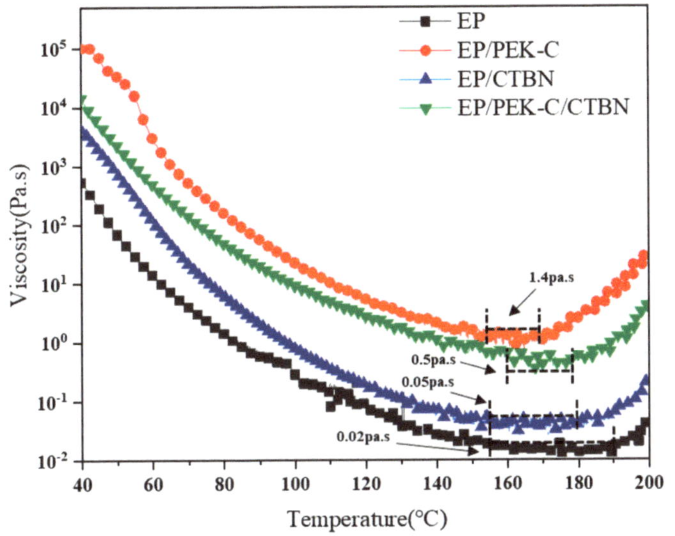

Figure 7. Viscosity−temperature curves of different toughening systems.

As shown in the graph, as the temperature rises, the viscosity of all four systems initially decreases, then tends to stabilize, and finally increases again. The initial stage of rapid viscosity reduction is primarily a physical change. As the temperature of the resin increases, the ratio of external heat energy to flow activation energy becomes larger and larger in the process of increasing temperature. The activity and migration ability of the molecular chain segments increase due to thermal activation, and thus the resin viscosity decreases sharply with increasing temperature. After increasing the temperature beyond a certain temperature, the EP and toughening agents are liquid. The viscosity of each blend system reaches a minimum, followed by the appearance of a low-viscosity plateau (the temperature range and viscosity corresponding to the plateau are shown in the marked part of Figure 7). This is because the temperature range has not reached the activation energy temperature required for the curing reaction of the epoxy resin system. The entire resin system has not undergone a curing cross-linking reaction, and its viscosity is at its lowest and most stable. As the temperature rises, the system undergoes a cross-linking reaction and quickly enters the gel state. It then forms a three-dimensional network structure, which limits the movement of the molecules, making the resin show a sharp increase in system viscosity on a macroscopic scale. In addition, the molecular weights of EP, CTBN, and PEK-C increase, with PEK-C having a significantly higher molecular weight than EP and CTBN. The larger the molecular weight, the longer the molecular chain and the higher the likelihood of entanglement. Once entanglement occurs, the flow resistance of the fluid increases and the viscosity increases. Therefore, in Figure 7, the order of gelation occurrence among the four systems is EP/PEK-C > EP/PEK-C/CTBN > EP/CTBN > EP. Figure 7 also shows that in the temperature range of 162.42–181.76 °C, the viscosity of the EP/PEK-C/CTBN synergistic toughening system is significantly higher than that of the untoughened EP system. Still, the viscosity of EP/PEK-C/CTBN is not high. In this low viscosity–temperature range, the preparation of prepreg and the selection of pressure time and other process parameters can be carried out effectively.

3.5. Microstructure of Modified Resin System

After curing, the microstructure determines the resin's performance under various external forces. Figure 8 shows the SEM fracture surfaces of untoughened EP and EP/CTBN toughening system samples after the impact test. Figure 8a shows that the fracture surface of the untoughened EP sample is smooth and flat, with few branching crack propagation paths and crack lines close to the coastline, exhibiting typical brittle fracture characteristics. The coastline-like morphology of the fracture surface suggests that the resistance encountered during crack propagation is small, and the energy required for crack propagation is low. Therefore, as stated previously, the untoughened EP system has the lowest impact strength and the poorest toughness.

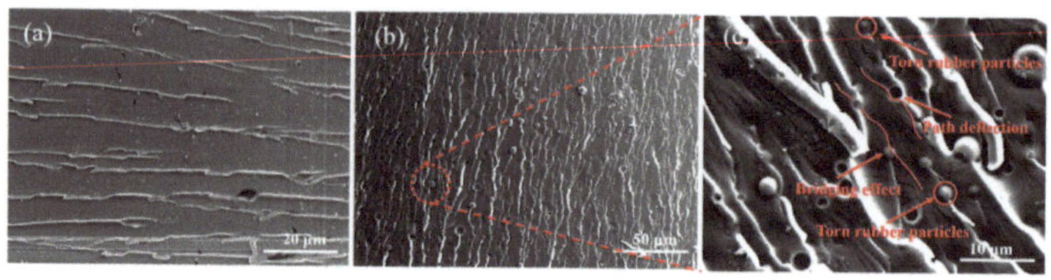

Figure 8. SEM images of (**a**) EP and (**b**,**c**) EP/CTBN toughening system.

Figure 8b shows that after the addition of CTBN, the fracture surface of the specimen becomes rough, with many folds and sawtooth-shaped cracks exhibiting obvious ductile fracture characteristics. In addition, Figure 8b shows that the rubber particles are uneven and distributed relatively uniformly, with sizes ranging from 1 to 5 μm. During the curing and precipitation process of CTBN, rubber particles that are close together will merge, resulting in larger sizes than those that have not merged. The larger particles can bridge the two sides of the crack (as indicated by the red line in Figure 8c) and prevent it from extending to a highly destructive level. On the other hand, smaller rubber particles induce shear bands through voids, which increases toughness. The uneven sizes of the particles can further enhance the toughening effect and improve the ability to relieve stress concentration [44–46]. In addition, Figure 8b shows that the rubber particles leave behind voids when detached from the resin matrix due to stress, resulting in a large surface area that consumes a significant amount of fracture energy. Figure 8c (the two red circles marked) shows that the rubber particles undergo stretching and tearing failure due to stress and cause the crack to diverge and turn due to the pinning effect; additionally, some rubber particles experience plastic deformation [47,48] and become irregular spheres under stress. All these changes caused by stress will consume a substantial amount of fracture energy, effectively preventing the crack from extending and reducing the material's fracture.

Figure 9 shows the SEM images of the EP/PEK-C toughened system before and after etching with tetrahydrofuran (THF). Figure 9a shows that the fracture surface morphology of the EP after toughening with PEK-C is completely different from that of the untoughened and CTBN-toughened systems. The surface of the fracture is smooth, without coastline or sawtooth-shaped cracks. Figure 9b shows that after THF etching, it can be found that the remaining epoxy-rich component after etching off PEK-C becomes blocky solids of varying sizes, resulting in a no longer smooth fracture surface. In addition, Figure 9c shows that PEK-C and EP actually become a bicontinuous phase structure of EP-coated PEK-C and PEK-C-coated EP, wherein the two phases are interlocked and continuous throughout the entire range, which is exactly the microstructure morphology controlled by the spinodal decomposition mechanism [49]. Compared with the untoughened system, which has only an epoxy-rich extension path, the double continuous phase structure requires the crack to propagate through many PEK-C-rich components. The plastic deformation capacity and energy dissipation of the PEK-C-rich phase are strong. The resistance required to overcome the crack propagation in this phase is also greater, which is why the toughness of the PEK-C toughened system is tougher than that of the untoughened EP system described earlier.

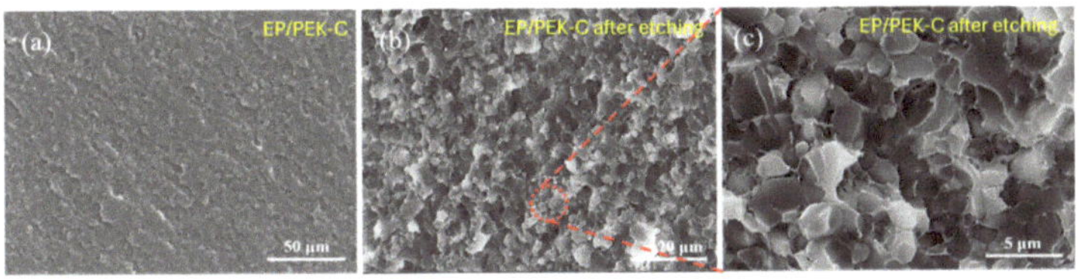

Figure 9. SEM images of EP/PEK-C toughening system before and after etching.

Figure 10 shows the SEM images of the EP/CTBN/PEK-C toughening system samples before and after THF etching. In Figure 10a,b, the fracture surface is rough, and there are many damaged surfaces, typical of toughness damage. It is also evident that many microvoids accompany the plastic stretching and cracking phenomena. This is because the rubber particles are subjected to static fluid tension during the curing and cooling process and the action of the triaxial stress field on the crack tip during loading. These two forces superimpose, causing the rupture of holes within the rubber particle or at the interface between the rubber particle and the matrix. The generation of these voids can alleviate the triaxial stress accumulated at the crack tip and on the other hand, will increase the stress concentration on the rubber particles, which makes the cavitation occur further. It also induces the local shear yielding of the matrix around the rubber particles, which in turn leads to the passivation of the crack tip [50], thereby further reducing the stress concentration in the matrix resin and preventing the occurrence of fracture. In addition, multiple stress-whitened regions can be observed in Figure 10a–c. The whitening of the stress is due to the scattering of visible light from the scattering center layer. In this case, it is void of the scattering center due to the cavitation of the CTBN particles [51]. This further proves that the shear yield induced by rubber cavitation is an important toughening mechanism of the EP/PEK-C/CTBN toughening system.

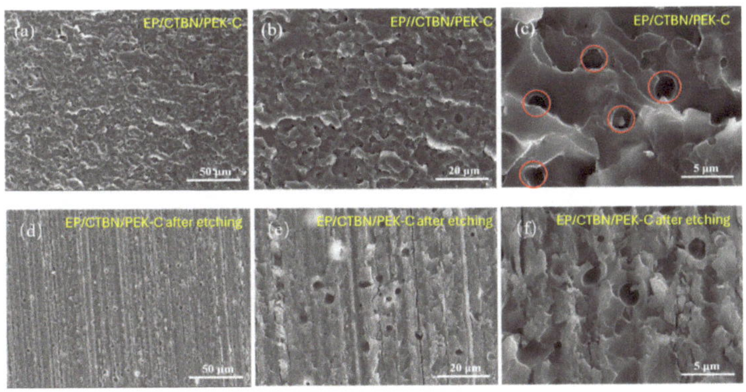

Figure 10. SEM images of EP/CTBN/PEK-C toughening system before and after etching.

Figure 10c shows that the fracture surface contains a large number of rubber particles (red circle mark) and many bow-shaped cracks of varying thickness, primarily around or through the rubber particles. This is because the rubber particles connect the two sides of the crack through bridging, which constrains and closes the crack propagation. In addition, the particle bridging not only restrains and limits the advancement of the crack leading edge, but the distributed bridging force also anchors the crack at the bridging point, giving the crack leading edge a wavy bow shape.

Figure 10d,e show that the microstructure of the EP/CTBN component left after removing the PEK-C is completely different from the morphology when either PEK-C or CTBN alone is toughened. No cracks propagating in multiple directions are visible in the two images—only linear cracks passing through rubber particles. Figure 10f shows that after the crack passes through the rubber particle, the rubber particle undergoes plastic deformation, transforming into an irregular sphere with numerous broken lumpy bumps or pits in the EP on both sides of the crack in the direction of crack expansion. The analysis suggests that after the EP/PEK-C/CTBN system cracks under external loading, the thermoplastic resin deforms, causing shear yielding. EP and PEK-C form a dual-phase interlocked co-continuous structure, which is closely linked, and the yielding effect of the thermoplastic resin will inevitably drive the changes in EP, thus leaving behind the deformed EP after etching away the PEK-C.

In addition, compared with the SEM image of CTBN alone, it can be observed that the rubber particles in the EP/PEK-C/CTBN synergistic toughening system are rarely exposed on the fracture surface and are primarily embedded in the thermoplastic/thermosetting resin. This indicates that the rubber particles cannot absorb fracture energy through debonding. Combining the above analysis with the current toughening theory of thermoplastic resins and rubber elastomers on EP, the following factors explain why the synergistic toughening effect of the PEK-C/CTBN system is significantly stronger than the individual toughening effects of PEK-C and CTBN: (1) During the CTBN and EP pre-polymerization stage, -COOH in CTBN and -CH(O)CH- in EP undergo an esterification reaction to form a flexible block polymer that can play a toughening role. (2) When CTBN is blended with the resin, reaction-induced phase separation occurs, and CTBN will precipitate and form a similar "sea-island" structure with the thermoplastic/thermoset resin system. When the system is subjected to external force after curing, the small rubber particles can produce shear bands through voids to improve the system's toughness. The large particles that undergo phase separation undergo tensile tearing and plastic deformation under stress [52–54], absorbing a large amount of fracture energy and slowing down material fracture. At the same time, the existence of a flexible block polymer makes the system less bound when it is deformed by force, which is easy to cause plastic deformation. The combined effect of the factors mentioned earlier increases the toughness of the cured material. (3) When encountering cracks, many dispersed-phase rubber particles will have a bridging effect which will play a restraining and closing role on crack expansion and prevent crack expansion to a highly destructive degree. In addition, the distributed bridging force exerts a pinning effect on the crack at the bridging point. The pinning effect causes the crack to deviate from its main plane and increases the surface area, resulting in a significant increase in the energy required for crack propagation. On the other hand, it also causes the coarser primary crack to branch into multiple finer secondary cracks [55]. The secondary cracks with changed expansion direction will inevitably encounter the thermoplastic PEK-C continuous phase. The toughness-rich PEK-C phase will induce greater cooperative deformation at the crack tip, triggering the yielding effect of the thermoplastic resin. The double continuous phase structure of EP and PEK-C will take advantage of the excellent interfacial adhesion to absorb the crack energy and to shear yield against the crack [56].

4. Conclusions

Dynamic rheological performance tests show that the EP/PEK-C/CTBN synergistic toughening system meets the requirements of conventional hot-melt prepreg preparation, and the toughening effect of the EP/PEK-C/CTBN synergistic toughening system is significantly higher than that of the single toughening system. The toughening mechanism combines the crack bridging effect, crack pinning effect, shear yielding, and plastic deformation of the rubber particle holes of a conventional single toughening system so that the synergistic system of thermoplastic resin and rubber particles shows an efficient synergistic energy dissipation effect under impact. According to the SEM image, the microstructure of the EP/PEK-C/CTBN synergistic toughening system is entirely distinct from the microstructures of the PEK-C and CTBN single toughening systems. The unique microstructure makes the thermoplastic resin and rubber particles exhibit efficient energy dissipation when subjected to crack damage, which ensures the synergistic toughening system has significantly higher toughness than the single toughening systems.

Author Contributions: Conceptualization, X.Y. and Y.X.; methodology, Z.W., X.Y. and Y.X.; validation, Y.L., X.Y., Z.W. and P.X.; formal analysis, Y.L., Z.W., X.Y. and P.X.; investigation, X.Y., Y.X. and J.M.; resources, Z.W.; data curation, Y.L., X.Y. and Z.W.; writing—original draft preparation, Y.L., Z.W., X.Y. and Y.X.; writing—review and editing, Z.W., X.Y. and Y.X.; visualization, X.Y., Y.X. and P.X.; supervision, X.Y., Y.X. and J.M.; project administration, X.Y.; funding acquisition, Y.X. All authors have read and agreed to the published version of the manuscript.

Funding: This research was funded by the Priority Academic Program Development of Jiangsu Higher Education Institutions (PAPD).

Institutional Review Board Statement: Not applicable.

Data Availability Statement: The raw data supporting the conclusions of this article will be made available by the authors on request.

Acknowledgments: Z.D. Wang and Y.C. Lai contributed equally to this work.

Conflicts of Interest: The authors declare no conflicts of interest.

References

1. Białkowska, A.; Bakar, M.; Kucharczyk, W.; Zarzyka, I. Hybrid Epoxy Nanocomposites: Improvement in Mechanical Properties and Toughening Mechanisms—A Review. *Polymers* **2023**, *15*, 1398. [CrossRef] [PubMed]
2. Wang, J.; Xue, Z.; Li, Y.; Li, G.; Wang, Y.; Zhong, W.-H.; Yang, X. Synergistically effects of copolymer and core-shell particles for toughening epoxy. *Polymer* **2018**, *140*, 39–46. [CrossRef]
3. Mansour, G.; Tsongas, K.; Tzetzis, D. Investigation of the dynamic mechanical properties of epoxy resins modified with elastomers. *Compos. Part B Eng.* **2016**, *94*, 152–159. [CrossRef]
4. Ricciardi, M.; Papa, I.; Langella, A.; Langella, T.; Lopresto, V.; Antonucci, V. Mechanical properties of glass fibre composites based on nitrile rubber toughened modified epoxy resin. *Compos. Part B Eng.* **2018**, *139*, 259–267. [CrossRef]
5. Thomas, R.; Durix, S.; Sinturel, C.; Omonov, T.; Goossens, S.; Groeninckx, G.; Moldenaers, P.; Thomas, S. Cure kinetics, morphology and miscibility of modified DGEBA-based epoxy resin—Effects of a liquid rubber inclusion. *Polymer* **2007**, *48*, 1695–1710. [CrossRef]
6. Gunwant, D.; Sah, P.L.; Zaidi, M. Morphology and micromechanics of liquid rubber toughened epoxies. *e-Polymers* **2018**, *18*, 511–527. [CrossRef]
7. Qu, C.; Zhang, X.; Wang, D.; Fan, X.; Li, H.; Liu, C.; Feng, H.; Wang, R.; Guo, K.; Tian, Y.; et al. Residual stress and thermal properties of rubber-modified epoxy systems for semiconductor package. *J. Appl. Polym. Sci.* **2022**, *139*, 51786. [CrossRef]
8. Sprenger, S. Nanosilica-toughened epoxy resins. *Polymers* **2020**, *12*, 1777. [CrossRef]
9. Mi, X.; Liang, N.; Xu, H.; Wu, J.; Jiang, Y.; Nie, B.; Zhang, D. Toughness and mechanism of epoxy resins. *Prog. Mater. Sci.* **2022**, *130*, 100977. [CrossRef]
10. Thomas, R.; Yumei, D.; Yuelong, H.; Le, Y.; Moldenaers, P.; Weimin, Y.; Czigany, T.; Thomas, S. Miscibility, morphology, thermal, and mechanical properties of a DGEBA based epoxy resin toughened with a liquid rubber. *Polymer* **2008**, *49*, 278–294. [CrossRef]
11. Dong, L.; Zhou, W.; Sui, X.; Wang, Z.; Cai, H.; Wu, P.; Zuo, J.; Liu, X. A Carboxyl-Terminated Polybutadiene Liquid Rubber Modified Epoxy Resin with Enhanced Toughness and Excellent Electrical Properties. *J. Electron. Mater.* **2016**, *45*, 3776–3785. [CrossRef]
12. Imanaka, M.; Narita, I.; Nakamura, Y.; Hisaka, S.; Yoshida, S.; Hara, K. Effect of matrix deformability on the fracture properties of epoxy resins modified with core-shell and cross-linked rubber particles. *J. Appl. Polym. Sci.* **2022**, *139*, 52316. [CrossRef]
13. Ali-Asgari Dehaghi, H.; Mazinani, S.; Zaarei, D.; Kalaee, M.; Jabari, H.; Sedaghat, N. Thermal and morphological characteristics of solution blended epoxy/NBR compound. *J. Therm. Anal. Calorim.* **2013**, *114*, 185–194. [CrossRef]
14. Ma, H.; Aravand, M.A.; Falzon, B.G. Phase morphology and mechanical properties of polyetherimide modified epoxy resins: A comparative study. *Polymer* **2019**, *179*, 121640. [CrossRef]
15. Szymańska, J.; Bakar, M.; Białkowska, A.; Kostrzewa, M. Study on the adhesive properties of reactive liquid rubber toughened epoxy-clay hybrid nanocomposites. *J. Polym. Eng.* **2018**, *38*, 231–238. [CrossRef]
16. Tripathi, G.; Srivastava, D. Effect of carboxyl-terminated poly(butadiene-co-acrylonitrile) (CTBN) concentration on thermal and mechanical properties of binary blends of diglycidyl ether of bisphenol-A (DGEBA) epoxy resin. *Mat. Sci. Eng. A Struct.* **2007**, *443*, 262–269. [CrossRef]
17. Chikhi, N.; Fellahi, S.; Bakar, M. Modification of epoxy resin using reactive liquid (ATBN) rubber. *Eur. Polym. J.* **2002**, *38*, 251–264. [CrossRef]
18. Wang, C.; Li, H.; Zhang, H.; Wang, H.; Liu, L.; Xu, Z.; Liu, P.; Peng, Z. Influence of addition of hydroxyl-terminated liquid nitrile rubber on dielectric properties and relaxation behavior of epoxy resin. *IEEE Trans. Dielectr. Electr. Insul.* **2016**, *23*, 2258–2269. [CrossRef]
19. Lee, S.-E.; Jeong, E.; Lee, M.Y.; Lee, M.-K.; Lee, Y.-S. Improvement of the mechanical and thermal properties of polyethersulfone-modified epoxy composites. *J. Ind. Eng. Chem.* **2016**, *33*, 73–79. [CrossRef]

20. Yang, G.; Zheng, B.; Yang, J.-P.; Xu, G.-S.; Fu, S.-Y. Preparation and cryogenic mechanical properties of epoxy resins modified by poly(ethersulfone). *J. Polym. Sci. Part A Polym. Chem.* **2008**, *46*, 612–624. [CrossRef]
21. Chen, D.; Li, J.; Yuan, Y.; Gao, C.; Cui, Y.; Li, S.; Wang, H.; Peng, C.; Liu, X.; Wu, Z. A new strategy to improve the toughness of epoxy thermosets by introducing the thermoplastic epoxy. *Polymer* **2022**, *240*, 124518. [CrossRef]
22. Ji, Y.; Zhang, Y.; Wang, P.; Li, Y.; Sui, J. Mechanical and thermal properties of epoxy resins modified by a novel thermoplastic-polyimide. *Fiber Polym.* **2021**, *22*, 205–212. [CrossRef]
23. Karthikeyan, L.; Robert, T.M.; Mathew, D.; Suma, D.D.; Thomas, D. Novel epoxy resin adhesives toughened by functionalized poly (ether ether ketone) s. *Int. J. Adhes. Adhes.* **2021**, *106*, 102816. [CrossRef]
24. Li, H.; Zhao, L.; Su, K.; Feng, H.; Wang, D.; Qu, C. A comparative study on the rheological, thermal, and mechanical performance of epoxy resin modified with thermoplastics. *J. Adhes. Sci. Technol.* **2021**, *35*, 1393–1403. [CrossRef]
25. Mathis, E.; Michon, M.-L.; Billaud, C.; Vergelati, C.; Clarke, N.; Jestin, J.; Long, D.R. Controlling the morphology in epoxy/thermoplastic systems. *ACS Appl. Polym. Mater.* **2022**, *4*, 2091–2104. [CrossRef]
26. Jung, H.-S.; Park, Y.; Nah, C.-W.; Lee, J.-C.; Kim, K.-Y.; Lee, C.S. Evaluation of the mechanical properties of polyether sulfone-toughened epoxy resin for carbon fiber composites. *Fiber Polym.* **2021**, *22*, 184–195. [CrossRef]
27. Zhou, S.; Chen, Z.; Tusiime, R.; Cheng, C.; Sun, Z.; Xu, L.; Liu, Y.; Jiang, M.; Zhou, J.; Zhang, H. Highly improving the mechanical and thermal properties of epoxy resin via blending with polyetherketone cardo. *Compos. Commun.* **2019**, *13*, 80–84. [CrossRef]
28. *GB/T 2567-2008*; Test Method for Impact Resistance of Resin Casting Body. Standards Press of China: Beijing, China, 2008.
29. *GB/T 2571-1995*; Test Method for Impact Resistance of Resin Casting Body. Standards Press of China: Beijing, China, 1995.
30. Yin, X.; Xie, Z.; Liu, Q.; Yuan, X.; Hou, X.; Zhao, J. Synergistic toughening of epoxy resin by CTBN and CM-β-CD. *J. Appl. Polym. Sci.* **2021**, *138*, 51248. [CrossRef]
31. Chen, H.; Zhu, Z.; Patil, D.; Bajaj, D.; Verghese, N.; Jiang, Z.; Sue, H.-J. Mechanical properties of reactive polyetherimide-modified tetrafunctional epoxy systems. *Polymer* **2023**, *270*, 125763. [CrossRef]
32. Ramos, V.D.; da Costa, H.M.; Soares, V.L.; Nascimento, R.S. Hybrid composites of epoxy resin modified with carboxyl terminated butadiene acrilonitrile copolymer and fly ash microspheres. *Polym. Test.* **2005**, *24*, 219–226. [CrossRef]
33. Tripathi, G.; Srivastava, D. Studies on the physico-mechanical and thermal characteristics of blends of DGEBA epoxy, 3, 4 epoxy cyclohexylmethyl, 3′, 4′-epoxycylohexane carboxylate and carboxyl terminated butadiene co-acrylonitrile (CTBN). *Mat. Sci. Eng. A Struct.* **2008**, *496*, 483–493. [CrossRef]
34. Akbari, R.; Beheshty, M.H.; Shervin, M. Toughening of dicyandiamide-cured DGEBA-based epoxy resins by CTBN liquid rubber. *Iran. Polym. J.* **2013**, *22*, 313–324. [CrossRef]
35. Nigam, V.; Setua, D.K.; Mathur, G.N. Characterization of liquid carboxy terminated copolymer of butadiene acrylonitrile modified epoxy resin. *Polym. Eng. Sci.* **1999**, *39*, 1425–1432. [CrossRef]
36. Januszewski, R.; Dutkiewicz, M.; Nowicki, M.; Szołyga, M.; Kownacki, I. Synthesis and properties of epoxy resin modified with novel reactive liquid rubber-based systems. *Ind. Eng. Chem. Res.* **2021**, *60*, 2178–2186. [CrossRef]
37. Ding, Y.; Liu, R.; Liu, H. Investigation on viscoelastic phase separation of phenolphthalein poly (ether ether ketone)/epoxy blends system. *Polym. Polym. Compos.* **2020**, *28*, 199–208. [CrossRef]
38. Mimura, K.; Ito, H.; Fujioka, H. Improvement of thermal and mechanical properties by control of morphologies in PES-modified epoxy resins. *Polymer* **2000**, *41*, 4451–4459. [CrossRef]
39. Son, B.T.; Trung, N.N.; Lim, D.-G.; Shin, S.; Bae, J.-Y. Improvements in thermal, mechanical, and dielectric properties of epoxy resin by chemical modification with a novel amino-terminated liquid-crystalline copoly (ester amide). *React. Funct. Polym.* **2012**, *72*, 542–548. [CrossRef]
40. Liu, X.-F.; Luo, X.; Liu, B.-W.; Zhong, H.-Y.; Guo, D.-M.; Yang, R.; Chen, L.; Wang, Y.-Z. Toughening epoxy resin using a liquid crystalline elastomer for versatile application. *ACS Appl. Polym. Mater.* **2019**, *1*, 2291–2301. [CrossRef]
41. Tiwari, S.N.; Agnihotri, P.K. Effect of crumb rubber addition on the deformation and fracture behavior of ductile epoxy matrix. *J. Appl. Polym. Sci.* **2023**, *140*, e53255. [CrossRef]
42. Ozturk, A.; Kaynak, C.; Tincer, T. Effects of liquid rubber modification on the behaviour of epoxy resin. *Eur. Polym. J.* **2001**, *37*, 2353–2363. [CrossRef]
43. Ahmad, H.; Shah, A.; Afaq, S.K.; Azad, M.; Arif, S.; Siddiqi, M.; Xie, L. Development and characterization of kevlar and glass fibers reinforced epoxy/vinyl ester hybrid resin composites. *Polym. Compos.* **2024**, *45*, 8133–8146. [CrossRef]
44. Dispenza, C.; Spadaro, G.; McGrail, P.T. The influence of the rubber carboxyl functionality on the processability and properties of solid rubber modified epoxy mixtures. *Macromol. Chem. Phys.* **2005**, *206*, 393–403. [CrossRef]
45. Gao, Z.; Yu, Y.; Xu, Y.; Li, S. Synthesis and characterization of a liquid crystalline epoxy containing azomethine mesogen for modification of epoxy resin. *J. Appl. Polym. Sci.* **2007**, *105*, 1861–1868. [CrossRef]
46. Morancho, J.; Salla, J. Relaxation in partially cured samples of an epoxy resin and of the same resin modified with a carboxyl-terminated rubber. *Polymer* **1999**, *40*, 2821–2828. [CrossRef]
47. Dadfar, M.; Ghadami, F. Effect of rubber modification on fracture toughness properties of glass reinforced hot cured epoxy composites. *Mater. Des.* **2013**, *47*, 16–20. [CrossRef]
48. Vijayan, P.P.; Pionteck, J.; Huczko, A.; Puglia, D.; Kenny, J.M.; Thomas, S. Liquid rubber and silicon carbide nanofiber modified epoxy nanocomposites: Volume shrinkage, cure kinetics and properties. *Compos. Sci. Technol.* **2014**, *102*, 65–73. [CrossRef]

49. Zhang, J.; Guo, Q.; Fox, B. Structural and material properties of a rapidly cured thermoplastic-toughened epoxy system. *J. Appl. Polym. Sci.* **2009**, *113*, 485–491. [CrossRef]
50. Xi, J.; Yu, Z. Toughening mechanism of rubber reinforced epoxy composites by thermal and microwave curing. *J. Appl. Polym. Sci.* **2018**, *135*, 45767. [CrossRef]
51. Bian, X.; Tuo, R.; Yang, W.; Zhang, Y.; Xie, Q.; Zha, J.; Lin, J.; He, S. Mechanical, thermal, and electrical properties of BN–epoxy composites modified with carboxyl-terminated butadiene nitrile liquid rubber. *Polymers* **2019**, *11*, 1548. [CrossRef]
52. Ocando, C.; Tercjak, A.; Mondragon, I. Nanostructured systems based on SBS epoxidized triblock copolymers and well-dispersed alumina/epoxy matrix composites. *Compos. Sci. Technol.* **2010**, *70*, 1106–1112. [CrossRef]
53. Ocando, C.; Tercjak, A.; Serrano, E.; Ramos, J.A.; Corona-Galván, S.; Parellada, M.D.; Fernández-Berridi, M.J.; Mondragon, I. Micro- and macrophase separation of thermosetting systems modified with epoxidized styrene-*block*-butadiene-*block*-styrene linear triblock copolymers and their influence on final mechanical properties. *Polym. Int.* **2008**, *57*, 1333–1342. [CrossRef]
54. Pearson, R.A.; Yee, A.F. Influence of particle size and particle size distribution on toughening mechanisms in rubber-modified epoxies. *J. Mater. Sci.* **1991**, *26*, 3828–3844. [CrossRef]
55. Klingler, A.; Bajpai, A.; Wetzel, B. The effect of block copolymer and core-shell rubber hybrid toughening on morphology and fracture of epoxy-based fibre reinforced composites. *Eng. Fract. Mech.* **2018**, *203*, 81–101. [CrossRef]
56. Gilbert, A.H.; Bucknall, C.B. Epoxy resin toughened with thermoplastic. *Makromol. Chemie. Macromol. Symp.* **1991**, *45*, 289–298. [CrossRef]

polymers

MDPI

Article

Bio-Sourced, High-Performance Carbon Fiber Reinforced Itaconic Acid-Based Epoxy Composites with High Hygrothermal Stability and Durability

Kaixuan Xiao [1,†], Yuan Fang [1], Zhaodi Wang [1,†], Nannan Ni [1], Ziqian Liu [2], Soochan Kim [3], Zongfu An [3], Zhiyi Lyu [4], Yahong Xu [1,*] and Xin Yang [1,*]

1 College of Materials Science and Engineering, Nanjing Tech University, Nanjing 211816, China; 202161203303@njtech.edu.cn (K.X.); 202261103096@njtech.edu.cn (Y.F.); 201910006628@njtech.edu.cn (Z.W.); 202010006705@njtech.edu.cn (N.N.)
2 Yangtze River Delta Carbon Fiber and Composites Innovation Center, Changzhou 213000, China; liuziqian@ccicyd.com
3 School of Chemical Engineering, Sungkyunkwan University, Suwon 16419, Republic of Korea; sc0916@skku.edu (S.K.); zfann@skku.edu (Z.A.)
4 Department of Physics, Sungkyunkwan University, 2066, Seobu-ro, Jangan-gu, Suwon 16419, Republic of Korea; louislv@skku.edu
* Correspondence: 201910006672@njtech.edu.cn (Y.X.); xin12.25@hotmail.com (X.Y.)
† These authors contributed equally to this work.

Abstract: Thermosetting polymers and composites are a class of high-performance materials with significant industrial applications. However, the widespread use of thermosets and their composites generates large quantities of waste and leads to serious economic and environmental problems, there is a critical need in the elaboration of sustainable composite materials. Here, we propose a method to prepare sustainable carbon fiber reinforced composites with different degrees of greenness by blending environmentally friendly EIA with DGEBA in different ratios, and the properties compared with a well-known commercial petroleum-based epoxy resin. The prepared carbon fiber reinforced polymer (CFRP) composites with different degrees of greenness had excellent dimensional stability under extreme hygrothermal aging. After aging, the green CFRP composite T700/EIA-30 has higher strength and performance retention than that of petroleum-based CFRP composites. The higher hygrothermal stability and durability of EIA-based epoxy resins as compared with BPA-based epoxy resins demonstrated significant evidence to design and develop a novel bio-based epoxy resin with high performance to substitute the petroleum-based epoxy resin.

Keywords: sustainable composite materials; itaconic acid epoxy resin; carbon fiber reinforced composites; hygrothermal aging

Citation: Xiao, K.; Fang, Y.; Wang, Z.; Ni, N.; Liu, Z.; Kim, S.; An, Z.; Lyu, Z.; Xu, Y.; Yang, X. Bio-Sourced, High-Performance Carbon Fiber Reinforced Itaconic Acid-Based Epoxy Composites with High Hygrothermal Stability and Durability. *Polymers* **2024**, *16*, 1649. https://doi.org/10.3390/polym16121649

Academic Editor: Yizhuo Gu

Received: 20 May 2024
Revised: 2 June 2024
Accepted: 7 June 2024
Published: 11 June 2024

1. Introduction

Epoxy resin (EP) is a thermoset resin with several advantages such as optimal properties, chemical stability, and electrical insulation properties. It is widely used in building materials, aerospace, coating adhesives, and polymer matrix composites [1–4]. In many high-tech fields, such as carbon fiber reinforced polymer (CFRP) composites, epoxy resins are a major component. However, the widespread use of EP and CFRP generates large quantities of waste and leads to serious economic and environmental problems. Currently, the global production of epoxy resins is expected to grow from 3.55 million tons in 2024 to 4.20 million tons in 2029, at a CAGR (Compound annual growth rate) of 3.41% during the forecast period (2023–2028), in which bisphenol A (BPA) epoxy resins dominate the market [5]. The dominant BPA epoxy resins in the market generate fossilized pollution during the production process. There is limited use of BPA epoxy resin worldwide due to its biotoxicity [6], which requires research and development of environmentally friendly green epoxy resins together with their composites.

It is commonly accepted that green resin composites are mainly fiber-reinforced resins made from bio-sourced materials. However, plant fibers as natural green reinforcements, are difficult to meet the requirements in performance since their hydrophilicity usually shows negative impact on composite properties [7]. Therefore, more attention has been paid to the development of bio-based epoxy resins. The bio-based epoxy resins that are widely studied and applied at present are mainly vegetable oils [8,9], such as soybean oil and hemp oil; lignin [10,11]; and other bio-based compounds that can be reacted with epoxy hydroxypropyl, such as rosin acid [12–14], cardanol [15,16], furan derivatives [17], gallic acid [18,19], isosorbide [20], itaconic acid [21], and other small molecules [22]. In recent years, scientists have developed bio-based feedstocks for epoxy resins, which resulted in a series of bio-based epoxy resins. However, epoxy vegetable oil is not suitable for application in the occasion of high requirements for mechanical properties due to the aliphatic long chain structure. Bio-based compounds possessing rigid groups, such as lignin and rosin acid, have been used to develop bio-based resins with high mechanical and thermal properties. Lignin-based epoxy resins result in slow curing and unstable properties due to the low mobility of large molecular substances and complex structures. The fused rings structure in rosin acid epoxy resin leads to an increase in brittleness [23]. Other bio-based compounds, such as cardanol, furan derivatives, gallic acid and isosorbide, still need to consider their cost, yield, and performance [17,24–26].

Among them, the development of itaconic acid epoxy resin is noteworthy. Itaconic acid is an unsaturated dicarboxylic acid usually produced by fermenting carbohydrates such as glucose or starch using aspergillus terreus [27]. It is an abundant and cost-effective raw material and has been selected by the U.S. Department of Energy as one of the twelve most promising bio-based platform chemicals [28–30]. To the best of our knowledge, itaconic acid has become an excellent raw material for industries such as synthetic resins, synthetic fibers, plastics, rubber, ion-exchange, surfactants, detergents [31]. Ma et al. obtained an itaconic acid epoxy resin (EIA) by reacting itaconic acid with epichlorohydrin, which was shown to have higher epoxide (0.625) and higher reactivity values than bisphenol A diglycidyl ether (DGEBA) [32]. It was found that the thermal and mechanical properties of the EIA were comparable to or better than those of DGEBA. Ma utilized carboxyl and itaconic acid double bonds to design and synthesize trifunctional itaconic acid epoxy resins (TEIA) with high epoxy value and low viscosity, which also has excellent mechanical property. Ma also used double bonds to achieve flame retardant structure and led to excellent flame retardant properties [21,32,33]. However, there is a lack of rigid groups in the crosslinked network of EIA or TEIA, they may be less resistant to solvents and heating. Liu et al. [34]. synthesized an epoxy network based on itaconic acid-based epoxy monomer, maleic anhydride, and glycerol, which possesses well-controlled degradability when combined with carbon fibers to manufacture composites, yet its lower mechanical properties and Tg limit its application.

The application of itaconic acid epoxy resin in carbon fiber-reinforced composites will have great prospects. Generally, applications of carbon fiber-reinforced composites are subject to aging due to ambient humidity or direct exposure to liquids [35]. It is generally believed that the diffusion of moisture in the resin matrix follows Fickian diffusion [36], where the resin undergoes swelling and the glass transition temperature decreases. It is reported that moisture may be located in an unbound state within the resin network or form hydrogen bonds with the resin network, making it difficult to remove [37,38]. It is also important to note that in carbon fiber reinforced composites, although the fibers do not readily absorb moisture, there are gaps between the fibers and the resin matrix, which may further lead to moisture ingress and reduced performance [39,40]. Temperature is a further factor in the aging of composites, as the resin may undergo network relaxation at high temperatures, which accelerates the ingress of moisture and leading to degradation of composite properties. Other factors could also impact composites performance, such as salinity [41], UV irradiation [42], corrosive environment [43] etc.

As an ester-containing bio-based epoxy resin, itaconic acid epoxy resin has low hygrothermal resistance [44]. In order to substitute the petroleum-based epoxy resin and improve the hygrothermal stability and durability of itaconic acid epoxy resin, we propose a method to prepare sustainable carbon fiber reinforced composites with different degrees of greenness by blending EIA with DGEBA in different ratios. The thermal and mechanical properties of the composites were investigated. The hygrothermal aging properties of the composites were also investigated. This paper used a commercial petroleum-based epoxy resin with DGEBA as the matrix of CFRP as control.

2. Materials and Methods

2.1. Materials

Dicyandiamide (DICY) was purchased from Taicang Changhe Polymer Co., Ltd. (Suzhou, China). Itaconic acid epoxy resin (EIA) with the epoxy value of 0.52 was purchased from Ningbo Institute of Materials Technology and Engineering (Ningbo, China). Diglycidyl ether bisphenol-A (DGEBA) epoxy resin with the epoxy value of 0.51 was purchased from Baling Petrochemical Co., Ltd. (Yueyang, China). Epoxy resin (E-20) was purchased from Baling Petrochemical Co., Ltd. (Yueyang, China). Replaces Urea (GLOC-500) was purchased from Suzhou Chenyang Polymer Material Co., Ltd. (Suzhou, China). Carbon fiber (T700) was purchased from Toray Inc., Tokyo, Japan. None of these chemicals was further purified before usage.

2.2. Preparation of the Cured Epoxy Resin

First, 20 wt% of E-20 was added to the beaker. After a high-speed stirring at 80 °C for 1 h, different proportions (0 wt%, 30 wt%, 40 wt%, 50 wt%) of EIA and different proportions (70 wt%, 40 wt%, 30 wt%, 20 wt%) of DGEBA were added, and stirred at 1000 r/min for 1 h. After adding 8 wt% of DICY and 2 wt% of GLOC-500, the mixture was heated up to 90 °C and stirred at a high speed of 1500 r/min for 30 min until the DICY and GLOC-500 were homogeneously dispersed in the resin. After vacuuming at 80 °C for half an hour, the product was poured into a polytetrafluoroethylene mold and cured at 80 °C for 1.5 h, 100 °C for 1.5 h, 110 °C for 2 h, and 130 °C for 2 h. Figure 1 and Table 1 show the specific experimental steps and experimental formulations of the products.

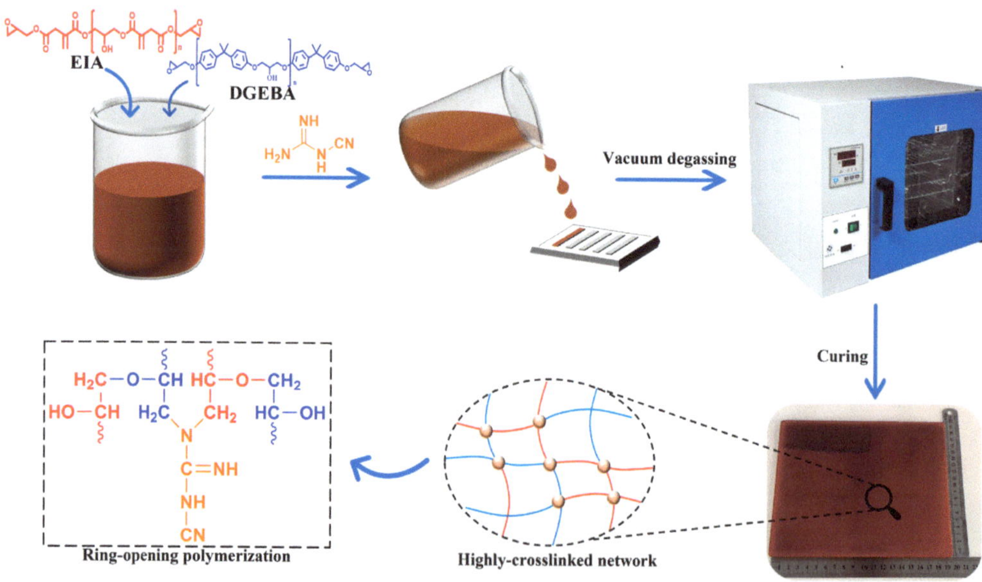

Figure 1. Schematic illustration of preparation of the cured epoxy resin.

Table 1. Formulations of epoxy resin with EIA.

Sample	EIA (wt%)	DGEBA (wt%)	E-20 (wt%)	DICY (wt%)	GLOC-500 (wt%)
EIA-0	0	70	20	8	2
EIA-30	30	50	20	8	2
EIA-40	40	40	20	8	2
EIA-50	50	30	20	8	2

2.3. Preparation of Carbon Fiber Composite Laminates

CFRP composite laminates of 320 mm × 320 mm were fabricated in an autoclave which allows good control of the curing process. The curing system was heated from room temperature at a heating rate of 3 °C/min to 70 °C for 30 min and then 130 °C for 120 min. This process was subject to a pressure of 3 MPa. The overall thickness of the CFRP laminates was 2 mm, and a total of 16 layers were laid, the other CFRP laminate had a total thickness of 1 mm and were laid with a total of 8 layers. There were 16 layers laid in the thickness direction with the sequence of $[0]_{16}$ and 8 layers laid in the thickness direction with the sequence of $[0]_8$, the prepreg average thickness of each layer was 0.13 mm. The density of composites was 1.55 g/cm^3 and the fiber volume fraction was measured at 60%.

2.4. Hygrothermal Aging

Hygrothermal aging tests were conducted on the laminates to observe the moisture absorption behavior of the composites and to determine the weight gain of the composite as a function of time. The size of the moisture absorption specimens was 50 mm × 50 mm × 2 mm, and the water absorption was determined according to ASTM D5229/D5229M standards with a number of five composite specimens in each group [45]. An analytical balance with an accuracy of 0.1 mg was used to measure the weight of each specimen hourly from the first day until the 12th hour, and at regular intervals (every 24 h) from the second day. The experiment was stopped when hygroscopic equilibrium (less than 0.01% change in weight) was reached.

All specimens were pre-treated in a drying oven at 70 °C before starting the experiments until they reached an engineered dry state (less than 0.02% change in weight), and then placed in a thermostatic water bath for hygrothermal aging. Mechanical properties of the composites were tested after 48 h of aging at 80 °C under water immersion. All samples were subsequently weighed to determine weight change. The weight gain was calculated according to

$$M = \frac{M_w - M_d}{M_d} \times 100\% \tag{1}$$

where M is the percentage of gained humidity, M_w is the weight of the wet sample, and M_d is the weight of the dry sample.

2.5. Characterization

Fourier Transform Infrared Spectroscopy (BRUKER, TENSOR II, Karlsruhe, Germany) was used to investigate the changes in the chemical structure of the resin in the composites before and after hygrothermal aging. FTIR spectra were recorded using a Bruker Tensor II spectrometer with a spectral range of 500 to 4000 cm^{-1} and a resolution of 4 cm^{-1}.

The static mechanical properties of the composites were tested using an Instron 3382 material testing machine. Tensile tests were based on ASTM D3039 [46], with each specimen measuring 250 mm × 15 mm × 1 mm, with reinforcing tabs affixed to the ends of each specimen, and with a loading rate of 2 mm/minute for tensile properties, with 6 specimens tested per composite system. Compression test according to ASTM D6641 [47], each specimen size 140 mm × 12 mm × 2 mm, 6 specimens per composite system. Flexural performance test according to ASTM D7264 [48], each specimen size 100 mm × 12.5 mm × 2 mm, flexural performance test loading rate 2 mm/min, 6 specimens per composite system. Interlaminar shear strength (ILSS) test according to ASTM D2344 [49], each specimen size

20 mm × 6 mm × 2 mm, 6 specimens per composite system. The mechanical test samples of composite laminates in this paper were all cut along the 0° direction of the fibers.

Dynamic thermodynamic analysis was carried out by a Dynamic Mechanical Analyzer (TA, DMA Q800, New Castle, DE, USA) in a nitrogen atmosphere. Samples with dimensions of 60 mm × 10 mm × 2 mm were heated from 30 °C to 230 °C at a heating rate of 3 °C/min in three-point bending mode at 1 Hz. 6 samples were tested for each composite system.

Thermogravimetric analysis (TGA) is carried out by means of a thermogravimetric analyzer (NETZSCH, TG 209F, Bavaria, Germany) under a nitrogen atmosphere, 20 mg of powdered solid were heated from 30 °C to 700 °C at a rate of 10 °C per minute.

Damage cross sections of CFRP specimens after flexural performance test were characterized using a TESCAN MIRA4 scanning electron microscope (TESCAN, MIRA4, Brno, Czech Republic). The specimens were gold-plated before SEM observation. The SEM had a working distance of 7.5–15.0 mm, a resolution of 10.0 um, and a voltage of 5 kV.

3. Results and Discussion

3.1. Moisture Uptake Behavior

Figure 2 shows the water absorption of the epoxy resin and the composite laminates for each system, and since the carbon fibers are essentially non-absorbent, the weight change is caused by the absorption of water by the resin matrix. While EIA lacks rigid groups, increased EIA content leads to increased flexibility of the resin network and easier entry of moisture into the resin. Pure EIA results in decreased dimensional stability of resin in a high-temperature water bath (Figure 2c). Detailed optical pictures are in Figures S1 and S2. However, this phenomenon is not a chemical degradation of the resin since there was no change in the FTIR spectra of the samples before and after hygrothermal aging (Figure S3).

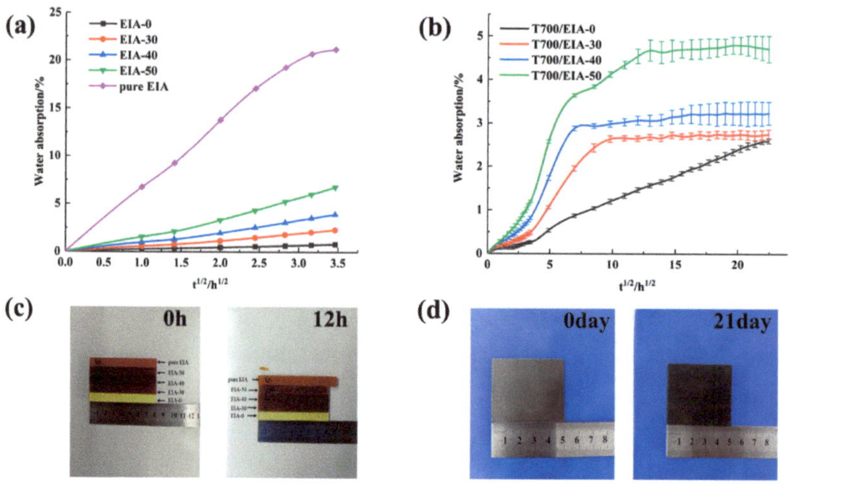

Figure 2. (**a**) Moisture absorption curves for the cured epoxy resin. (**b**) Moisture absorption curves for the composites. (**c**) Representative images of cured epoxy resin after hygrothermal aging. (**d**) Representative images of composites after hygrothermal aging.

As shown in Figure 2b, for bio-based composite laminates, the water uptake seems to follow a two-stage diffusion response, with the first stage showing a roughly linear relationship between the weight change and $t^{1/2}$, and the diffusion of water following Fickian responses [50]. Whereas, over a longer period, the slow increase in water absorption and equilibrium is reached. As shown in Figure 3, the main reason for the moisture absorption of the resin matrix is the presence of microcracks and the pores on the surface, water can easily enter the composite through these pores, so the water diffusion rate is very fast at the beginning. With longer testing time, the rate of moisture diffusion becomes lower and reaches equilibrium due to the relaxation of the glassy epoxy network [51] and the filling of voids and debonding zones with water by wicking. In Figure 3, the resin in the fiber surface are significantly reduced after hygrothermal aging, but the fibers are still bonded tightly besides the T700/EIA-50. This is because of the higher content of EIA in the T700/EIA-50, the resin matrix in the high temperature water immersion condition is more likely to absorb water, resulting in obvious cracks between the fibers.

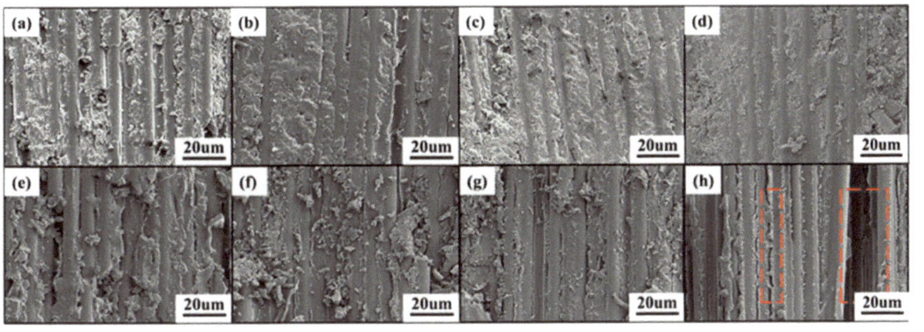

Figure 3. SEM images of the unaged and the aged specimens (**a**) T700/EIA-0 unaged, (**b**) T700/EIA-0 aged, (**c**) T700/EIA-30 unaged, (**d**) T700/EIA-30 aged, (**e**) T700/EIA-40 unaged, (**f**) T700/EIA-40 aged (**g**) T700/EIA-50 unaged, (**h**) T700/EIA-50 aged.

As shown in Figure 2b, the composites exhibit a strong tendency to absorb moisture across all systems. This is attributed to the crosslinked network of dicyandiamide cured resin that contains polar groups such as imine, amide, and tertiary amine. With an increase in EIA content, the composites display a higher initial diffusion rate and absorb more moisture at saturation. This is due to the hydrophilic hydroxyl groups of EIA, which facilitate water diffusion [52].

For T700/EIA-0, the benzene ring's presence and the hydrogen bonds formed between water molecules and the amide or imine impede water diffusion in petroleum-based resin composites. As a result, the rate of water diffusion is lower, and it takes longer to reach water absorption saturation. Nonetheless, the high density of polar groups in these composites results in a higher hydrophilic water absorption rate.

3.2. FT-IR Analysis of Composites before and after Hygrothermal Aging

Figure 4 illustrates the changes in the infrared spectrogram of the composite EIA-30 before and after hygrothermal aging. the band assigned to the -OH stretching is visible at 3400–3300 cm^{-1} for all the prepared specimens. At 2950 cm^{-1} and 2870 cm^{-1} the asymmetric and symmetric C-H stretch of the methyl and methylene groups were observed. The stretching vibration peak at 1725 cm^{-1} was attributed to -C=O, and the characteristic absorption peaks of C=C was appeared at 1642 cm^{-1}, both of which are considered to be the typical structures of EIA [21]. The absorption peak at 2195 cm^{-1} was attributed to -C≡N stretch. It is also possible to observe the characteristic peak at 1604 cm^{-1} and 1506 cm^{-1} were assigned to the benzene ring, and the stretching vibration peaks at 1241 cm^{-1} and 1020 cm^{-1} were assigned to the ether bond. While EIA belongs to aliphatic ether, and the peak of ether bond is reflected at 1090 cm^{-1}. There is no increase or decrease in the number of peaks before and after aging, especially the absorption peak at 910 cm^{-1} was attributed to the epoxy group before and after treatment. Indicating that all the specimens have been completely cured during the preparation, and there is no post-curing phenomenon during the aging process [53].

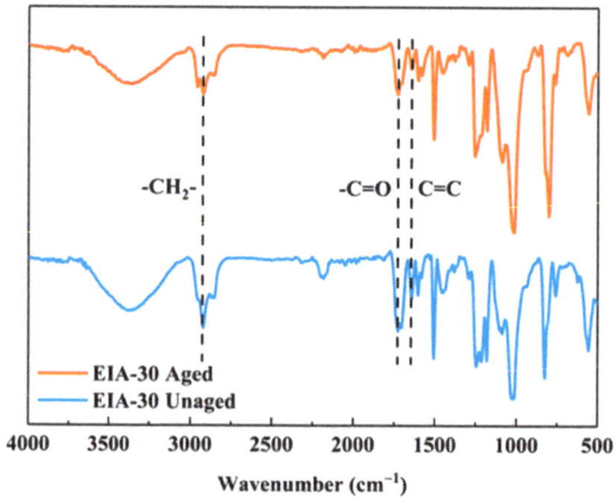

Figure 4. FTIR spectra of unaged EIA-30 resin and aged EIA-30 resin.

3.3. Mechanical Properties of the CFRP Composite

Figure 5 shows the basic properties of composite systems with different itaconic acid contents. The detailed data curves can be found in Figures S4–S7. Flexural strength and interlaminar shear strength were significantly increased with the addition of EIA compared to the T700/EIA-0 composite. This is probably attributed to the fact that the resin matrix is more strongly bonded to the carbon fibers, and there is greater resistance to prevent the adjacent layers from undergoing relative displacements. The decline in tensile strength from 3064.70 MPa to 2748.62 MPa with the increase in EIA content is due to the gradual decrease in network rigidity and the ease of displacement and deformation of molecular chain segments under load. There was no significant change in compressive, which may be related to the fracture of fibers unable to contain the stress.

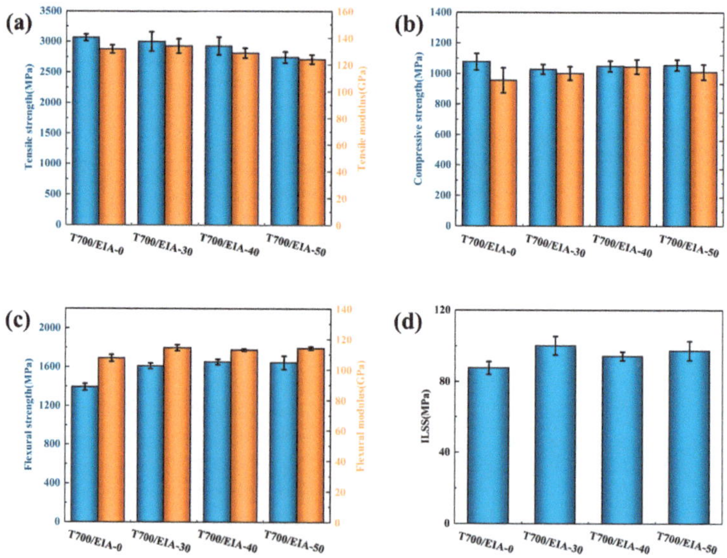

Figure 5. (**a**) Tensile strength and tensile modulus of composites, (**b**) Compressive strength and compressive modulus of composites, (**c**) Flexural strength and flexural modulus of composites, (**d**) Interlaminar shear strength of composites.

The EIA molecular chain contains an unsaturated double bond structure and contains ester bonds, and exhibits chain flexibility. A limited amount of EIA, the crosslinking network is still dominated by petroleum-based epoxy resin, which will not significantly affect the crosslinking density but can make the crosslinking network toughness increase, and improve the degree of adhesion between the resin and the fiber [54]. While excess EIA is added to the crosslinked network, the relative molecular mass between the crosslinked points is increased after curing and the crosslink density of the system is reduced. The short network chain in the system is prone to become a stress concentration point and would fracture at first so that the material tensile strength was reduced.

Figure 6 and Table 2 shows the changes in the properties of the composite system before and after hygrothermal aging. The detailed data curves can be found in Figures S8 and S9. When T700/EIA-30 was immersed in deionized water at 80 °C for 2 days, the bending strength was reduced from 1609.67 MPa to 1369.23 MPa, and the retention rate was 85.06%; the ILSS decreased from 100.11 MPa to 82.42 MPa with a retention rate of 82.33%. As for T700/EIA-0 composites under the same conditions, their flexural strength decreased from 1393.22 MPa to 1175.89 MPa, with a retention rate of 84.40%; ILSS decreased from 86.03 MPa to 64.98 MPa, with a retention rate of 75.53%. When the ratio of EIA to DGEBA is 50/50, the composite has higher strength and performance retention after wet and thermal aging than T700/EIA-0. However, as the percentage of EIA increases, the performance of the composite system decreases more and more after treatment.

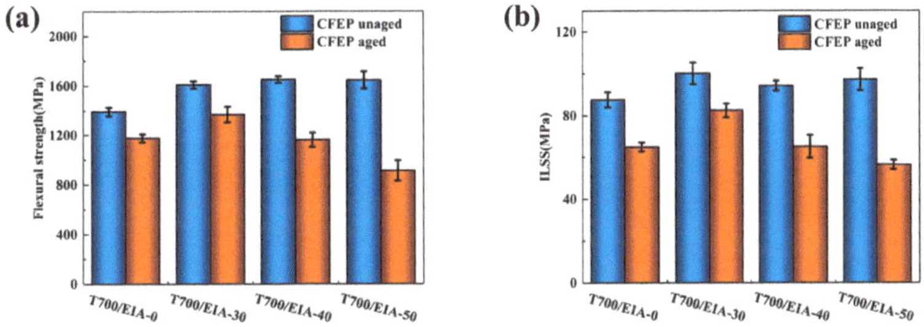

Figure 6. (**a**) Comparing the average flexural strength of the unaged and the aged specimens, (**b**) Comparing the average ILSS of the unaged and the aged specimens.

Table 2. List of mechanical properties of the unaged and the aged specimens.

Sample		Tensile Strength/MPa	Compressive Strength/MPa	Flexural Strength/MPa	Retention Rate of Flexural Strength/%	ILSS/MPa	Retention Rate of ILSS/%
T700/EIA-0	Unaged	3064.70	1077.79	1393.22	100	87.54	100
	aged	-	-	1175.89	84.40	64.98	74.23
T700/EIA-30	Unaged	3000.74	1077.79	1609.67	100	100.11	100
	aged	-	-	1369.23	85.06	82.42	82.33
T700/EIA-40	Unaged	2927.52	1048.90	1651.34	100	94.17	100
	aged	-	-	1162.78	70.41	65.17	69.20
T700/EIA-50	Unaged	2748.62	1055.77	1644.89	100	97.24	100
	aged	-	-	912.53	55.48	56.41	58.01

The performance degradation of CFRP can be attributed to several factors: temperature, humidity, and load. In the process of moisture absorption, due to the swelling and plasticizing effect of moisture on the resin matrix, the performance of the resin matrix decreases, the composite material would generate new micro-cracks and pores, and the original defects continue to expand. Moisture present within the interface layer of a composite laminate can create internal stresses that diminish interlayer adhesion. In more severe instances, this can cause the resin and fiber to separate, resulting in yield deformation of the fiber and a significant decrease in its strength [55]. The high temperature will accelerate the process of moisture absorption of the laminate so that the composite material will reach the service life and face the scrap. In Table 2, T700/EIA-30 under the condition of high temperature and high humidity can maintain high performance since the EIA improved the adhesion strength of resin and fiber interface, in which preventing moisture from entering. With the increased amount of EIA, the ratio of rigid groups was decreased which led composites have less resistance and make it easier for the system to absorb water, resulting in lower performance.

As shown in the figures (Figure 7a,c,e,g), the fracture morphology of the unaged CFRP specimens was observed by SEM. It can be seen that the fracture surface is uneven, a large amount of resin remains on the fiber surface, and the resin is tightly bonded to the fibers. The main reason for the failure of the composite is the brittle fracture of the fiber and resin. Compare to T700/EIA-0, the adhesion strength of resin and fiber interface was increased after the incorporation of EIA, which improved the mechanical properties of composites. However, with the increased amount of EIA, the mechanical properties of matrix would decrease due to the flexibility of EIA and decreased of crosslink density. Figure 7b,d,f,h show the fracture morphology of CFRP specimens after hygrothermal aging. The attached debris on the fracture surface is obviously reduced, with a neat texture and relatively smooth fiber surface. Resin-fiber debonding occurs and elevated with the increase of EIA content. This is because the EIA molecular chain lacks rigid groups, and the increase in its content resulted in an increase in the water absorption of the composite material.

Figure 7. SEM images of the unaged and the aged specimens with fractures (**a**) T700/EIA-0 unaged, (**b**) T700/EIA-0 aged, (**c**) T700/EIA-30 unaged, (**d**) T700/EIA-30 aged, (**e**) T700/EIA-40 unaged, (**f**) T700/EIA-40 aged (**g**) T700/EIA-50 unaged, (**h**) T700/EIA-50 aged.

3.4. Thermodynamic Properties of Composites

The dynamic mechanical properties of composite laminates of different systems were investigated using DMA and the results are shown in Figure 8 and Table 3. Figure 8a shows that the initial energy storage modulus (E′) value of T700/EIA-0 composite is greater than that of the composite with EIA addition. As the temperature increases from room temperature, the energy storage modulus retention values of the composites tend to decrease. The higher the EIA content, the lower the temperature at which the energy storage modulus starts to decrease. This result may be due to the thermal movement of the resin chain segments of the composites, while the EIA molecular configuration lacks rigid groups and is worse heat resistant compared to DGEBA. Continuing to increase the temperature, the resin in the composite changes from a glassy state to a rubbery state, allowing the chain segments in the "frozen state" to move freely, thus keeping the energy storage modulus at its sharply decreasing value.

Figure 8b shows the loss modulus (E″) of the composites for each system. In the initial environment, T700/EIA-0 contains a higher loss modulus, which is due to the large proportion of benzene ring content in the resin network structure. At the same time, the addition of EIA reduces the spatial hindrance and chain stiffness, which not only reduces the friction between the polymer chain segments, but also reduces the friction between the resin and the carbon fibers, and less energy is lost. When raised to a certain temperature, the molecular chains are further thawed, the molecular thermal movement is intensified, the friction between the chains as well as the stretching of the molecular chains themselves is intensified, and the loss modulus is increased.

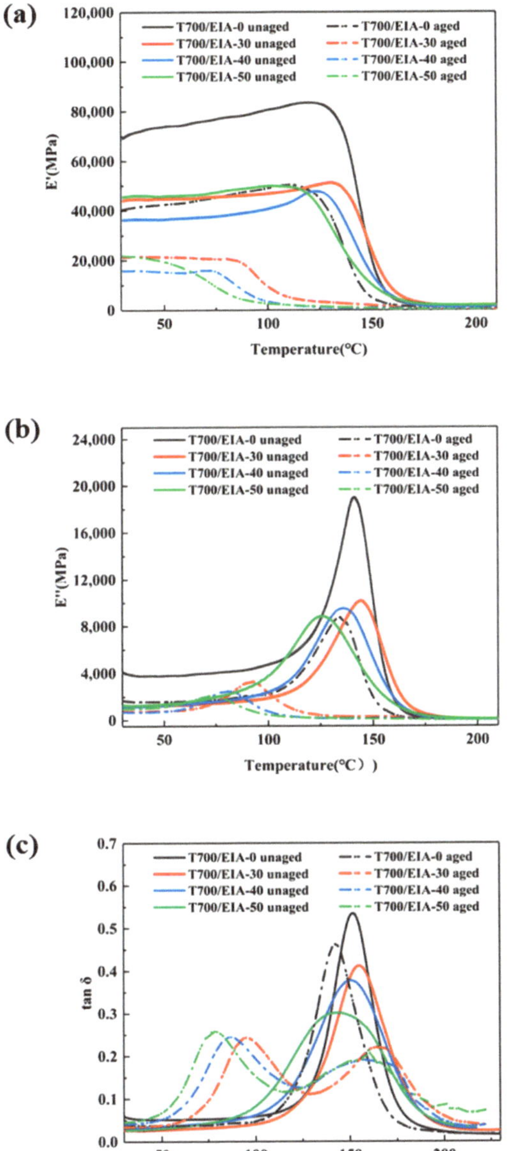

Figure 8. (**a**) Storage modulus, (**b**) loss modulus, and (**c**) loss factor of the CFRP composites with different addition levels (wt%) of EIA.

Table 3. List of thermodynamic properties of the unaged and the aged specimens.

Sample		Initial Storage Modulus/GPa	Initial Loss Modulus/GPa	T_g/°C	$T_{d5\%}$/°C
T700/EIA-0	Unaged	69.27	4.12	151.91	318.62
	aged	40.56	1.68	142.57	319.87
T700/EIA-30	Unaged	44.05	1.20	155.12	322.18
	aged	21.59	0.72	94.88	322.62
T700/EIA-40	Unaged	36.29	1.17	150.44	305.63
	aged	15.81	0.68	86.77	308.81
T700/EIA-50	Unaged	45.48	1.28	143.63	293.73
	aged	21.60	0.97	80.98	300.55

The loss factor is the ratio of loss modulus to energy storage modulus, and Figure 8c shows the loss factor curves of the composites for each resin system. The temperature corresponding to the peak loss factor is the glass transition temperature (T_g), and the glass transition temperature of T700/EIA-30 (T_g = 155.12 °C) is slightly higher than that of T700/EIA-0 (T_g = 151.91 °C). As the percentage of EIA content continues to increase, their glass transition temperatures gradually decrease and slightly lower than the T700/EIA-0. This may be due to the fact that low content of EIA increase the cross-linking density of the resin, while when the EIA content is higher, the relative molecular mass between the cross-linking points increases after curing, and the cross-linking density of the system decreases, thus lowering the glass transition temperature of the materials.

The thermomechanical properties of the composites after the hygrothermal aging treatment are also shown in Figure 8. The initial energy storage modulus and initial loss modulus values of the treated composites decrease compared to the untreated values, and the decrease in energy storage modulus is more pronounced for the composites with the addition of EIA. The energy storage modulus retention values start to decrease at lower temperatures, and the higher the percentage of EIA content, the lower the temperature required for the energy storage modulus to start decreasing. The main reason is that the increase of EIA content makes the water absorption rate of the composite material accelerated, and the water enters into the interior of the resin matrix to make the resin swollen, which destroys the network structure of the polymer, and the network is more likely to be thawed out at lower temperatures and transformed from the glassy state to the rubbery state. At the same time, there is a clear relationship between the loss modulus and water absorption, the higher water absorption leads to lower loss modulus. It indicates that the entry of moisture reduces the friction between chain segments, and the collision friction between chain segments may appear as the phenomenon of slipping, which reduces the energy loss. In Figure 8c, the loss factor curves of the composites which contains EIA displayed two peaks after hydrothermal aging, the temperature corresponding to the peak in front is the real glass transition temperature of the resin. The coupling effect of moisture and temperature makes the glass transition temperature shift forward with respect to that of the unaged composites. In contrast, the peak in the back is due to the drying effect in the testing process [56,57].

3.5. *Thermal Degradation Behavior of the Composites*

The TGA curves of the composites under nitrogen are shown in Figure 9. The values of initial degradation temperature at 5% weight loss ($T_{d5\%}$) are shown in Table 3. The initial decomposition temperature ($T_{d5\%}$) of the T700/EIA-30 system is slightly higher than that of the T700/EIA-0 system. When the ratio of EIA in the resin network is higher, the heat resistance of the composites is inferior, because EIA has a large number of easily cleaved ester bonds, it is more susceptible to thermal decomposition than DGEBA. This result is similar to the previous findings of Ma et al. [21]. Another phenomenon is that the initial decomposition temperature ($T_{d5\%}$) of each composite system after hygrothermal aging is almost unchanged, suggesting that such treatment does not affect the thermal stability of the material.

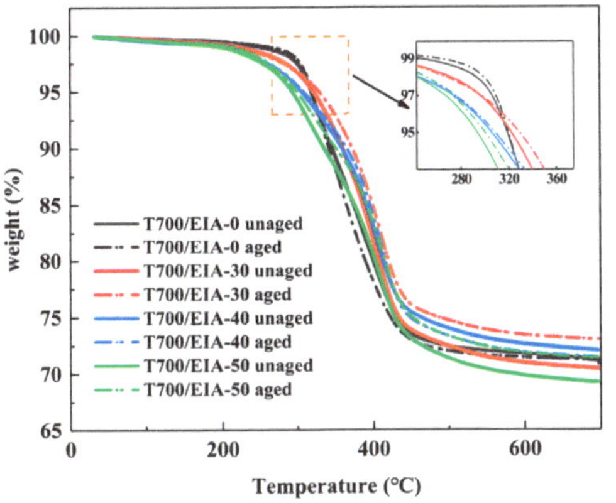

Figure 9. TGA curves of composites with different contents of EIA.

4. Conclusions

In conclusion, several sustainable carbon fiber reinforced composites with different degrees of greenness by blending EIA with DGEBA in different ratios were prepared. The mechanical and thermodynamic properties of green resin composites were compared with the properties of T700/EIA-0 composites. The effects of hygrothermal aging on the properties of these composites were investigated.

The water absorption tests showed that the prepared specimens had excellent dimensional stability under extreme hygrothermal conditions. Elevated EIA content in the resin leads to faster moisture absorption equilibrium and higher water absorption of the specimens. FTIR spectroscopy showed no evident effect of hygrothermal aging chemical structure of these composites. The mechanical properties of CFRP prepared by each system were comparable. The flexural strengths and ILSS are better than those of T700/EIA-0 in the unaged condition, and the differences in thermomechanical properties were not significant. After 48 h of hygrothermal aging, the water absorption of T700/EIA-30 was higher than that of T700/EIA-0, but the flexural strength, ILSS, and the corresponding retention rate were higher than that of T700/EIA-0, and the thermal stability is comparable to that of T700/EIA-0. It can be seen that the EIA-based epoxy resins as compared with BPA-based epoxy resins demonstrated significant evidence to design and develop a novel bio-based epoxy resin with high performance to substitute the petroleum-based epoxy resin.

Supplementary Materials: The following supporting information can be downloaded at: https://www.mdpi.com/article/10.3390/polym16121649/s1, Figure S1: The digital photos of EP samples. (a) EIA-0, (b) EIA-30, (c) EIA-40, (d) EIA-50, (e) Pure EIA; Figure S2: The digital photos of the hygrothermal aging process of EP samples; Figure S3: (a) FTIR spectra of unaged EIA-0 resin and aged EIA-0 resin, (b) FTIR spectra of unaged EIA-40 resin and aged EIA-40 resin, (c) FTIR spectra of unaged EIA-50 resin and aged EIA-50 resin, (d) FTIR spectra of unaged Pure EIA resin and aged Pure EIA resin; Figure S4: Tensile stress-strain curves for CFRP composites. (a) specimens of T700/EIA-0, (b) specimens of T700/EIA-30, (c) specimens of T700/EIA-40, (d) specimens of T700/EIA-50; Figure S5: Compressive stress-strain curves for CFRP composites. (a) specimens of T700/EIA-0, (b) specimens of T700/EIA-30, (c) specimens of T700/EIA-40, (d) specimens of T700/EIA-50; Figure S6: Flexural stress-strain curves for CFRP composites. (a) specimens of T700/EIA-0, (b) specimens of T700/EIA-30, (c) specimens of T700/EIA-40, (d) specimens of T700/EIA-50; Figure S7: ILSS stress-strain curves for CFRP composites. (a) specimens of T700/EIA-0, (b) specimens of T700/EIA-30, (c) specimens of T700/EIA-40, (d) specimens of T700/EIA-50; Figure S8: Flexural stress-strain curves for aged CFRP composites. (a) specimens of T700/EIA-0, (b) specimens of T700/EIA-30, (c) specimens of T700/EIA-40, (d) specimens of T700/EIA-50; Figure S9: ILSS stress-strain curves for aged CFRP composites. (a) specimens of T700/EIA-0, (b) specimens of T700/EIA-30, (c) specimens of T700/EIA-40, (d) specimens of T700/EIA-50.

Author Contributions: Conceptualization, K.X., X.Y. and Y.X.; methodology, K.X., X.Y. and Y.X.; validation, X.Y., N.N. and Z.W.; formal analysis, K.X. and Y.F.; investigation, K.X., X.Y. and Y.X.; resources, Z.W., Z.L. (Zhiyi Lyu) and N.N.; data curation, X.Y., S.K. and Z.A.; writing—original draft preparation, K.X., Y.F. and X.Y.; writing—review and editing, K.X., Z.L. (Ziqian Liu) and X.Y.; visualization, X.Y. and Y.X.; supervision, X.Y. and Y.X.; project administration, X.Y.; funding acquisition, Y.X. All authors have read and agreed to the published version of the manuscript.

Funding: This research was funded by The Priority Academic Program Development of Jiangsu Higher Education Institutions (PAPD).

Institutional Review Board Statement: Not applicable.

Data Availability Statement: The raw data supporting the conclusions of this article will be made available by the authors on request.

Conflicts of Interest: The authors declare no conflict of interest.

References

1. Mazzon, E.; Habas-Ulloa, A.; Habas, J.-P. Lightweight rigid foams from highly reactive epoxy resins derived from vegetable oil for automotive applications. *Eur. Polym. J.* **2015**, *68*, 546–557. [CrossRef]
2. Diez-Barcenilla, G.; Salvador, A.A.; Zubieta, K.G.; Zuza, E. Epoxy matrix composites from waste for manufacturing tools for energy efficient moulding process. *Clean Technol. Environ. Policy* **2022**, *24*, 1793–1810. [CrossRef]
3. Stefan, W.; Petra, V.; Peter, M.; Nerea, M. Investigations on thermoforming of carbon fiber reinforced epoxy vitrimer composites. *Compos. Part A Appl. Sci. Manuf.* **2021**, *154*, 106791. [CrossRef]
4. Carter, J.T.; Emmerson, G.T.; Faro, C.L.; McGrail, P.T.; Moore, D.R. The development of a low temperature cure modified epoxy resin system for aerospace composites. *Compos. Part A Appl. Sci. Manuf.* **2003**, *34*, 83–91. [CrossRef]
5. Mordor Intelligence Research & Advisory. Epoxy Resin Market Size & Share Analysis—Growth Trends & Forecasts (2024–2029). Available online: https://www.mordorintelligence.com/industry-reports/global-epoxy-resin-market-industry (accessed on 31 May 2024).
6. Ďurovcová, I.; Kyzek, S.; Fabová, J.; Makuková, J.; Gálová, E.; Ševčovičová, A. Genotoxic potential of bisphenol A: A review. *Environ. Pollut.* **2022**, *306*, 119346. [CrossRef] [PubMed]
7. Wang, Q.; Chen, T.; Wang, X.; Zheng, Y.; Zheng, J.; Song, G.; Liu, S. Recent Progress on Moisture Absorption Aging of Plant Fiber Reinforced Polymer Composites. *Polymers* **2023**, *15*, 4121. [CrossRef] [PubMed]
8. Fernandes, F.C.; Kirwan, K.; Lehane, D.; Coles, S.R. Epoxy resin blends and composites from waste vegetable oil. *Eur. Polym. J.* **2017**, *89*, 449–460. [CrossRef]
9. Marriam, F.; Irshad, A.; Umer, I.; Asghar, M.A.; Atif, M. Vegetable oils as bio-based precursors for epoxies. *Sustain. Chem. Pharm.* **2023**, *31*, 100935. [CrossRef]
10. Nikafshar, S.; Wang, J.; Dunne, K.; Sangthonganotai, P.; Nejad, M. Choosing the Right Lignin to Fully Replace Bisphenol A in Epoxy Resin Formulation. *ChemSusChem* **2021**, *14*, 1184–1195. [CrossRef]
11. Nicastro, K.H.; Kloxin, C.J.; Epps, T.H., III. Potential Lignin-Derived Alternatives to Bisphenol A in Diamine-Hardened Epoxy Resins. *ACS Sustain. Chem. Eng.* **2018**, *6*, 14812–14819. [CrossRef]
12. Mantzaridis, C.; Brocas, A.-L.; Llevot, A.; Cendejas, G.; Auvergne, R.; Caillol, S.; Carlotti, S.; Cramail, H. Rosin acid oligomers as precursors of DGEBA-free epoxy resins. *Green Chem.* **2013**, *15*, 3091–3098. [CrossRef]
13. Wang, H.; Liu, B.; Liu, X.; Zhang, J.; Xian, M. Synthesis of biobased epoxy and curing agents using rosin and the study of cure reactions. *Green Chem.* **2008**, *10*, 1190–1196. [CrossRef]
14. Tserpes, K.; Tzatzadakis, V.; Katsiropoulos, C. Effect of hygrothermal ageing on the interlaminar shear strength of carbon fiber-reinforced rosin-based epoxy bio-composites. *Compos. Struct.* **2019**, *226*, 111211. [CrossRef]
15. Mora, A.-S.; Decostanzi, M.; David, G.; Caillol, S. Cardanol-Based Epoxy Monomers for High Thermal Properties Thermosets. *Eur. J. Lipid Sci. Technol.* **2019**, *121*, 1800421. [CrossRef]
16. Kinaci, E.; Can, E.; Scala, J.J.L.; Palmese, G.R. Influence of Epoxidized Cardanol Functionality and Reactivity on Network Formation and Properties. *Polymers* **2020**, *12*, 1956. [CrossRef] [PubMed]
17. Eid, N.; Ameduri, B.; Boutevin, B. Synthesis and Properties of Furan Derivatives for Epoxy Resins. *ACS Sustain. Chem. Eng.* **2021**, *9*, 8018–8031. [CrossRef]
18. Xu, J.; Liu, X.; Fu, S. Bio-based epoxy resin from gallic acid and its thermosets toughened with renewable tannic acid derivatives. *J. Mater. Sci.* **2022**, *57*, 9493–9507. [CrossRef]
19. Hou, G.; Li, N.; Han, H.; Run, M.; Gao, J. Preparation and thermal properties of bio-based gallic acid epoxy/carbon nanotubes composites by cationic ring-opening reaction. *Polym. Compos.* **2016**, *37*, 3093–3102. [CrossRef]

20. Hong, J.; Radojcic, D.; Ionescu, M.; Petrović, Z.; Eastwood, E. Advanced Materials from Corn: Isosorbide-Based Epoxy Resins. *Polym. Chem.* **2014**, *5*, 5360–5368. [CrossRef]
21. Ma, S.; Liu, X.; Jiang, Y.; Tang, Z.; Zhang, C.; Zhu, J. Bio-based epoxy resin from itaconic acid and its thermosets cured with anhydride and comonomers. *Green Chem.* **2013**, *15*, 245–254. [CrossRef]
22. Ramon, E.; Sguazzo, C.; Moreira, P. A Review of Recent Research on Bio-Based Epoxy Systems for Engineering Applications and Potentialities in the Aviation Sector. *Aerospace* **2018**, *5*, 110. [CrossRef]
23. Deng, L.; Wang, Z.; Qu, B.; Liu, Y.; Qiu, W.; Qi, S. A Comparative Study on the Properties of Rosin-Based Epoxy Resins with Different Flexible Chains. *Polymers* **2023**, *15*, 4246. [CrossRef] [PubMed]
24. Kathalewar, M.; Sabnis, A. Epoxy resin from cardanol as partial replacement of bisphenol-A-based epoxy for coating application. *J. Coat. Technol. Res.* **2014**, *11*, 601–618. [CrossRef]
25. Dutta, K.; Karak, N. Bisphenol-A free bio-based gallic acid amide epoxy thermosets. *J. Appl. Polym. Sci.* **2023**, *140*, e54306. [CrossRef]
26. Łukaszczyk, J.; Janicki, B.; Kaczmarek, M. Synthesis and properties of isosorbide based epoxy resin. *Eur. Polym. J.* **2011**, *47*, 1601–1606. [CrossRef]
27. Saha, B.C.; Kennedy, G.J.; Qureshi, N.; Bowman, M.J. Production of itaconic acid from pentose sugars by Aspergillus terreus. *Biotechnol. Prog.* **2017**, *33*, 1059–1067. [CrossRef] [PubMed]
28. Choi, S.; Song, C.W.; Shin, J.H.; Lee, S.Y. Biorefineries for the production of top building block chemicals and their derivatives. *Metab. Eng.* **2015**, *28*, 223–239. [CrossRef] [PubMed]
29. Diankristanti, P.A.; Ng, I.S. Microbial itaconic acid bioproduction towards sustainable development: Insights, challenges, and prospects. *Bioresour. Technol.* **2023**, *384*, 129280. [CrossRef] [PubMed]
30. Rebolledo-Leiva, R.; Moreira, M.T.; González-García, S. Environmental assessment of the production of itaconic acid from wheat straw under a biorefinery approach. *Bioresour. Technol.* **2022**, *345*, 126481. [CrossRef]
31. Magalhães, A.I.; de Carvalho, J.C.; Medina, J.D.C.; Soccol, C.R. Downstream process development in biotechnological itaconic acid manufacturing. *Appl. Microbiol. Biotechnol.* **2017**, *101*, 1–12. [CrossRef]
32. Ma, S.; Liu, X.; Fan, L.; Jiang, Y.; Cao, L.; Tang, Z.; Zhu, J. Synthesis and Properties of a Bio-Based Epoxy Resin with High Epoxy Value and Low Viscosity. *ChemSusChem* **2014**, *7*, 555–562. [CrossRef] [PubMed]
33. Ma, S.; Liu, X.; Jiang, Y.; Fan, L.; Feng, J.; Zhu, J. Synthesis and properties of phosphorus-containing bio-based epoxy resin from itaconic acid. *Sci. China-Chem.* **2014**, *57*, 379–388. [CrossRef]
34. Liu, Y.; Wang, B.; Ma, S.; Yu, T.; Xu, X.; Li, Q.; Wang, S.; Han, Y.; Yu, Z.; Zhu, J. Catalyst-free malleable, degradable, bio-based epoxy thermosets and its application in recyclable carbon fiber composites. *Compos. Part B Eng.* **2021**, *211*, 108654. [CrossRef]
35. Almudaihesh, F.; Holford, K.; Pullin, R.; Eaton, M. The influence of water absorption on unidirectional and 2D woven CFRP composites and their mechanical performance. *Compos. Part B Eng.* **2020**, *182*, 107626. [CrossRef]
36. Wong, T.C.; Broutman, L.J. Water in epoxy resins Part II. Diffusion mechanism. *Polym. Eng. Sci.* **1985**, *25*, 529–534. [CrossRef]
37. Zhou, J.; Lucas, J.P. Hygrothermal effects of epoxy resin. Part I: The nature of water in epoxy. *Polymer* **1999**, *40*, 5505–5512. [CrossRef]
38. Mijović, J.; Zhang, H. Local Dynamics and Molecular Origin of Polymer Network−Water Interactions as Studied by Broadband Dielectric Relaxation Spectroscopy, FTIR, and Molecular Simulations. *Macromolecules* **2003**, *36*, 1279–1288. [CrossRef]
39. Korkees, F.; Alston, S.; Arnold, C. Directional diffusion of moisture into unidirectional carbon fiber/epoxy Composites: Experiments and modeling. *Polym. Compos.* **2018**, *39*, E2305–E2315. [CrossRef]
40. Bone, J.E.; Sims, G.D.; Maxwell, A.S.; Frenz, S.; Ogin, S.L.; Foreman, C.; Dorey, R.A. On the relationship between moisture uptake and mechanical property changes in a carbon fibre/epoxy composite. *J. Compos. Mater.* **2022**, *56*, 2189–2199. [CrossRef]
41. Beura, S.; Chakraverty, A.P.; Pati, S.N.; Pradhan, D.D.; Thatoi, D.N.; Mohanty, U.K. Effect of salinity and strain rate on sea water aged GFRP composite for marine applications. *Mater. Today Commun.* **2023**, *34*, 105056. [CrossRef]
42. Shi, Z.; Zou, C.; Zhou, F.; Zhao, J. Analysis of the Mechanical Properties and Damage Mechanism of Carbon Fiber/Epoxy Composites under UV Aging. *Materials* **2022**, *15*, 2919. [CrossRef] [PubMed]
43. Aslan, A.; Salur, E.; Düzcükoğlu, H.; Sinan Şahin, Ö.; Ekrem, M. The effects of harsh aging environments on the properties of neat and MWCNT reinforced epoxy resins. *Constr. Build. Mater.* **2021**, *272*, 121929. [CrossRef]
44. Shen, M.; Almallahi, R.; Rizvi, Z.; Gonzalez-Martinez, E.; Yang, G.; Robertson, M.L. Accelerated hydrolytic degradation of ester-containing biobased epoxy resins. *Polym. Chem.* **2019**, *10*, 3217–3229. [CrossRef]
45. *ASTM D5229/D5229M*; Standard Test Method for Moisture Absorption Properties and Equilibrium Conditioning of Polymer Matrix Composite Materials. ASTM International: West Conshohocken, PA, USA, 2020.
46. *ASTM D3039*; Standard Test Method for Tensile Properties of Polymer Matrix Composite Materials. ASTM International: West Conshohocken, PA, USA, 2017.
47. *ASTM D6641*; Standard Test Method for Compressive Properties of Polymer Matrix Composite Materials Using a Combined Loading Compression (CLC) Test Fixture. ASTM International: West Conshohocken, PA, USA, 2023.
48. *ASTM D7264*; Standard Test Method for Flexural Properties of Polymer Matrix Composite Materials. ASTM International: West Conshohocken, PA, USA, 2021.
49. *ASTM D2344*; Standard Test Method for Short-Beam Strength of Polymer Matrix Composite Materials and Their Laminates. ASTM International: West Conshohocken, PA, USA, 2022.

50. Korkees, F. Moisture absorption behavior and diffusion characteristics of continuous carbon fiber reinforced epoxy composites: A review. *Polym.-Plast. Technol. Mater.* **2023**, *62*, 1789–1822. [CrossRef]
51. Zhavoronok, E.S.; Senchikhin, I.N.; Roldughin, V.I. Physical aging and relaxation processes in epoxy systems. *Polym. Sci. Ser. A* **2017**, *59*, 159–192. [CrossRef]
52. McConnell, B.K.; Pethrick, R.A. Dielectric studies of water absorption and desorption in epoxy resins: Influence of cure process on behaviour. *Polym. Int.* **2008**, *57*, 689–699. [CrossRef]
53. Li, K.; Wang, K.; Zhan, M.-S.; Xu, W. The change of thermal–mechanical properties and chemical structure of ambient cured DGEBA/TEPA under accelerated thermo-oxidative aging. *Polym. Degrad. Stab.* **2013**, *98*, 2340–2346. [CrossRef]
54. Kumar, S.; Samal, S.K.; Mohanty, S.; Nayak, S.K. Curing kinetics of bio-based epoxy resin-toughened DGEBA epoxy resin blend: Synthesis and characterization. *J. Therm. Anal. Calorim.* **2019**, *137*, 1567–1578. [CrossRef]
55. Behera, A.; Vishwakarma, A.; Thawre, M.M.; Ballal, A. Effect of hygrothermal aging on static behavior of quasi-isotropic CFRP composite laminate. *Compos. Commun.* **2020**, *17*, 51–55. [CrossRef]
56. Xian, G.; Karbhari, V.M. DMTA based investigation of hygrothermal ageing of an epoxy system used in rehabilitation. *J. Appl. Polym. Sci.* **2007**, *104*, 1084–1094. [CrossRef]
57. Xian, G.; Karbhari, V.M. Segmental relaxation of water-aged ambient cured epoxy. *Polym. Degrad. Stab.* **2007**, *92*, 1650–1659. [CrossRef]

polymers

MDPI

Article

Effects of Carbon Black on Mechanical Properties and Oil Resistance of Liquid Silicone Rubber

Beom-Joo Lee and Hyeong-Min Yoo *

School of Mechanical Engineering, Korea University of Technology and Education (KOREATECH), Cheonan 31253, Republic of Korea; gadc336@koreatech.ac.kr
* Correspondence: yhm2010@koreatech.ac.kr; Tel.: +82-41-560-1255; Fax: +82-41-560-1253

Abstract: Liquid silicone rubber (LSR) garners attention across a diverse range of industries owing to its commendable fluidity and heat resistance. Nonetheless, its mechanical strength and oil resistance fall short compared to other rubbers, necessitating enhancement through the incorporation of a suitable filler. This research focuses on reinforcing LSR using carbon black (CB) particles as a filler, evaluating the mechanical properties and oil resistance of neat LSR, and LSR containing up to 3 wt% of CB filler. CB was added in powder form to investigate its effect on LSR. When LSR was impregnated with oil, the deterioration of rubber was noticeably observed under high-temperature conditions compared to room-temperature conditions. Consequently, the mechanical properties and oil resistance, excluding the permanent compression reduction rate, tended to increase as the filling content of CB increased compared to the unfilled state. Notably, in the specimen with 2 wt% CB filler, the tensile modulus increased significantly by 48% and the deterioration rate was reduced by about 50% under accelerated deterioration conditions. Additionally, the swelling rate in oil decreased by around 14%. This validates a notable improvement in both mechanical properties and oil resistance. Based on the identified mechanism for properties enhancement in this study, CB/LSR composite is expected to have a wide range of applications in fields such as gaskets, oil seals, and flexible sensors.

Keywords: liquid silicone rubber; carbon black; mechanical properties; swelling; oil resistance

Citation: Lee, B.-J.; Yoo, H.-M. Effects of Carbon Black on Mechanical Properties and Oil Resistance of Liquid Silicone Rubber. *Polymers* **2024**, *16*, 933. https://doi.org/10.3390/polym16070933

Academic Editors: Xiaoquan Cheng, Wenjun Huang and Qian Zhang

Received: 19 February 2024
Revised: 23 March 2024
Accepted: 26 March 2024
Published: 28 March 2024

1. Introduction

Liquid silicone rubber (LSR) has excellent biocompatibility and thermal compatibility, and its good fluidity makes it easy to implement into products, so it can be used even in places where complex shapes and strict deviations are required [1–5]. Due to these characteristics, LSR is widely used and attracts attention in various applications, such as medicine [6], electronics [7,8], and automobiles [9]. However, LSR's mechanical strength and oil resistance are relatively inferior to those of other rubbers, which are directly related to product performance and limit its widespread application. In particular, when LSR is used in products usually exposed to oil, such as gaskets, oil seals, and O-rings, compression set behavior that maintains elastic properties even under long-term compressive stress, temperature, and swelling ratio for oil must be considered. Corrosion or deterioration which occur due to the oil exposure reduce the performance and lifespan of the products, so research on processing the silicone rubber with improved mechanical properties and oil resistance is required [10–14].

LSR is mainly composed of long divinyl polydimethylsiloxane chains and short SiH-multifunctionalized copolymers, along with silica, platinum catalysts, and cure retarders. It is usually supplied in two parts, Part A and Part B, one containing the catalyst and the other without, and is used in a 1:1 ratio. The curing process of LSR involves hydrosilylation, wherein Si-H bonds are added to the carbon double bonds of the polydimethylsiloxane chain. For stability and compatibility with silicon, a platinum catalyst is employed, and a cure retarder is used to prevent premature crosslinking of the polymer at room temperature.

The platinum-mediated crosslinking reaction then takes place at elevated temperatures. Since LSR products are processed from A and B solutions, it has the advantage that reinforcing materials can be added during the process. Therefore, Consequently, fillers like silica, clay, graphene oxide, carbon nanotubes (CNTs), and carbon black can be readily added to enhance the properties of LSR [2–4].

Carbon-based reinforcement materials such as carbon black (CB), CNT, and graphene have high thermal conductivity and chemical stability, so they are most commonly used to produce various elastomer composites for the purpose of reinforcing conductivity and mechanical properties. Among them, CB in particular has the advantage of a low price, and it is known to have a large specific surface area and a chemically active surface in a complex microstructure, which has a rich reaction with rubber molecular chains [15–18]. ZH Li et al. [15] added five types of carbon black with different surface areas to EPDM rubber and confirmed that physical and chemical crosslinking occurred most actively in CB with the largest surface area compared to other CBs. In addition, it had the best hardness, tensile strength, and 300% elastic modulus, but showed relatively low elongation. N. Stübler et al. [16] confirmed the mechanical stress of CB-containing Elastomer (EPDM, SBR, NR) through uniaxial deformation cycle tests and, simultaneously, DC (direct current) resistance of composite samples was investigated under deformation conditions. The effects of strain and relaxation on the carbon black network of the elastomer were confirmed through the investigated stress and piezo resistance. A. Mostafa et al. [10] investigated the mechanical properties and oil resistance performance of CB-reinforced SBR and NBR rubber according to time, temperature, and CB content. It was revealed that as the CB content increases, the swelling rate with exposure to the oil decreases, and as the exposure time and temperature to the oil increase, the oil penetrates between the rubber chains and the swelling rate increases. In most existing studies, CB is added in micro and nano forms to ensure uniform dispersion. When adding CB, if the amount is excessive, agglomeration between fillers may occur and the mechanical properties of the material would not be uniform, so it is necessary to consider the amount of filler for even dispersion [19–24].

Also, the addition of carbon-based reinforcements enhances the electrical properties of flexible materials, making them applicable to the electrical industry. As the electronics and electrical equipment industry evolves, LSR composites with CB are utilized in a variety of applications for sensing or electrical detection purposes, act as shielding for electromagnetic interface (EMI), and improve the durability of medical and automotive components where LSRs have traditionally been used. Pin Liu et al. [25] designed a pressure-/temperature-sensitive conductive silicone rubber filled with different volume fractions of CB, prepared by a liquid mixing method. Pan Song et al. [26] designed flexible sensors made of silicone rubber/graphene/modified CB composites, which are highly likely to be applied in the wearable electronic devices field. Javad Jeddi et al. [27] investigated the electrical conductivity and EMI shielding properties of silicon rubber, CB, nanographite, and polyurethane foam hybrid composites.

Previous studies on CB/LSR composites have focused on specific properties depending on the application, such as thermal conductivity, thermal stability, conductivity, and electrical resistance [28–31]. Zhang et al. [28] measured the activation energy through thermal decomposition by adding different weights of conductive CB to silicone rubber and confirmed that conductive silicone rubber had superior thermal stability compared to general silicone rubber, and Ding et al. [29] produced a CB-filled silicone rubber composite, and confirmed that both the stress and electrical resistance of the specimen increased immediately after compression, and that the stress relaxed and gradually decreased, resulting in the use of a flexible sensor that can measure compressive stress relief.

Although research is actively being conducted on other characteristics of LSR containing CB, there is a lack of quantitative research on oil resistance. Considering the diverse applications of LSR, there is still a need to investigate the oil resistance and extensive mechanical properties of CB-reinforced LSR. Accordingly, in this study, we fabricated LSR reinforced with CB of various concentrations, and performed a compression test, tensile test, swelling test, and compression set test to determine the effects of CB on the oil resistance of LSR.

2. Experimental Section

2.1. Preparation of CB/LSR Composite Specimen

KCC Corp's SL7270 (Seoul, Republic of Korea) product was used as LSR, and Orion Engineered Carbon's No. 0924795 (Senningerberg, Luxembourg) product was used for CB as a rubber reinforcement. To produce specimens, CB was dispersed in the LSR part B solution, which included vinyl polymer, silica, H polymer, and a curing retardant. The part B solution was then mixed with the LSR part A solution for 10 min by using a mechanical stirrer [23,24], consisting of vinyl polymer, silica, and platinum catalyst, in a mixing ratio of 1:1. Subsequently, using the Automatic Heating Plate, QM900A from Qmesys Corp (Uiwang-si, Republic of Korea), a specimen was manufactured by pressurizing it at 20 bar for 10 min at a temperature of 170 °C (Figure 1). The curing parameters were optimized according to the KCC LSR 7270 data sheet.

LSR (A + B)	23°C_Dens (g/cm^3)	1wt% (g)	2wt% (g)	3wt% (g)
LSR	1.14	200	200	200
Carbon black	1.81	2.02	4.11	6.19

Figure 1. The preparation process of the CB/LSR composite sample.

2.2. Measurements

2.2.1. Tensile and Compression Tests

The tensile test adhered to the ASTM D412 standard [32], and a 1-ton UTM (TestOne, T0-102, Siheung-si, Republic of Korea) was employed for the testing. The crosshead's test speed was set at 500 mm/min. The specimen size was 40 × 25 mm, and appropriate gripping was applied on both sides of the gripper in the equipment to prevent sample slipping. Tensile tests were conducted in two cases. First, tests were conducted according to the CB content at room temperature, and second, specimens were deteriorated for 72 h in an oil at 130 °C and then measured to determine their CB content.

The compression test also utilized the same UTM as the tensile test, with testing procedures referencing the ASTM D1229 standard [33]. The crosshead test speed was set at 20 mm/min and conducted at room temperature.

2.2.2. Swelling Test

Deterioration needs to be taken into account in the lifespan of rubber products such as gaskets and seals when exposed to an environment containing oil. The deterioration can be quantified as the swelling ratio of rubber exposed to oil under specific temperature and time conditions. In this study, motor oil was chosen and used to simulate the environment surrounding gaskets and seals. The swelling test followed the ASTM D471 standard [34] and was conducted using motor oil at room-temperature and high-temperature conditions (100, 130 °C). The specimen size was 25 × 50 × 2 mm (Figure 2). The specimen underwent immersion for 3 days at room temperature and in an oven (Labtech, LDO-150F, Namyangjusi, Republic of Korea) at 100 °C and 130 °C, and the change in swelling ratio over time was examined. When measuring, the specimen was washed in acetone, wiped with filter paper, and then weighed. In the case of the high-temperature swelling tests, the specimen at high temperature was transferred to the room-temperature motor oil for cooling, followed by washing in acetone and wiping with filter paper. The formula for calculating the swelling ratio is as follows.

$$Q\% = \frac{M_t - M_0}{M_0} \times 100 \tag{1}$$

$$(M_t = Mass\ after\ swelling,\ \ M_0 = initial\ mass)$$

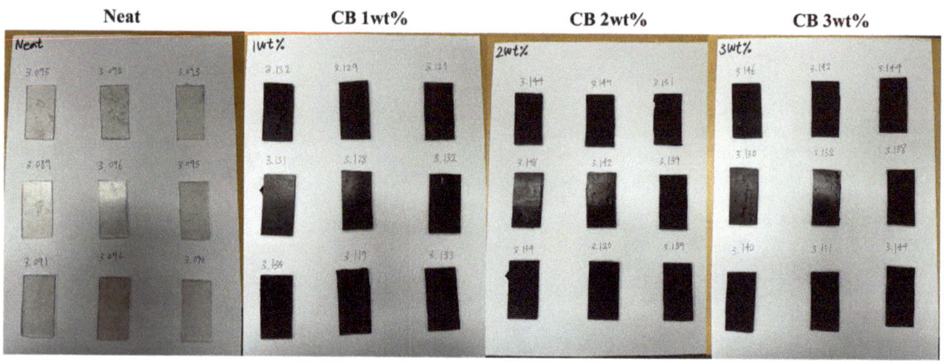

Figure 2. Swelling test specimens.

2.2.3. Scanning Electron Microscopy (SEM) Measurement

The SEM measurement was conducted using COXEM's EM-30N series SEM, and multiple images were presented on a computer through field emission scanning electrons. In this experiment, the LSR samples, considering LSR's status as a non-conductive material incapable of reflecting electrons, were coated with platinum. High-quality SEM images were achieved by directing electrons onto a thin platinum film applied to the sample, and images were captured at 7 kV with a working distance ranging from 4.0 mm to 5.0 mm.

2.2.4. Compression Set Test

The ability of rubber compounds to retain elasticity under prolonged compressive stress is crucial for products demanding long-term performance, such as gaskets and seals. This characteristic can be assessed through a compression set test. The formula for calculating the permanent compression set rate (C%) is as follows.

$$C\% = \frac{T_0 - T_1}{T_0 - T_s} \times 100 \tag{2}$$

(T_0 = *Initial thickness of Specimen*, T_1 =*Thickness of specimen after clamp removal*, T_s = *Thickness between clamps*).

As evidenced by the above equation, the lower the compression set, the more resistant the material is to permanent deformation at a given deflection and temperature. The compression set test adhered to the ASTM D395 standard [35], employing a cylindrical compression-molded specimen with a diameter of 25 mm and a thickness of 12 mm. As shown in Figure 3, a specimen was inserted between the tools, and the bolt was tightened to narrow the space of the plates. The space between the plates was adjusted to achieve a compression percentage of 25% of the original thickness of the specimen. The final thickness was measured 30 min after loosening the bolt and removing the clamp. The tests were conducted at room temperature and 130 °C.

Figure 3. Compression set test jig and specimens.

3. Results and Discussion

3.1. Tensile and Compression Properties

Through tensile and compressive strength tests, the impact of temperature and the addition of CB on the mechanical properties of LSR was determined. According to the compression test results in Figure 4a, the compressive modulus of the initial elastic deformation section was 40.86 MPa in neat LSR, 44.23 MPa in 1 wt% of CB, 50.72 MPa in 2 wt% of CB, and 54.21 MPa in 3 wt% of CB. The compression modulus of LSR tended to increase with the increasing CB content. From the graphs, it is evident that the compressive behavior of each composite is not consistently uniform throughout. It is apparent that they initially exhibit linear elasticity, but as the deformation progresses beyond a critical value, there is an irregular nonlinear deformation regime, likely attributable to local network destruction. Nevertheless, it has been noted that this phenomenon does not significantly alter the initial slope of the graph, namely, the modulus [31,36,37]. In Figure 4b, the result of the tensile test revealed that as the CB content increased, the elongation of LSR decreased, but the tensile modulus tended to increase. Both the compression and tensile test results indicate that CB contributes to the enhancement of modulus, consistent with previous studies [16–18]. In Figure 4c, the specimens were exposed to the same accelerated deterioration condition (130 °C, 72 h) to observe deterioration according to CB content. The specimens with CB had relatively better resistance to deterioration compared to the neat specimen, which shows that the deterioration of mechanical properties may progress more slowly under extreme conditions due to the added CB. The deterioration rates of mechanical properties, specifically ultimate tensile strength and tensile modulus, for each specimen are summarized and presented in Tables 1 and 2. Notably, for the specimen with 2 wt% CB added, the ultimate tensile strength was shown to be 50% less and the modulus to be 67% less deteriorated compared to the neat LSR specimen. Figure 4d shows that when the specimens were immersed in oil, the decrease in tensile strength and elongation of the specimen accelerated and proceeded under high-temperature conditions compared to low-temperature conditions.

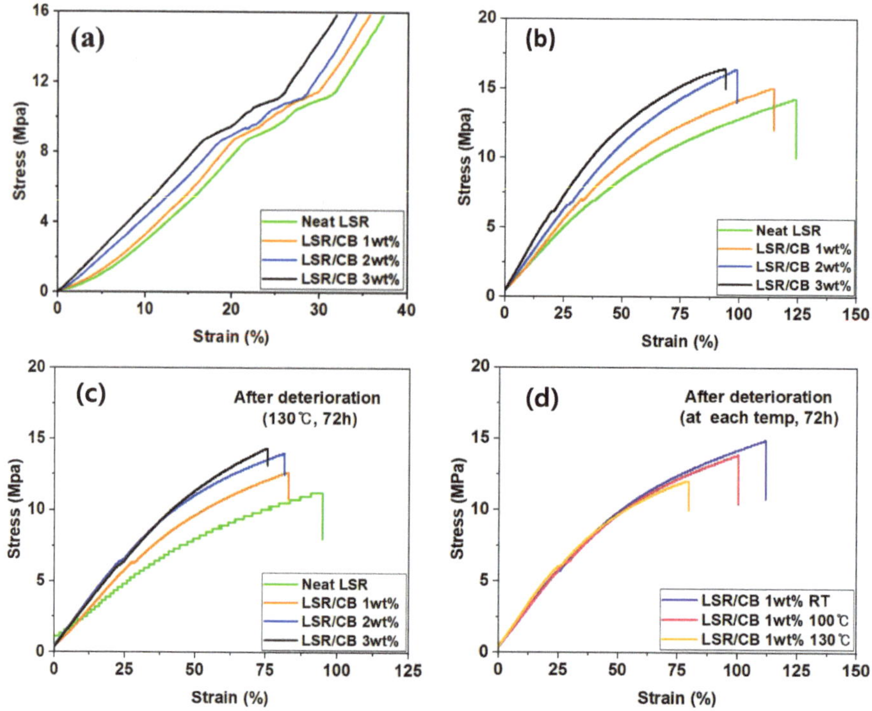

Figure 4. (**a**) Compression curve of CB/LSR, (**b**) tensile curve of CB/LSR, (**c**) tensile curve of CB/SLR after accelerated deterioration (130 °C, 72 h), (**d**) tensile curve of CB/SLR after accelerated deterioration (1 wt%, 72 h) with different temperatures.

Table 1. Deterioration rate of CB/LSR after accelerated deterioration with ultimate tensile strength (130 °C, 72 h).

Material	Ultimate Tensile Strength-before Deterioration (MPa)	Ultimate Tensile Strength-after Deterioration (MPa)	Deterioration Rate (%)
Neat	14.22	11.21	21.67
CB 1 wt%	14.99	12.62	15.81
CB 2 wt%	16.34	14.57	10.83
CB 3 wt%	16.35	14.32	12.42

Table 2. Deterioration rate of CB/LSR after accelerated deterioration with tensile modulus (130 °C, 72 h).

Material	Tensile Modulus—before Deterioration (MPa)	Tensile Modulus—after Deterioration (MPa)	Deterioration Rate (%)
Neat	16.53	13.35	19.24
CB 1 wt%	19.79	17.72	10.46
CB 2 wt%	24.05	22.48	6.53
CB 3 wt%	29.03	23.69	18.39

3.2. Swelling Characteristics

Swelling tests were conducted at various oil exposure times and temperatures to determine the effect of CB content on the swelling behavior of LSR. Table 3 shows the change in weight corresponding to the oil exposure time of LSR containing 0 to 3 wt% CB at various temperatures of RT, 100 °C, and 130 °C, respectively. As time passed, the oil permeated the specimen, increasing its weight. The change in swelling rate was also calculated using Equation (1) in Section 2.2.2, as shown in Figure 5 and Table 4. Each specimen exhibited a generally similar pattern, wherein the oil absorption initially increases rapidly but subsequently decreases over time, leading to a reduction in the slope of the graphs. The swelling tests revealed a swelling ratio of 5.21% at room temperature, 9.02% at 100 °C, and 11.11% at 130 °C for neat LSR. Compared to the neat LSR results, the swelling ratio decreased as the CB content in LSR increased up to 2 wt%, and this effect was particularly pronounced at high temperatures. The findings show that the CB has a mitigating effect on oil-induced deterioration in LSR. However, as the CB content increased, the swelling ratio decreased, and it was observed that the swelling ratio increased slightly when 3 wt% of CB was added, which is presumed to be due to the agglomeration of the CB [38,39].

Table 3. Weight ($M_t(g)$) change over oil exposure time of CB/LSR.

Time (h)		0	8	16	24	32	40	48	56	64	72
Neat	RT	3.095	3.112	3.133	3.146	3.164	3.201	3.226	3.250	3.252	3.256
	100	3.092	3.135	3.182	3.223	3.262	3.302	3.334	3.353	3.364	3.371
	130	3.093	3.167	3.229	3.286	3.326	3.365	3.401	3.417	3.431	3.437
CB 1 wt%	RT	3.132	3.149	3.163	3.184	3.194	3.229	3.254	3.285	3.289	3.294
	100	3.129	3.289	3.204	3.244	3.276	3.315	3.343	3.348	3.377	3.389
	130	3.127	3.196	3.258	3.314	3.352	3.392	3.424	3.437	3.444	3.449
CB 2 wt%	RT	3.144	3.160	3.171	3.194	3.200	3.241	3.271	3.289	3.292	3.302
	100	3.149	3.182	3.297	3.240	3.289	3.306	3.354	3.373	3.391	3.397
	130	3.151	3.209	3.272	3.327	3.369	3.411	3.442	3.452	3.459	3.462
CB 3 wt%	RT	3.146	3.163	3.178	3.198	3.206	3.240	3.273	3.302	3.306	3.307
	100	3.142	3.168	3.199	3.246	3.294	3.335	3.370	3.390	3.396	3.397
	130	3.144	3.211	3.274	3.326	3.367	3.409	3.427	3.462	3.467	3.471

Table 4. Swelling ratio with different temperature of CB/LSR.

Material			**RT 72 h**		**100 °C 72 h**		**130 °C 72 h**	
Neat	M_0 (g)	M_t (g)	3.095	3.256	3.092	3.371	3.093	3.437
	swelling ratio (%)		**5.202**		**9.238**		**11.121**	
CB 1 wt%	M_0 (g)	M_t (g)	3.132	3.294	3.129	3.389	3.127	3.449
	swelling ratio (%)		**5.172**		**8.309**		**10.297**	
CB 2 wt%	M_0 (g)	M_t (g)	3.144	3.302	3.149	3.397	3.151	3.462
	swelling ratio (%)		**5.025**		**7.876**		**9.858**	
CB 3 wt%	M_0 (g)	M_t (g)	3.146	3.307	3.142	3.397	3.144	3.471
	swelling ratio (%)		**5.118**		**8.116**		**10.407**	

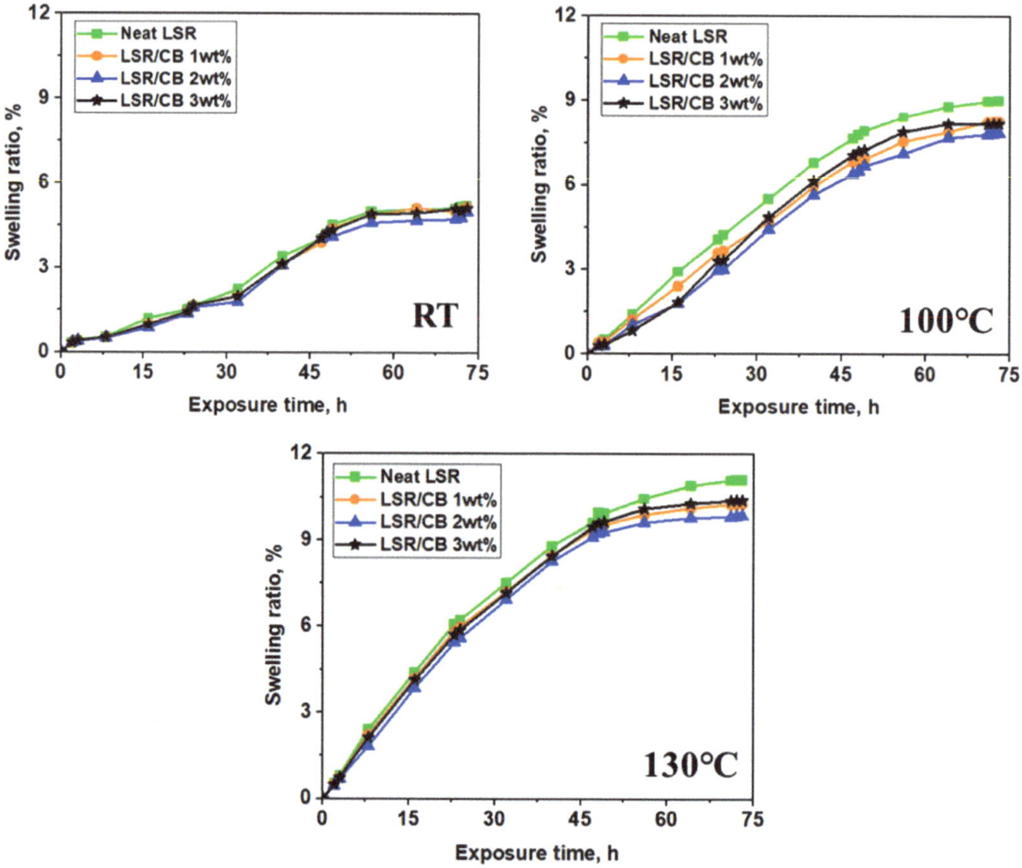

Figure 5. Swelling ratio with time for different temperature of CB/LSR.

Furthermore, it was verified that the variation in swelling rate, dependent on the CB content, becomes more pronounced at elevated temperatures compared to room temperature. This observation indicates that with increasing temperature, the mobility of the rubber chains increases within the viscoelastic behavior of the rubber. Consequently, the available space for oil diffusion between rubber molecules increases, facilitating a more pronounced rubber expansion.

3.3. Morphological and Surface Characterization

SEM is a commonly employed method for observing the morphology of rubber/CB composites. The SEM measurement results are shown in Figure 6. We captured SEM images of the surfaces of both Neat LSR and the CB/CLSR composites after cutting them with a razor blade. The CB particles were observed on the surface of the specimen with a CB content ranging from 1 wt% to 3 wt%. Most CB was well distributed in the LSR matrix. However, the aggregation pattern varied depending on the content of CB. In particular, in the specimen with a CB content of 3 wt%, uneven particle distribution was noted compared to the specimens with CB content of 1 wt% and 2 wt%, and some agglomeration was also observed. This accounts for the slight increase in the swelling ratio observed in the CB content of 3 wt%, as mentioned in Section 3.2. Consequently, in the CB/LSR tensile test shown in Figure 4b, the tensile strength increased with the rising CB content but experienced a slight decrease at 3 wt%, which was attributed to the observed agglomeration [38,39].

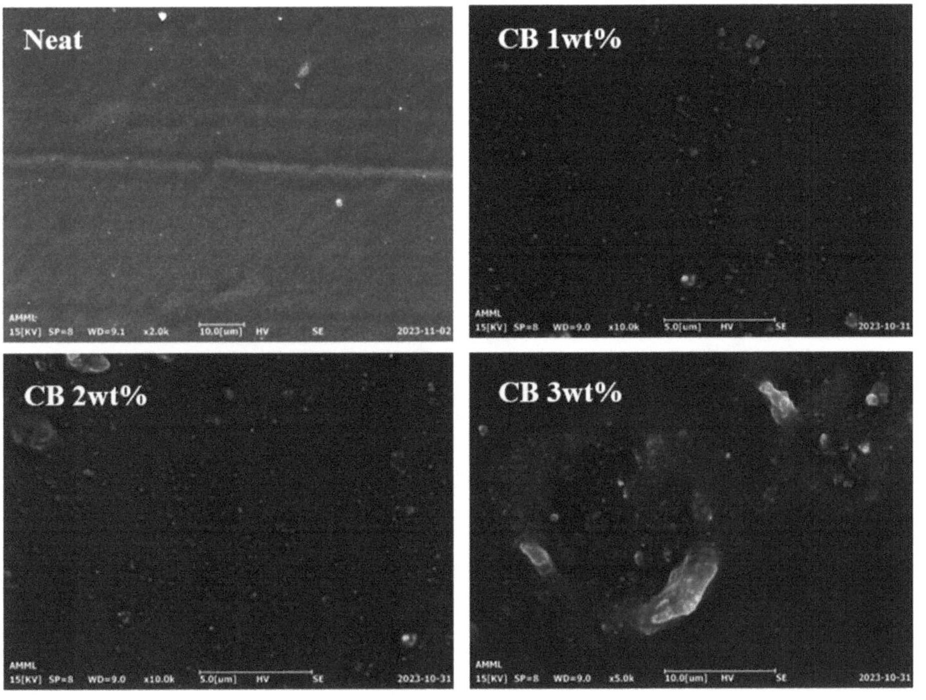

Figure 6. SEM images of CB/LSR specimens.

3.4. Compression Set Property

As can be seen in Figure 7 and Table 5, the results of the compression set test exhibit minimal differences at room temperature. However, at 130 °C, an increase in CB content leads to a reduction in the rate of elastic recovery against compression, resulting in a slight increase in the compression set. Specifically, the compression set value at 24 h was 5.43% in neat LSR and 6.81% in CB 3 wt% of LSR. This phenomenon is attributed to the increased crosslinking density of the LSR as CB content rises, actively participating in the crosslinking of the rubber chain. Simultaneously, the movement of the silicone rubber chain is constrained, leading to an increase in stiffness. Consequently, under a continuous load, the number of crosslinks resisting deformation rises, but so does the number of crosslinks broken during the resistance. This results in a lowered crosslinking ratio responsible for deformation recovery, preventing the full recovery of the original thickness. Therefore, the compression set value increases, albeit subtly, with the increment in CB content.

Table 5. Compression set with different temperature of CB/LSR.

	Material		**RT 72 h**		**130 °C 72 h**	
Neat	t_0 (cm)	t_1 (cm)	12.471	12.450	12.469	12.444
	Compression set (%)		**4.44**		**5.43**	
CB 1 wt%	t_0 (cm)	t_1 (cm)	12.454	12.433	12.452	12.426
	Compression set (%)		**4.62**		**5.83**	
CB 2 wt%	t_0 (cm)	t_1 (cm)	12.478	12.455	12.481	12.452
	Compression set (%)		**4.72**		**6.12**	
CB 3 wt%	t_0 (cm)	t_1 (cm)	12.499	12.474	12.503	12.469
	Compression set (%)		**5.01**		**6.81**	

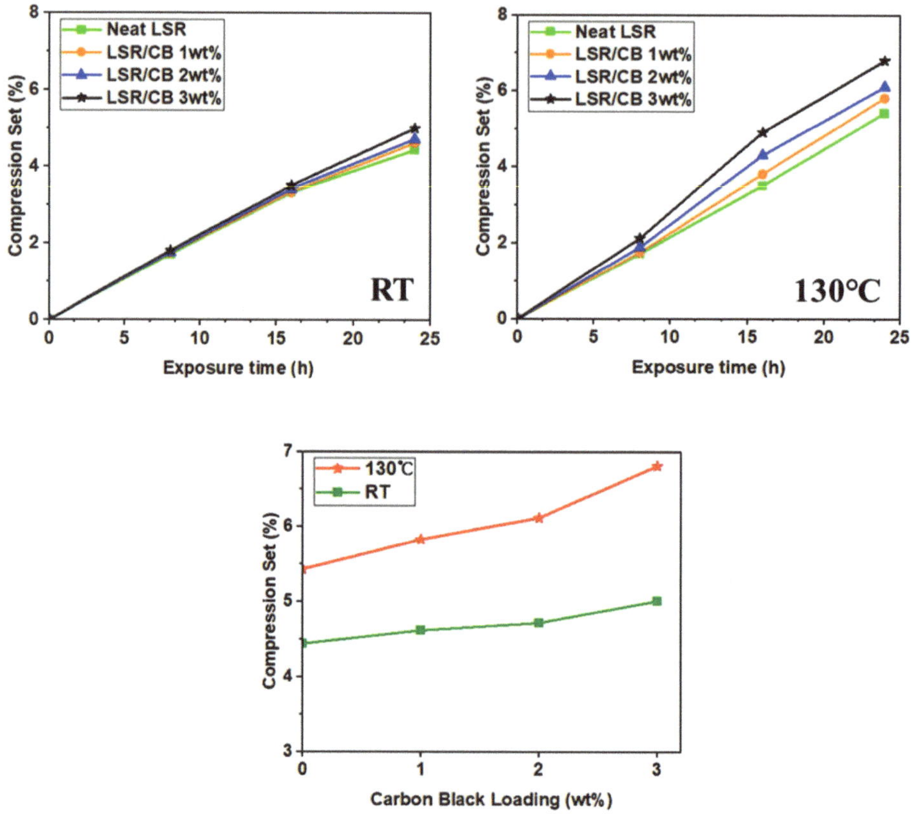

Figure 7. The effect of CB loading on the compression set test of CB/LSR.

3.5. The Mechanism of CB Effect on LSR

In the tensile tests, we confirmed that with the increase in CB content added to LSR, the mechanical property (modulus) increases and the strain rate decreases. Additionally, in a tensile test under deteriorated conditions, as the CB content increased, we observed that LSR composites showed an improvement in resistance to deteriorated conditions. From the results of the swelling test, it was observed that when CB was added to LSR, a low swelling ratio was observed at an appropriate content (2 wt%), demonstrating excellent oil resistance performance. Furthermore, in the compression set test, as the content of CB increased, there was in increase in the cross-linking of the rubber chain. This led to an increase in the number of broken cross-links and a reduction in the cross-linking ratio responsible for deformation recovery, thereby hindering the recovery of the original thickness.

The mechanism behind the excellent mechanical properties and low swelling ratio of CB-added LSR can be elucidated by the physical interaction between the surface of CB and the LSR chain, as well as the chemical bonding in which the OH group of CB reacts with the hydrogen of LSR (Figure 8) [40]. The chemical bonding results in the secure attachment of CB to the LSR chain. The carbon black chemically bonded with LSR can then surround the rubber molecular chain and cause physical crosslinking, roughly as shown in Figure 9a [41]. Essentially, the tightly bound CB, integrated into the crosslinking of LSR, enhances the rigidity of CB/LSR and restricts the mobility of the rubber chain, thereby impeding the penetration of oil between the rubber chains, as illustrated in Figure 9. This, in turn, increases the resistance to oil.

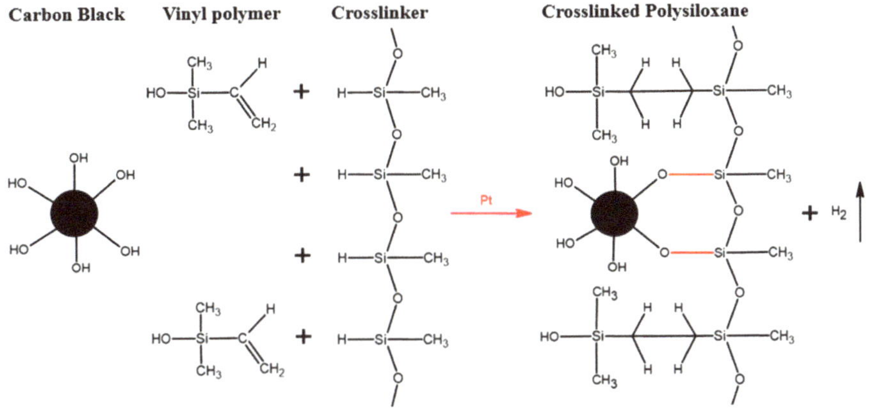

Figure 8. The chemical reaction mechanism between LSR and CB.

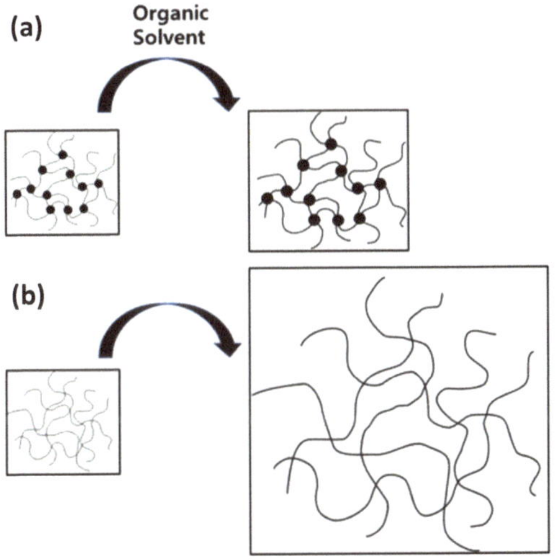

Figure 9. Scheme of the physical network structure LSR swelling with (**a**) and without (**b**) the addition of carbon black.

4. Conclusions

In this study, we investigated the mechanical properties and swelling characteristics of the CB and LSR composite, considering variables such as CB content, time, and temperature. The mechanical property test results demonstrated that with an increase in CB content, both the tensile modulus and deterioration resistance improved, which was attributed to the strong bonding between CB and the LSR chain effectively delaying the deterioration caused by oil. In the same vein, the restricted mobility of the chain led to a reduction in the swelling ratio for oil. Although there was a slight increase in the compression set with the addition of CB, it was not significant. These findings suggest the potential utility of CB in LSR products exposed to the oil, such as gaskets and seals. However, it is crucial to note that at a relatively high CB content (3 wt%), mechanical properties and oil resistance slightly decreased due to the dispersion and agglomeration of the CB. Therefore, future efforts should focus on improving the surface treatment and dispersion methods of CB during the production of CB/LSR composites to enhance their acceptability.

Author Contributions: Conceptualization, H.-M.Y.; Methodology, H.-M.Y.; Investigation, B.-J.L.; Data curation, B.-J.L.; Writing—original draft, H.-M.Y. All authors have read and agreed to the published version of the manuscript.

Funding: This paper was supported by the Education and Research promotion program of KOREATECH in 2022 and Cheonan Institute of Science & Technology Platform (2023 Enabling tech development project).

Institutional Review Board Statement: Not applicable.

Data Availability Statement: Data are contained within the article.

Conflicts of Interest: The authors declare no conflicts of interest.

References

1. Etienne, D.; Ganachaud, F. Looking over Liquid Silicone Rubbers: (1) Network Topology vs. Chemical Formulations. *ACS Appl. Mater. Interfaces* **2012**, *4*, 3340–3352.
2. Matthew, B.; Barry, C.; Johnston, S. A review of liquid silicone rubber injection molding: Process variables and process molding. *Polym. Eng. Sci.* **2021**, *61*, 331–347.
3. Haberstroh, E.; Michaeli, W.; Henze, E. Simulation of the Filling and Curing Phase in Injection Molding of Liquid Silicone Rubber (LSR). *J. Compos. Mater.* **2002**, *21*, 461–471. [CrossRef]
4. Delebecq, E.; Hermeline, N.; Flers, A.; Ganachaud, F. Looking over Liquid Silicone Rubbers: (2) Mechanical Properties vs. Network Topology. *ACS Appl. Mater. Interfaces* **2012**, *4*, 3353–3363. [CrossRef] [PubMed]
5. Shit, S.C.; Shah, P. A Review on Silicone Rubber. *Natl. Acad. Sci. Lett.* **2013**, *36*, 355–365. [CrossRef]
6. Li, P.; Jiang, S. Biomaterial characteristics and application of silicone rubber and PVA hydrogels mimicked in organ groups for prostate brachytherapy. *J. Mech. Behav. Biomed. Mater.* **2015**, *49*, 220–234. [CrossRef] [PubMed]
7. Yun, Y.J.; Ju, J.; Lee, J.H.; Moon, S.H.; Park, S.J.; Kim, Y.H.; Hong, W.G.; Ha, D.H.; Jang, H.; Lee, G.H. Highly elastic graphenebased electronics toward electronic skin. *Adv. Funct. Mater.* **2017**, *27*, 1701513. [CrossRef]
8. Min, S.H.; Lee, G.Y.; Ahn, S.H. Direct printing of highly sensitive, stretchable, and durable strain sensor based on silver nanoparticles/multi-walled carbon nanotubes composites. *Compos. B Eng.* **2019**, *161*, 395–401. [CrossRef]
9. Fink, J.K. *Liquid Silicone Rubber: Chemistry, Materials, and Processing*; John Wiley & Sons: Hoboken, NJ, USA, 2019.
10. Mostafa, A.; Abouel-Kasem, A.; Bayoumi, M.R.; El-Sebaie, M.G. Effects of carbon black loading on the swelling and compression set behavior of SBR and NBR rubber compounds. *Mater. Des.* **2009**, *30*, 1561–1568. [CrossRef]
11. Pehilivan-Davis, S.; Clarke, J.; Armour, S. Comparison of accelerated aging of silicone rubber gasket material with aging in a fuel cell environment. *J. Appl. Polym. Sci.* **2013**, *129*, 1446–1454. [CrossRef]
12. Widodo, W.S.; Soenoko, R.; Choiron, M.A.; Sonief, A.A. Sealing performance analysis of composite gaskets made of silicone rubber filled with ramie natural fibers. *J. Mech. Eng. Sci.* **2019**, *13*, 6178–6194.
13. Nakano, M.; Sasaki, H.; Hanashima, K. The prediction of long-term and emergency sealability of silicone and EPDM gaskets. In Proceedings of the 18th International Conference on Nuclear Engineering, Xi'an, China, 17–21 May 2010; Volume 49330, pp. 309–316.
14. Jie, Z.; Feng, S. Temperature effects of electrical resistivity of conductive silicone rubber filled with carbon blacks. *J. Appl. Polym. Sci.* **2003**, *90*, 3889–3895.
15. Li, Z.H.; Zhang, J.; Chen, S.J. Effects of carbon blacks with various structures on vulcanization and reinforcement of filled ethylene-propylene-diene rubber. *Express Polym. Lett.* **2008**, *2*, 695–704. [CrossRef]
16. Stübler, N.; Fritzsche, J.; Klüppel, M. Mechanical and electrical analysis of carbon black networking in elastomers under strain. *Polym. Eng. Sci.* **2011**, *51*, 1206–1217. [CrossRef]
17. Kumar, V.; Alam, M.N.; Manikkavel, A.; Song, M.; Lee, D.J.; Park, S.S. Silicone rubber composites reinforced by carbon nanofillers and their hybrids for various applications: A review. *Polymers* **2021**, *13*, 2322. [CrossRef]
18. Song, P.; Song, J.; Zhang, Y. Stretchable conductor based on carbon nanotube/carbon black silicone rubber nanocomposites with highly mechanical, electrical properties and strain sensitivity. *Compos. B Eng.* **2020**, *191*, 107979. [CrossRef]
19. Boonstra, B.B.; Medalia, A.I. Effect of carbon black dispersion on the mechanical properties of rubber vulcanizates. *Rubber Chem. Technol.* **1963**, *36*, 115–142. [CrossRef]
20. Rwei, S.P.; Ku, F.H.; Cheng, K.C. Dispersion of carbon black in a continuous phase: Electrical, rheological, and morphological studies. *Colloid. Polym. Sci.* **2002**, *280*, 1110–1115.
21. Sau, K.P.; Khastgir, D.; Chaki, T.K. Electrical conductivity of carbon black and carbon fibre filled silicone rubber composites. *Macromol. Mater. Eng.* **1998**, *258*, 11–17. [CrossRef]
22. Grunert, F.; Wehmeier, A.; Blume, A. New insights into the morphology of silica and carbon black based on their different dispersion behavior. *Polymers* **2020**, *12*, 567. [CrossRef]
23. Shivashankar, H.; Sangamesh, R.; Kulkarni, S.M. Processing and investigation of mechanical characteristics on the polydimethylsiloxane/carbon black composites. *Mater. Res. Express* **2019**, *6*, 105340. [CrossRef]

24. Schubert, D.W.; Werner, S.; Hahn, I.; Solovieva, V. Effect of particle size and size distribution on the permeability of soft magnetic liquid silicone rubber composites. *Compos. Sci. Technol.* **2019**, *177*, 26–33. [CrossRef]
25. Liu, P.; Liu, C.; Huang, Y.; Wang, W.; Fang, D.; Zhang, Y.; Ge, Y. Transfer function and working principle of a pressure/temperature sensor based on carbon black/silicone rubber composites. *J. Appl. Polym. Sci.* **2016**, *133*, 42979. [CrossRef]
26. Song, P.; Wang, G.; Zhang, Y. Enhanced positive temperature coefficient effect by crosslinking reaction for silicone rubber/carbon black composites with high pressure sensitivity. *J. Appl. Polym. Sci.* **2022**, *139*, 51682. [CrossRef]
27. Jeddi, J.; Katbab, A.A. The Electrical Conductivity and EMI Shielding Properties of Polyurehane Foam/Silicone Rubber/Carbon Black/Nanographite Hybrid Composites. *Polym. Compos.* **2018**, *39*, 3452–3460. [CrossRef]
28. Jie, Z.; Feng, S.; Ma, Q. Kinetics of the thermal degradation and thermal stability of conductive silicone rubber filled with conductive carbon black. *J. Appl. Polym. Sci.* **2003**, *89*, 1548–1554.
29. Ding, T.; Wang, L. Changes in electrical resistance of carbon-black-filled silicone rubber composite during compression. *J. Polym. Sci. B Polym. Phys.* **2007**, *45*, 2700–2706. [CrossRef]
30. Hussain, R.; Moxter, J.; Hinrichsen, V. Influence of temperature and electric field strength on the space charge behavior of liquid silicone rubber with carbon black nanofillers. *IEEE Trans. Dielectr. Electr. Insul.* **2020**, *27*, 427–432. [CrossRef]
31. Wang, S.; Wang, P.; Ding, T. Resistive viscoelasticity of silicone rubber/carbon black composite. *Polym. Compos.* **2011**, *32*, 29–35. [CrossRef]
32. *ASTM D412-16*; Standard Test Methods for Vulcanized Rubber and Thermoplastic Elastomer Tension. ASTM International: West Conshohocken, PA, USA, 2021.
33. *ASTM D1229-03*; Standard Test Methods for Rubber Property-Compresion Set at Low Temperatures. ASTM International: West Conshohocken, PA, USA, 2019.
34. *ASTM D471-16a*; Standard Test Methods for Rubber Property-Effect of Liquids. ASTM International: West Conshohocken, PA, USA, 2021.
35. *ASTM D395-18*; Standard Test Methods for Rubber Property-Compreession Set. ASTM International: West Conshohocken, PA, USA, 2018.
36. Omnes, B.; Thuillier, S.; Pilvin, P.; Grohens, Y.; Gillet, S. Effective properties of carbon black filled natural rubber: Experiments and modeling. *Compos. Part. A Appl. Sci. Manuf.* **2008**, *39*, 1141–1149. [CrossRef]
37. Herd, C.R.; McDonald, G.C.; Hess, W.M. Morphology of Carbon-Black Aggregates: Fractal Versus Euclidean Geometry. *Rubber Chem. Technol.* **1992**, *65*, 107–129. [CrossRef]
38. Dou, Y.; Sun, S.; Lu, S.; Yao, W.; Guan, D. Preparation of carbon black/silicone rubber composites with large-area-homogeneous-low electrical-resistance used as electroplating matrix. *RSC Adv.* **2012**, *12*, 32448. [CrossRef] [PubMed]
39. Zhu, X.; Li, W. Research on Dynamic Compression Testing of Silicone Rubber under Different Temperatures. *J. Phys. Conf. Ser.* **2021**, *2002*, 012047. [CrossRef]
40. Luis, E.; Pan, H.M.; Sing, S.L.; Bastola, A.K.; Goh, G.D.; Goh, G.L.; Tan, H.K.J.; Bajpai, R.; Song, J.; Yong, W.Y. Silicone 3D printing: Process optimization, product biocompatibility, and reliability of silicone meniscus implants. *3D Print. Addit. Manuf.* **2019**, *6*, 319–332. [CrossRef]
41. VLitvinov, V.M.; Steeman, P.A.M. EPDM-Carbon Black Interactions and the Reinforcement Mechanisms, As Studied by Low-Resolution 1H NMR. *Macromolecules* **1999**, *32*, 8476–8490. [CrossRef]

MDPI AG
Grosspeteranlage 5
4052 Basel
Switzerland
Tel.: +41 61 683 77 34

Polymers Editorial Office
E-mail: polymers@mdpi.com
www.mdpi.com/journal/polymers

www.ingramcontent.com/pod-product-compliance
Lightning Source LLC
Chambersburg PA
CBHW041405210926
51850CB00013B/3181